Artificial Intelligence and Deep Learning in Sensors and Applications

Artificial Intelligence and Deep Learning in Sensors and Applications

Editors

Shyan-Ming Yuan
Zeng-Wei Hong
Wai-Khuen Cheng

Basel • Beijing • Wuhan • Barcelona • Belgrade • Novi Sad • Cluj • Manchester

Editors

Shyan-Ming Yuan
Department of
Computer Science
National Yang Ming Chiao
Tung University
Hsinchu
Taiwan

Zeng-Wei Hong
Department of Information
Engineering and
Computer Science
Feng Chia University
Taichung
Taiwan

Wai-Khuen Cheng
Faculty of Information and
Communication Technology
Universiti Tunku
Abdul Rahman
Kampar
Malaysia

Editorial Office
MDPI AG
Grosspeteranlage 5
4052 Basel, Switzerland

This is a reprint of articles from the Special Issue published online in the open access journal *Sensors* (ISSN 1424-8220) (available at: https://www.mdpi.com/journal/sensors/special_issues/YI50W8MD78).

For citation purposes, cite each article independently as indicated on the article page online and as indicated below:

Lastname, A.A.; Lastname, B.B. Article Title. *Journal Name* **Year**, *Volume Number*, Page Range.

ISBN 978-3-7258-1451-0 (Hbk)
ISBN 978-3-7258-1452-7 (PDF)
doi.org/10.3390/books978-3-7258-1452-7

© 2024 by the authors. Articles in this book are Open Access and distributed under the Creative Commons Attribution (CC BY) license. The book as a whole is distributed by MDPI under the terms and conditions of the Creative Commons Attribution-NonCommercial-NoDerivs (CC BY-NC-ND) license.

Contents

About the Editors vii

Shyan-Ming Yuan, Zeng-Wei Hong and Wai-Khuen Cheng
Artificial Intelligence and Deep Learning in Sensors and Applications
Reprinted from: *Sensors* **2024**, *24*, 3258, doi:10.3390/s24103258 1

Joana Vale Sousa, Pedro Matos Francisco Silva, Pedro Freitas, Hélder P. Oliveira and Tania Pereira
Single Modality vs. Multimodality: What Works Best for Lung Cancer Screening?
Reprinted from: *Sensors* **2023**, *23*, 5597, doi:10.3390/s23125597 6

Sanghoon Park, Jihun Kim, Han-You Jeong, Tae-Kyoung Kim and Jinwoo Yoo
C2RL: Convolutional-Contrastive Learning for Reinforcement Learning Based on Self-Pretraining for Strong Augmentation
Reprinted from: *Sensors* **2023**, *23*, 4946, doi:10.3390/s23104946 17

Hookyung Lee, Jaeseung Jeon, Seokjin Hong, Jeesu Kim and Jinwoo Yoo
TransNet: Transformer-Based Point Cloud Sampling Network
Reprinted from: *Sensors* **2023**, *23*, 4675, doi:10.3390/s23104675 28

Zheng Xu, Yumeng Yang, Xinwen Gao and Min Hu
DCFF-MTAD: A Multivariate Time-Series Anomaly Detection Model Based on Dual-Channel Feature Fusion
Reprinted from: *Sensors* **2023**, *23*, 3910, doi:10.3390/s23083910 40

Gawon Lee and Jihie Kim
MTGEA: A Multimodal Two-Stream GNN Framework for Efficient Point Cloud and Skeleton Data Alignment
Reprinted from: *Sensors* **2023**, *23*, 2787, doi:10.3390/s23052787 60

Ivan Grubišić, Marin Oršić and Siniša Šegvić
Revisiting Consistency for Semi-Supervised Semantic Segmentation †
Reprinted from: *Sensors* **2023**, *23*, 940, doi:10.3390/s23020940 73

Ren-Hung Hwang, Jia-You Lin, Sun-Ying Hsieh, Hsuan-Yu Lin and Chia-Liang Lin
Adversarial Patch Attacks on Deep-Learning-Based Face Recognition Systems Using Generative Adversarial Networks
Reprinted from: *Sensors* **2023**, *23*, 853, doi:10.3390/s23020853 99

Yuang Li, Yuntao Wang, Xin Liu, Yuanchun Shi, Shwetak Patel and Shao-Fu Shih
Enabling Real-Time On-Chip Audio Super Resolution for Bone-Conduction Microphones
Reprinted from: *Sensors* **2023**, *23*, 35, doi:10.3390/s23010035 128

Jehn-Ruey Jiang and Yan-Ting Lin
Deep Learning Anomaly Classification Using Multi-Attention Residual Blocks for Industrial Control Systems
Reprinted from: *Sensors* **2022**, *22*, 9084, doi:10.3390/s22239084 147

Yupeng Wei and Hongrui Liu
Convolutional Long-Short Term Memory Network with Multi-Head Attention Mechanism for Traffic Flow Prediction
Reprinted from: *Sensors* **2022**, *22*, 7994, doi:10.3390/s22207994 161

Abdur Rahim Mohammad Forkanb, Ahsan Morshedc, Prem Prakash Jayaramanb and Muhammad Shoaib Siddiqui
Weed Detection Using Deep Learning: A Systematic Literature Review
Reprinted from: *Sensors* **2023**, *23*, 3670, doi:10.3390/s23073670 . . . . **176**

Ruey-Kai Sheu and Mayuresh Sunil Pardeshi
A Survey on Medical Explainable AI (XAI): Recent Progress, Explainability Approach, Human Interaction and Scoring System
Reprinted from: *Sensors* **2022**, *22*, 8068, doi:10.3390/s22208068 . . . . **221**

About the Editors

Shyan-Ming Yuan

Shyan-Ming Yuan received a Ph.D. degree in computer science from the University of Maryland, College Park, in 1989. In 1990, he became an Associate Professor with the Department of Computer and Information Science, National Chiao Tung University, Hsinchu, Taiwan. Since 1995, he has been a Professor with the Department of Computer Science, National Yang Ming Chiao Tung University, where he is also the Director of Library. His current research interests include distance learning, internet technologies, distributed computing, cloud computing, and blockchain technology.

Zeng-Wei Hong

Zeng-Wei Hong received B.S., M.S., and Ph.D degrees in computer science from the Department of Information Engineering and Computer Science, Feng-Chia University, Taichung, Taiwan. He acted as an Assistant Professor and Associate Professor in Asia University, Taiwan, from August 2007 to July 2015. He was also an Assistant Professor in the Faculty of Information and Communication Technology, UTAR, Kampar, Malaysia, from August 2015 to July 2020. He is now an Associate Professor at the Department of Information Engineering and Computer Science, Feng Chia University, Taiwan. His current research interests include e-learning and software engineering.

Wai-Khuen Cheng

Wai-Khuen Cheng received his Bachelor of Computer Science (Hons.) and Doctor of Philosophy degrees from Universiti Sains Malaysia in 2004 and 2009, respectively. Cheng is currently serving as an Associate Professor and Deputy Dean of the Faculty of Information and Communication Technology at Universiti Tunku Abdul Rahman (UTAR). He has been a Professional Technologist with the Malaysia Board of Technologists (MBOT) since 2018, a Senior Member of the ACM since 2013, a Member of the IEEE since 2022, and an Associate Fellow of the ASEAN Academy of Engineering & Technology (AAET) since 2023. He has published more than 50 articles related to cloud computing, multi-agent systems, recommender systems, the Internet of Things, financial technology, and artificial intelligence. With over 20 years of experience in distributed and high-performance computing, he has also served as a Guest Editor for the journals Mathematics and Sensors. Dr. Cheng was a recipient of the Malaysia Digital Economy Corporation's (MDEC) Premier Digital Tech Institution (PDTI) Best Faculty Member Award in 2021 and 2022, the Alibaba Cloud MVP Award in 2022, and the Best Academic Staff Award in 2022.

Editorial

Artificial Intelligence and Deep Learning in Sensors and Applications

Shyan-Ming Yuan [1,*], Zeng-Wei Hong [2] and Wai-Khuen Cheng [3]

1. Department of Computer Science, National Yang Ming Chiao Tung University, Hsinchu 300093, Taiwan
2. Department of Information Engineering and Computer Science, Feng Chia University, Taichung 40724, Taiwan; zwhong@fcu.edu.tw
3. Faculty of Information and Communication Technology, Department of Computer Science, Universiti Tunku Abdul Rahman, Kampar 31900, Malaysia; chengwk@utar.edu.my
* Correspondence: smyuan@cs.nycu.edu.tw

Citation: Yuan, S.-M.; Hong, Z.-W.; Cheng, W.-K. Artificial Intelligence and Deep Learning in Sensors and Applications. *Sensors* **2024**, *24*, 3258. https://doi.org/10.3390/s24103258

Received: 27 April 2024
Accepted: 6 May 2024
Published: 20 May 2024

Copyright: © 2024 by the authors. Licensee MDPI, Basel, Switzerland. This article is an open access article distributed under the terms and conditions of the Creative Commons Attribution (CC BY) license (https:// creativecommons.org/licenses/by/ 4.0/).

1. Introduction

To effectively solve the increasingly complex problems experienced by human beings, the latest development trend is to apply a large number of different types of sensors to collect data in order to establish effective solutions based on deep learning and artificial intelligence [1–4]. This not only creates a huge demand for sensors, providing business opportunities, but also creates new challenges for the development of sensor devices and their related applications [5,6]. These technological developments that combine AI and sensors are being actively used in various application fields such as healthcare, manufacturing, agriculture and fisheries, transportation, construction, environmental monitoring, etc.

For instance, in environmental monitoring, sensors integrated with deep learning and AI algorithms possess the capability to swiftly analyze extensive datasets, identifying patterns, anomalies, and trends in real-time [7,8]. Consider weather forecasting, where sensors driven by AI can gather data from various sources like satellites, weather stations, and drones, enabling more precise predictions of weather patterns. Through deep learning models, sensors can dynamically adjust and incorporate new data, thereby refining their predictive accuracy over time. Additionally, in industrial settings, sensors enhanced with AI play a crucial role in optimizing manufacturing operations by monitoring equipment health, forecasting potential failures, and scheduling maintenance preemptively [9–12]. This approach reduces operational downtime and enhances overall efficiency.

In this context, the Special Issue "Artificial Intelligence and Deep Learning in Sensors and Applications" collected high-quality original contributions on new developments in AI (specifically deep learning) and sensor technology in various fields, as well as sharing ideas, designs, data-driven applications, and production and deployment experiences and challenges. The call for papers for this Special Issue included topics such as applications and sensors for manufacturing, machinery, and semiconductors; smart applications and sensors for architecture, construction, buildings, e-learning; recommendation systems; applications and sensors for autonomous vehicles, traffic monitoring, and transportation; object recognition, image classification, object detection, speech processing, human behavior analysis; and other related sensing applications [13,14].

2. Overview of Published Papers

All submissions were evaluated based on their technical excellence, resulting in the selection of ten research articles, and two reviews on weed detection using deep learning and medical XAI, for inclusion in this Special Issue. Below, all of the contributions are listed, followed by a short summary of each contribution.

In contribution 1, the author contributes to the field by exploring the integration of AI and deep learning for enhancing lung cancer screening through sensors, such as CT scans, enriched with clinical data. By comparing single-modality (imaging data alone) and multimodality approaches (combining imaging with clinical data), the study underlines the importance of leveraging multiple data types for improved diagnostic accuracy. Specifically, it showcases the application of ResNet18 for 3D CT imaging analysis and random forest algorithms for clinical data evaluation, demonstrating a significant leap in performance with multimodal data fusion. This work aligns with the journal's focus by illustrating how AI and deep learning can optimize sensor data utilization in healthcare applications, pushing the boundaries of early cancer detection capabilities.

In contribution 2, the innovative C2RL framework introduces a convolutional-contrastive learning approach that integrates with reinforcement learning to enhance the generalization capabilities of agents in varied environments, especially those captured through high-dimensional image sensors. This paper's approach addresses the challenge of leveraging strongly augmented sensor data without compromising the reinforcement learning process. By demonstrating its effectiveness in the DeepMind Control suite, the paper contributes to the journal's theme by showcasing how AI and deep learning techniques can enhance the interpretability and utility of sensor data in complex, dynamic systems, paving the way for more robust AI agents that can operate in diverse and unpredictable real-world scenarios.

In contribution 3, TransNet leverages transformer architecture to introduce a novel approach for point cloud sampling, addressing the computational complexities associated with direct usage of dense point cloud data from sensors. This contribution is particularly relevant to the journal's interests, as it illustrates the application of AI and deep learning in optimizing sensor data processing. By implementing self-attention mechanisms, TransNet not only reduces computational demands, but also enhances precision in tasks requiring detailed spatial analysis, such as autonomous driving and robotic navigation, highlighting the potential of advanced AI models in improving the efficiency and accuracy of sensor-based applications.

In contribution 4, DCFF-MTAD presents a dual-channel feature fusion model for anomaly detection in multivariate time-series data, a common output from various sensors. This paper's approach, combining spatial STFT and graph attention networks, exemplifies the integration of AI to enhance sensor data analytics for predictive maintenance and monitoring of complex systems. It aligns with the journal's focus by demonstrating the potential of deep learning for extracting and fusing features from sensor-generated data to identify anomalies, offering valuable insights for applications in industrial IoT, environmental monitoring, and more, where accurate and timely anomaly detection is crucial.

In contribution 5, MTGEA addresses the challenge of aligning sparse point cloud data from radar sensors with skeleton data from Kinect sensors for human activity recognition, a critical task in smart home systems. The paper's development of a multimodal two-stream GNN framework exemplifies innovative use of deep learning to enhance sensor data compatibility and application efficiency. This work is highly relevant to the journal, as it showcases the fusion of data from diverse sensors through AI to improve accuracy in privacy-preserving human activity recognition, contributing to the advancement of smart home technologies and healthcare monitoring systems.

In contribution 6, the author explores semi-supervised learning techniques for semantic segmentation, leveraging perturbed unlabeled inputs to enforce consistency in dense prediction tasks. By focusing on one-way consistency and introducing a novel perturbation model, this research advances the application of AI in processing and interpreting sensor data, especially in real-time and low-power scenarios. The findings are pertinent to the journal's scope, as they highlight how deep learning can reduce reliance on extensively labelled datasets in dense prediction, facilitating more efficient and scalable implementations of AI in semantic segmentation tasks using data from visual sensors for applications in autonomous driving, surveillance, and robotics.

In contribution 7, the author contributes significantly to the realm of AI and sensor applications by addressing the vulnerabilities of deep-learning-based face recognition systems to adversarial patch attacks. By using Generative Adversarial Networks (GANs), the study proposes a method that successfully implements black-box attacks, enhancing the realism and applicability of security assessments in systems that rely on facial recognition sensors. This contribution is vital for developing robust face recognition technologies that are increasingly integrated into various security and personal identification applications, ensuring they can resist real-world adversarial threats.

In contribution 8, the introduction of a real-time, on-chip audio super-resolution system tailored for bone-conduction microphones marks a notable advancement in sensor technology and AI applications. This paper's development of lightweight deep-learning models for enhancing audio quality directly addresses the limitations of existing sensor technology in noisy environments. By optimizing these models for low-power, real-time processing on embedded systems, the study provides a practical solution that can be integrated into next-generation hearing aids and communication devices, leveraging AI to significantly improve the user's experience.

In contribution 9, the research enhances the application of AI in industrial sensors by introducing a deep learning framework capable of detecting anomalies in the network traffic of industrial control systems. The use of multi-attention residual blocks to refine the feature extraction process represents a forward leap in improving the accuracy and reliability of anomaly detection in critical infrastructure. This paper's methodology contributes to safer, more secure industrial operations by leveraging deep learning to interpret complex sensor data streams effectively.

In contribution 10, the author contributes to the field of AI and sensor applications through their innovative use of convolutional LSTM networks combined with a multi-head attention mechanism, specifically designed to enhance traffic flow predictions. By effectively analyzing and predicting traffic conditions using vast arrays of sensor data, this approach helps with optimizing traffic management and urban planning. The integration of spatial and temporal data analyses in a unified deep learning model exemplifies how AI can harness sensor data to facilitate smarter, more efficient city infrastructure.

In contribution 11, this systematic literature review aggregates and synthesizes the use of AI in agricultural sensors for weed detection, presenting an overview of the state-of-the-art deep learning techniques applied to this problem. By reviewing various approaches and evaluating their effectiveness, this paper contributes to the broader application of AI in precision agriculture. It guides future research and practical implementations that can help farmers reduce crop loss and manage fields more efficiently through advanced sensor technology.

In contribution 12, this survey paper explores the critical role of explainable AI (XAI) within the medical field, emphasizing the integration of AI with medical sensor data to improve diagnostics and patient care. By discussing recent progress and proposing new frameworks for XAI, the paper contributes to the ongoing development of transparent, understandable AI applications that can work alongside healthcare professionals. This work is especially important for enhancing the trustworthiness and efficacy of AI systems in interpreting complex medical sensor data, thus supporting better clinical decisions and patient outcomes.

3. Conclusions

The intersection of artificial intelligence (AI) and deep learning with sensor technologies presents an evolving frontier with profound implications across a myriad of applications, from healthcare diagnostics to autonomous systems and beyond. This collection of twelve papers contributes significant insights and innovations at this intersection, showcasing the potential of AI and deep learning in extracting, processing, and analyzing sensor data to solve complex real-world problems.

We appreciate all contributors of this Special Issue, as well as the reviewers of the submitted papers. Their dedicated efforts and expertise in providing thorough reviews greatly helped with the completion of this successful publication.

Author Contributions: Conceptualization, all authors; writing—original draft preparation, Z.-W.H. and W.-K.C.; writing—review and editing, all authors. All authors have read and agreed to the published version of the manuscript.

Conflicts of Interest: The authors declare no conflicts of interest.

List of Contributions:

1. Sousa, J.V.; Matos, P.; Silva, F.; Freitas, P.; Oliveira, H.P.; Pereira, T. Single Modality vs. Multimodality: What Works Best for Lung Cancer Screening? *Sensors* **2023**, *23*, 5597.
2. Park, S.; Kim, J.; Jeong, H.-Y.; Kim, T.-K.; Yoo, J. C2RL: Convolutional-Contrastive Learning for Reinforcement Learning Based on Self-Pretraining for Strong Augmentation. *Sensors* **2023**, *23*, 4946.
3. Lee, H.; Jeon, J.; Hong, S.; Kim, J.; Yoo, J. TransNet: Transformer-Based Point Cloud Sampling Network. *Sensors* **2023**, *23*, 4675.
4. Xu, Z.; Yang, Y.; Gao, X.; Hu, M. DCFF-MTAD: A Multivariate Time-Series Anomaly Detection Model Based on Dual-Channel Feature Fusion. *Sensors* **2023**, *23*, 3910.
5. Lee, G.; Kim, J. MTGEA: A Multimodal Two-Stream GNN Framework for Efficient Point Cloud and Skeleton Data Alignment. *Sensors* **2023**, *23*, 2787.
6. Grubišić, I.; Oršić, M.; Šegvić, S. Revisiting Consistency for Semi-Supervised Semantic Segmentation. *Sensors* **2023**, *23*, 940.
7. Hwang, R.-H.; Lin, J.-Y.; Hsieh, S.-Y.; Lin, H.-Y.; Lin, C.-L. Adversarial Patch Attacks on Deep-Learning-Based Face Recognition Systems Using Generative Adversarial Networks. *Sensors* **2023**, *23*, 853.
8. Li, Y.; Wang, Y.; Liu, X.; Shi, Y.; Patel, S.; Shih, S.-F. Enabling Real-Time On-Chip Audio Super Resolution for Bone-Conduction Microphones. *Sensors* **2023**, *23*, 35.
9. Jiang, J.-R.; Lin, Y.-T. Deep Learning Anomaly Classification Using Multi-Attention Residual Blocks for Industrial Control Systems. *Sensors* **2022**, *22*, 9084.
10. Wei, Y.; Liu, H. Convolutional Long-Short Term Memory Network with Multi-Head Attention Mechanism for Traffic Flow Prediction. *Sensors* **2022**, *22*, 7994.
11. Murad, N.Y.; Mahmood, T.; Forkan, A.R.M.; Morshed, A.; Jayaraman, P.P.; Siddiqui, M.S. Weed Detection Using Deep Learning: A Systematic Literature Review. *Sensors* **2023**, *23*, 3670.
12. Sheu, R.-K.; Pardeshi, M.S. A Survey on Medical Explainable AI (XAI): Recent Progress, Explainability Approach, Human Interaction and Scoring System. *Sensors* **2022**, *22*, 8068.

References

1. Ramanujam, E.; Thinagaran, P.; Padmavathi, S. Human Activity Recognition with Smartphone and Wearable Sensors Using Deep Learning Techniques: A Review. *IEEE Sens. J.* **2021**, *21*, 13029. [CrossRef]
2. Zhang, S.; Li, Y.; Zhang, S.; Shahabi, F.; Xia, S.; Deng, Y.; Alshurafa, N. Deep Learning in Human Activity Recognition with Wearable Sensors: A Review on Advances. *Sensors* **2022**, *22*, 1476. [CrossRef] [PubMed]
3. Ji, I.H.; Lee, J.H.; Kang, M.J.; Park, W.J.; Jeon, S.H.; Seo, J.T. Artificial Intelligence-Based Anomaly Detection Technology over Encrypted Traffic: A Systematic Literature Review. *Sensors* **2024**, *24*, 898. [CrossRef] [PubMed]
4. Shajari, S.; Kuruvinashetti, K.; Komeili, A.; Sundararaj, U. The Emergence of AI-Based Wearable Sensors for Digital Health Technology: A Review. *Sensors* **2023**, *23*, 9498. [CrossRef] [PubMed]
5. Kadhim, I.; Abed, F.M. A Critical Review of Remote Sensing Approaches and Deep Learning Techniques in Archaeology. *Sensors* **2023**, *23*, 2918. [CrossRef] [PubMed]
6. Ma, W.; Sun, Y.; Qi, X.; Xue, X.; Chang, K.; Xu, Z.; Li, M.; Wang, R.; Meng, R.; Li, Q. Computer-Vision-Based Sensing Technologies for Livestock Body Dimension Measurement: A Survey. *Sensors* **2024**, *24*, 1504. [CrossRef] [PubMed]
7. Yuan, Q.; Shen, H.; Li, T.; Li, Z.; Li, S.; Jiang, Y.; Zhang, L. Deep Learning in Environmental Remote Sensing: Achievements and Challenges. *Remote Sens. Environ.* **2020**, *241*, 111716. [CrossRef]
8. Jin, X.-B.; Zheng, W.-Z.; Kong, J.-L.; Wang, X.-Y.; Zuo, M.; Zhang, Q.-C.; Lin, S. Deep-Learning Temporal Predictor via Bidirectional Self-Attentive Encoder–Decoder Framework for IOT-Based Environmental Sensing in Intelligent Greenhouse. *Agriculture* **2021**, *11*, 802. [CrossRef]
9. Wang, J.; Ma, Y.; Zhang, L.; Gao, R.X.; Wu, D. Deep Learning for Smart Manufacturing: Methods and Applications. *J. Manuf. Syst.* **2018**, *48*, 144. [CrossRef]
10. Sarivan, I.M.; Greiner, J.N.; Álvarez, D.D.; Euteneuer, F.; Reichenbach, M.; Madsen, O.; Bøgh, S. Enabling Real-Time Quality inspection in smart manufacturing through wearable smart devices and deep learning. *Procedia Manuf.* **2020**, *51*, 373–380. [CrossRef]

11. Andronie, M.; Lăzăroiu, G.; Iatagan, M.; Uță, C.; Ștefănescu, R.; Cocoșatu, M. Artificial Intelligence-Based Decision-Making Algorithms, Internet of Things Sensing Networks, and Deep Learning-Assisted Smart Process Management in Cyber-Physical Production Systems. *Electronics* **2021**, *10*, 2497. [CrossRef]
12. Lăzăroiu, G.; Andronie, M.; Iatagan, M.; Geamănu, M.; Ștefănescu, R.; Dijmărescu, I. Deep Learning-Assisted Smart Process Planning, Robotic Wireless Sensor Networks, and Geospatial Big Data Management Algorithms in the Internet of Manufacturing Things. *ISPRS Int. J. Geo-Inf.* **2022**, *11*, 277. [CrossRef]
13. Mo, S.; Shi, Y.; Yuan, Q.; Li, M. A Survey of Deep Learning Road Extraction Algorithms Using High-Resolution Remote Sensing Images. *Sensors* **2024**, *24*, 1708. [CrossRef] [PubMed]
14. Rafique, S.H.; Abdallah, A.; Musa, N.S.; Murugan, T. Machine Learning and Deep Learning Techniques for Internet of Things Network Anomaly Detection—Current Research Trends. *Sensors* **2024**, *24*, 1968. [CrossRef] [PubMed]

Disclaimer/Publisher's Note: The statements, opinions and data contained in all publications are solely those of the individual author(s) and contributor(s) and not of MDPI and/or the editor(s). MDPI and/or the editor(s) disclaim responsibility for any injury to people or property resulting from any ideas, methods, instructions or products referred to in the content.

Article

Single Modality vs. Multimodality: What Works Best for Lung Cancer Screening?

Joana Vale Sousa [1,2,*], Pedro Matos [2], Francisco Silva [1,3], Pedro Freitas [1,2], Hélder P. Oliveira [1,3] and Tania Pereira [1]

1. Institute for Systems and Computer Engineering, Technology and Science (INESC TEC), 4200-465 Porto, Portugal
2. Faculty of Engineering (FEUP), University of Porto, 4200-465 Porto, Portugal
3. Faculty of Science (FCUP), University of Porto, 4169-007 Porto, Portugal
* Correspondence: joana.v.sousa@inesctec.pt

Abstract: In a clinical context, physicians usually take into account information from more than one data modality when making decisions regarding cancer diagnosis and treatment planning. Artificial intelligence-based methods should mimic the clinical method and take into consideration different sources of data that allow a more comprehensive analysis of the patient and, as a consequence, a more accurate diagnosis. Lung cancer evaluation, in particular, can benefit from this approach since this pathology presents high mortality rates due to its late diagnosis. However, many related works make use of a single data source, namely imaging data. Therefore, this work aims to study the prediction of lung cancer when using more than one data modality. The National Lung Screening Trial dataset that contains data from different sources, specifically, computed tomography (CT) scans and clinical data, was used for the study, the development and comparison of single-modality and multimodality models, that may explore the predictive capability of these two types of data to their full potential. A ResNet18 network was trained to classify 3D CT nodule regions of interest (ROI), whereas a random forest algorithm was used to classify the clinical data, with the former achieving an area under the ROC curve (AUC) of 0.7897 and the latter 0.5241. Regarding the multimodality approaches, three strategies, based on intermediate and late fusion, were implemented to combine the information from the 3D CT nodule ROIs and the clinical data. From those, the best model—a fully connected layer that receives as input a combination of clinical data and deep imaging features, given by a ResNet18 inference model—presented an AUC of 0.8021. Lung cancer is a complex disease, characterized by a multitude of biological and physiological phenomena and influenced by multiple factors. It is thus imperative that the models are capable of responding to that need. The results obtained showed that the combination of different types may have the potential to produce more comprehensive analyses of the disease by the models.

Keywords: deep learning; multimodality; feature fusion; lung cancer; CT scan; clinical data

Citation: Sousa, J.V.; Matos, P.; Silva, F.; Freitas, P.; Oliveira, H.P.; Pereira, T. Single Modality vs. Multimodality: What Works Best for Lung Cancer Screening? *Sensors* **2023**, *23*, 5597. https://doi.org/10.3390/s23125597

Academic Editors: Guillermo Villanueva, Shyan-Ming Yuan, Zeng-Wei Hong and Wai-Khuen Cheng

Received: 9 February 2023
Revised: 13 June 2023
Accepted: 13 June 2023
Published: 15 June 2023

Copyright: © 2023 by the authors. Licensee MDPI, Basel, Switzerland. This article is an open access article distributed under the terms and conditions of the Creative Commons Attribution (CC BY) license (https://creativecommons.org/licenses/by/4.0/).

1. Introduction

Lung cancer is the leading cause of cancer-related deaths, being responsible for approximately over 2 million new cases and 1.8 millions deaths in 2020 [1]. Despite the increasing risk of developing cancer related with age, tobacco consumption persists as the main contributor for all major histological types of lung cancer, accounting for about 80% of cases [2–4]. Nevertheless, there are other risk factors that can have a key role as well in the development of this condition, such as exposure to air pollution and second-hand smoke, occupational exposure, a diet poor in nutrients, alcohol consumption, genetic susceptibility and positive family history of lung cancer [3,4]. Given the lack of clear and distinct symptoms at early stages, when this condition begins to manifest itself in a more evident manner, by the time patients are diagnosed, lung cancer is usually in an advanced stage, and, as result, the 5-year survival rate is low, around 19%. On the contrary, if the disease is detected at earlier stages, the 5-year survival rate can increase up to 54% [5], reinforcing, for this reason, the urgent need for screening and prevention measures.

In the clinical practice, it is common for physicians to take into account the information obtained from multiple sources, namely imaging findings, clinical and demographic data, and family history, in order to give an accurate diagnosis for the patient. Through the visual inspection of medical images, such as computed tomography (CT) scans, radiologists search for evidence of lung cancer, and in case there is a suspicion of the presence of malignant nodules, patients are submitted to biopsy, an invasive procedure with associated risks. However, very often, false positives are identified, leading to unnecessary procedures in patients that are cancer-free. Furthermore, given the great amount of medical images to analyze and because physicians cannot overlook them, this task becomes time demanding, exhausting and human-error prone [6]. Artificial intelligence (AI) based methods can assist the practitioners in the correct classification of these nodules, helping to decrease the high rates of false positives and negatives, and give more accurate diagnoses for these patients. Nonetheless, the vast majority of available methods use a single modality for the task of classification, mainly imaging modalities [2], which may put constraints on the learning process of the models, as they are limited to a single type of information [6].

Motivated by the variation in the size and morphology of lung nodules, Lyu et al. [7] introduced a multi-level cross ResNet that includes three sets of parallel residual blocks, each with a specific convolutional kernel size, in order to extract features at different scales. Data from the Lung Image Database Consortium (LIDC) [8] dataset were retrieved and because they contain nodules that can fall into three malignancy categories—benign, malignant and indeterminate—the authors conducted experiments for a ternary classification and a binary classification (that only considers benign and malignant nodules). Accuracies of 0.85 and 0.92 were obtained for the former and the latter experiments. Calheiros et al. [9] presented a work that studied the importance of the perinodular area for the malignancy classification of lung nodules. Radiomic features were extracted from the perinodular and intranodular regions of the 3D CT images from the LIDC database, and different combinations of the extracted features were made. The authors tested six different machine learning methods, namely decision tree, logistic regression, random forest, k-nearest neighbor (kNN), support vector machine (SVM) and extreme gradient boosting (XGBoost), with a total of 15 models, as a result of the combination of various hyper-parameters. The overall best performance was obtained with SVM trained with the set of features pertaining to the nodule, margin sharpness and the perinodular zone, having achieved an area under the receiver operating characteristics curve (AUC) of 0.91 ± 0.031. In addition, from the feature ranking analysis of the tree-based models, the results demonstrated that 6 of the 20 top features were extracted from the perinodular region, thus highlighting its relevance for the classification task. In [10], the authors developed a 3D axial-attention network for the classification of CT lung nodules, and data were retrieved from the LIDC dataset. The model presented an AUC, accuracy, precision and sensitivity of 0.96, 0.92, 0.92 and 0.92, respectively. The authors in [11] extracted features from CT images using the convolutional neural network, histogram of oriented gradients (HOG), extended HOG and local binary pattern, and tested four different algorithms: SVM, kNN, random forest and decision trees. The LIDC dataset was, once again, used for development and evaluation, and the best performance model presented an accuracy of 0.95. Liu et al. [12] proposed an architecture denominated as Res-trans networks that combines residual and transformer blocks for the lung cancer classification of CT nodules. The method is assessed in the LIDC dataset and presents an AUC of 0.96 and an accuracy of 0.93. With the aim of studying the relationship between chronic obstructive pulmonary disease (COPD), pulmonary nodules and the risk of lung cancer, Uthoff et al. [13] explored the idea of fusing clinical features (that include the data and clinical history of the patients, the diameter of the nodules, and four pulmonary function tests) with automatically extracted features from CT images (such as measurements from the whole pulmonary parenchyma, the lobe that contained nodules, and the airways). Three approaches were implemented to study the impact of these features on the developed models: using clinical features only; using imaging features only; and combining both. Mutual information optimization (IO) and least absolute shrinkage and selection operator (LASSO) were applied for feature selection, and LASSO and an ensemble neural network (ENN) were chosen as classification models. For training and evaluation, data were collected from the COPD Genetic Epidemiology Study (COPDGene) [14], Inflammation, Health, and Lung Epidemiology Study (INHALE) [15] and National Lung Screening Trial (NLST) [16] databases, and only patients with pulmonary nodules ≥ 4 mm were selected, for a total of 327 individuals. The highest performance metric, an AUC of 0.79, was achieved with the ENN when trained with both clinical and imaging features, selected with the IO method. Motivated by the possible complementarity between the information of CT images and serum biomarkers, Jing et al. [17] developed two malignancy classification algorithms for lung cancer, one for each modality, and then studied the combination of the predictions of those two algorithms to output a final one. CT scans and serum biomarkers were collected with a total of 173 patients used. For all pulmonary nodules, the malignancy was confirmed with a biopsy. A total of 78 quantitative features were extracted from the CT segmented nodules and given to a SVM classifier. Five serum biomarkers were investigated (squamous cell carcinoma antigen (SCC); carcinoembryonic antigen (CEA), cytokeratin fragment 21-1 (CYFRA21-1); cancer antigen 15-3 (CA15-3); and carbohydrate antigen 19-19 (CA19-9)) and also given to a SVM. As for the combination of predictions of the two algorithms, three fusion methods were studied: minimum score between the two predictions; maximum score between the two predictions; and an weighted average of the two, in which the weights assigned vary between 0.1 and 0.9. The imaging model demonstrated higher performance metrics than the biomarker model, and the maximum AUC, 0.85 ± 0.03, was obtained by combining the predictions of the two models with weight factors of 0.3 and 0.7, respectively.

As mentioned above, in a clinical setting, data from a variety of sources are considered for lung cancer diagnosis. On the other hand, a great number of current AI approaches makes use of a single data modality, with the LIDC dataset being one of the most commonly used datasets for the development of image-based models [7,9–12], as it includes labeled nodules, yet no other data modalities are provided. In more recent years, multimodal approaches applied to the biomedical field have emerged, and often deep fusion methods surpass the performance of unimodal strategies [18]. Lung cancer is a complex disease, characterized by a multitude of biological and physiological phenomena and influenced by multiple factors. Multimodality data represent the possibility of developing learning models that are capable of responding to that need. With that in mind, the goal of this work was to study and compare lung cancer classification models that are dependent on a single modality with models that translate the clinical context by integrating information from different modalities, and with that, ascertain if improvements are registered when a broader view and analysis of the patients are taken into account. Furthermore, experiments were conducted with the NLST dataset [16] since it allows the combination of those modalities and it contains more challenging cases (as seen by the results obtained in [13]) which may enable the development of a more comprehensive analysis by the learning models.

2. Materials and Methods

In this section, the data used and the methods implemented in this work are described. Section 2.1 gives a detailed description of the dataset used and the pre-processing steps applied, whereas Section 2.2 describes the methodologies implemented, namely the single-modality approaches in Section 2.2.1 and the multimodality approaches in Section 2.2.2.

2.1. Dataset

2.1.1. National Lung Screening Trial

The NLST [16] was a clinical trial conducted in partnership between the Lung Screening Study group and the American College of Radiology Imaging Network, with the aim of ascertaining whether the use of low-dose helical CT for lung cancer screening in high-risk patients would reduce mortality in comparison to chest radiography. For that reason, individuals with ages between 55 and 74 and considered high-risk (current or former smokers with 30 years or more of cigarette pack smoking history) were randomly assigned to one of two possible study arms: one in which participants were scanned with chest radiography, and another in which CT was used as a screening imaging modality. Given the scope of this work, the focus was on participants who were screened with CT, and, as such, data regarding those participants were retrieved (representing a subset of the entire dataset), which included CT images, abnormalities annotations and lung cancer screening results, as well as participant data.

2.1.2. CT Scans

The CT scans provided were acquired with different equipment and scanning protocols, which resulted in differences in slice thickness and pixel spacing. For this reason, and to ensure homogeneity across all images, resampling was applied that set the pixel spacing to 1 mm in axes x, y and z. Afterwards, images were resized to a dimension of 128 × 128 pixels and submitted to a *min–max* normalization, with −1000 and 400 Hounsfield Units (HU) defined as lower and upper limits, to transform the original range of HU intensities to a range of [0, 1]. In the end, each scan had a dimension of 128 × 128 ×s, in which s represents the number of slices for that scan. This dataset does not provide the segmentation masks of the identified nodules; thus, 20 × 50 × 50 bounding boxes containing the nodule in their center were manually created, with a total of 1079 3D nodule regions of interest (ROI) obtained, from which 655 were of the benign class and 424 of the malignant class. For some of the patients, more than one nodule was identified, and thus, the 1079 cases represented, in fact, a total of 1005 patients. Examples of the bounding boxes of CT slices in axial view are presented in Figure 1.

  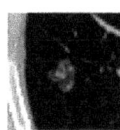 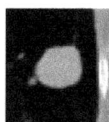 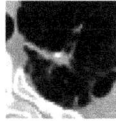 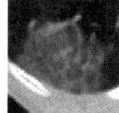

Figure 1. Example of bounding boxes of the CT slices of the NLST dataset. From left to right, the first three images correspond to malignant nodules, whereas the last three images correspond to benign nodules [16].

2.1.3. Clinical Features

The NLST dataset also provides participant data with regards to the study in which they were enrolled; participant identifier demographics (such as age, height, weight and education); smoking habits; screening; invasive procedures and possible complications; lung cancer results; last contact; death; occupational exposure to pollutants and prevention measures; medical history; cancer history; family history of lung cancer; alcohol habits; and lung cancer progression. Given the fact that some of these features were related to lung cancer screening results and further outcomes, they were discarded in the feature selection process, in order not to introduce bias during the learning of the models, and, as a result, a total of 136 features, out of the original 324, were selected, under the following tags: demographic, smoking, work history, disease history, personal cancer history, family history, and alcohol.

2.1.4. Summary

As explained above, the number of participants differs from the number of CT volumes of nodules obtained since for some patients, there was more than one nodule identified; hence, the distribution of classes benign and malignant of the CT scans and clinical data is different as presented in Table 1.

Table 1. Class distribution for imaging and clinical modalities.

Data Modality	Class		Task
	# Benign	# Malignant	
CT Scans	522	339	Train
	133	85	Test
Clinical Features	463	337	Train
	121	84	Test

2.2. Methodology

Firstly, each data modality, the CT scans and the clinical features, was analyzed separately with the purpose of investigating its individual effect on the classification task. Afterwards, three different strategies that combine both modalities were implemented to study whether joining information from different sources is beneficial and complementary to the learning of the models. An overview of the pipeline implemented is depicted in Figure 2.

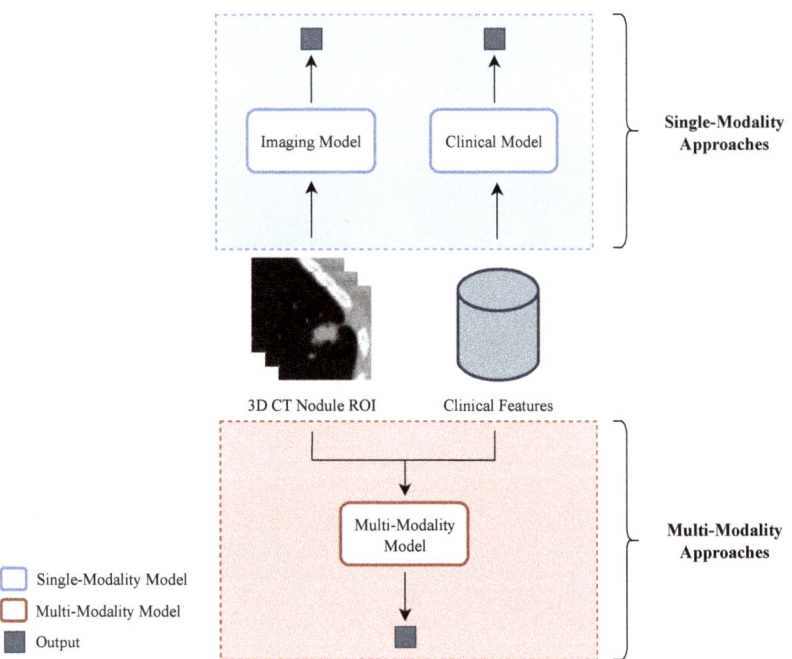

Figure 2. Overview of the pipeline implemented for study and comparison of the single- and multimodality strategies for lung cancer classification. Concerning the single-modality approaches, a classification model was developed for each of the data types utilized: an imaging model for the 3D CT nodule regions of interest and a clinical model for the clinical data. In the multimodality approaches, there is a fusion of the information from the two modalities.

In all experiments, for the division of the data into training and evaluation, the identifiers associated with the nodules were considered, with 80% used as training data and the remaining 20% for testing. As for the clinical data, their division was made by taking into consideration the task previously assigned to the respective nodule(s), see Table 1. With the goal of identifying the best combination of hyper-parameters, 5-fold cross validation was implemented, using 80% of the data assigned for training. In this implementation, for each combination of hyper-parameters, the 80% was divided 5-fold.

Four were used for training (64% of complete data), whereas the remaining was used for evaluation (16% of complete data). The process was repeated five times, and an average AUC was obtained. After all combinations were evaluated, the optimal parameters were selected as the ones that obtained the highest AUC. At last, the network was trained with the selected optimal set of parameters and using the 80% of the data assigned for training. AUC was used as a performance metric [19], and binary cross entropy (BCE) was used as the loss function.

2.2.1. Single-Modality Aproaches

With respect to the imaging data, a 3D ResNet-18 architecture was chosen, given its proven efficiency in classification tasks. In the search for an optimal combination of hyper-parameters, a 5-fold cross-validation was performed. The values used for the optimizer—learning rate, batch size, dropout, and weight decay—are presented in Table 2. When employing the 5-fold cross validation, it was ensured that nodules belonging to the same patient were assigned to the same fold and that no data leakage occurred. The models were trained for 50 epochs.

Table 2. Hyper-parameters used for the development of the imaging and intermediate fusion models.

Hyper-Parameter	Value
Optimizer	Adam, SGD
Learning rate	0.01, 0.001, 0.0001
Weight Decay	0.01, 0.001, 0.0001
Batch size	16, 32, 64
Dropout	0.3, 0.4, 0.5, 0.6

As for the clinical data, the random forest algorithm was chosen since it allows the identification of the features to which more importance was given by the models. A grid search with a 5-fold cross-validation strategy was implemented, using the AUC as a scoring metric, and the parameters and respective values analyzed are presented in Table 3. After assessing the impurity-based feature ranking produced by the highest-performing model, the scope of features was narrowed down to 42. These features are as follows: demographic (age, educat, ethnic, height, marital, race, and weight); smoking (age_quit, cigar, pkyr, smokeage, smokeday, and smokeyr); work history (yrsasbe, yrsbutc, yrschem, yrscott, yrsfarm, yrsfoun, yrspain, and yrssand); disease history (ageadas, agechas, agechro, agecopd, agediab, ageemph, agehear, agehype, agepneu, agestro, diagchas, diagchro, and diagpneu); personal cancer history (ageoral and cancoral); and alcohol (acrin_drink24h, acrin_drinknum_curr, acrin_drinknum_form, acrin_drinkyrs_curr, acrin_drinkyrs_form, and lss_alcohol_num).

Table 3. Hyper-parameters used for the development of clinical models.

Hyper-Parameter	Value
# Estimators	200, 300, 400, 500, 600
Criterion	gini, entropy
Max features	sqrt, log2
Maximum depth	3–9
Class weight	None, balanced

2.2.2. Multimodality Approaches

Regarding the fusion of the two modalities, there are three main strategies that can be implemented: early fusion, in which the raw data from two or more modalities are combined and given to a single model; intermediate fusion, in which features from each modality are extracted, concatenated, and given to a single model; and late fusion, in which the final classification output is a combination of the outputs given by each modality model [20]. In order to better exploit the information inherent to each modality and because the CT volumes and the clinical features present distinct formats, the early fusion was discarded, and priority was given to the intermediate and late fusion approaches. Figure 3 depicts the pipeline implemented for the three multimodality strategies.

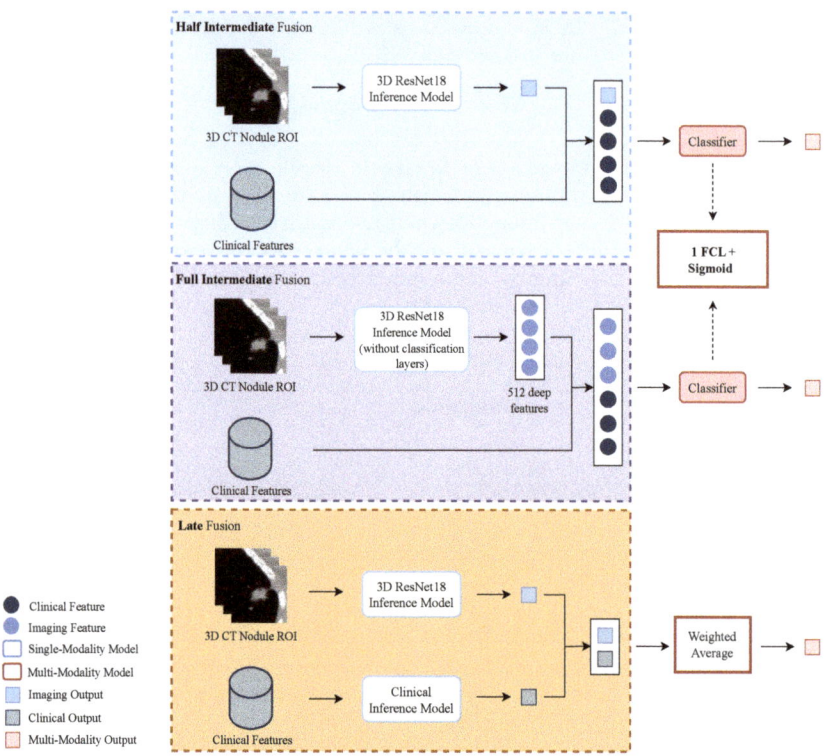

Figure 3. Overview of the pipeline implemented for the multi-modalities strategies. From top to bottom: half intermediate fusion (HIF), with the fusion of the imaging output and clinical features; full intermediate fusion (FIF) with the fusion of deep imaging features and clinical features; and late fusion (LF) with the fusion of the outputs given by the imaging and clinical models. For the HIF and FIF approaches, the lung cancer classification is given by a classifier constituted by one fully connected layer (FCL). In the LF approach, the classification is a weighted average of the predictions of the single-modality models.

In relation to the intermediate fusion, two methods were studied: one denominated half-intermediate fusion (HIF), in which the malignancy probability of the volumes of the nodules, given by an inference model (the ResNet18 imaging model that achieved the highest AUC), was fused with the clinical features; and full intermediate fusion (FIF), in which 512 deep imaging features of the volumes of the nodules, given by the last layer prior to the classification layer of that same inference model of the HIF, are fused with the clinical features. In both, the concatenated features are fed to one fully connected layer (FCL), followed by a sigmoid activation layer that outputs the final probability. Furthermore, with respect to the clinical features used, two different sets were tested: one with the original 136, and another with the selected 42, as described above. A 5-fold cross validation was performed in the search for the optimal parameters. The hyper-parameters implemented are presented in Table 2. The models were trained for 200 epochs.

As for the late fusion (LF) approach, the weighted average of the outputs of the imaging model and the clinical model was computed and used to estimate the malignancy. The weight assigned to each output ranged between 0.1 and 0.9.

3. Results and Discussion

This section includes the results obtained for the strategies developed and further discussion.

Table 4 presents the results of the models that demonstrated the best performance for each one of the five methods studied, as well as the number of features used, in the cases in which they were necessary. From the results of the 5-fold cross-validation implementation, for the single-modality approaches, the mean AUC and standard deviation obtained for the image and clinical models were, respectively, 0.7227 ± 0.0311 and 0.5924 ± 0.0188. As for the intermediate fusion approaches, mean AUC and standard deviation of 0.9195 ± 0.0029 and 0.8750 ± 0.0129 were obtained for the half intermediate fusion and full intermediate fusion models, respectively. Table 5 presents the hyper-parameters for three of these models, namely the imaging model and both intermediate fusion models. As for the clinical model, the set of parameters that achieved the best performance was as follows: 300 estimators with a maximum depth of 7 and the maximum number of features given by $log2$. The weight of the classes is balanced, and *gini* was used to measure the quality of the splits. The hyper-parameters of the single-modality models of the LF approach were formerly described. The result presented in Table 4 corresponds to an image output weight of 0.8 and respective clinical model output weight of 0.2, which is the combination of weight factors that achieved the highest AUC.

Table 4. Results obtained for the five methods implemented. The highest performance metric, highlighted in bold, is obtained for the Full Intermediate Fusion approach.

	Approach	# Clinical Features	AUC
Single-Modality	Image Model	-	0.7897
	Clinical Model	136	0.5241
Multimodality	HIF	42	0.7934
	FIF	42	**0.8021**
	LF	136	0.7911

Table 5. Hyper-parameters of models with the highest performance metric for the image-only and intermediate fusion approaches.

	Approach	Optimizer	Learning Rate	Weight Decay	Batch Size	Dropout
Single Modality	Image Model	SGD	0.0001	0.001	32	0.4
Multimodality	HIF	Adam	0.01	0	16	0.4
	FIF	Adam	0.0001	0	64	0.5

It is possible to observe that the multimodality approaches are the ones that present the highest performance metric, which can indicate that combining information from different sources has the potential to improve the performance of the models, particularly in comparison with the clinical model. Nonetheless, these improvements are minimal when compared to the value obtained for the imaging model. Effectively, when analyzing the results obtained by the imaging model, one can see that the CT volumes containing the nodules lead to a higher capability to distinguish cancer from non-cancer diagnosis.

One possible explanation could reside in the fact that the clinical features used may not bring enough relevance to the learning, as made evident by the poor results obtained by the clinical model. These results are also in agreement with what one would expect since in a clinical context, the lung cancer diagnosis is not based solely on the characteristics of the patient, pertaining to personal information and medical history. Similarly, considering that the LF approach combines the predictions of the single-modality models and given the results of the clinical model, it was likely that it would present the lowest AUC among the three multimodality methodologies. Those insights are reflected as well in the results of the intermediate fusion approaches, for which the attention of the network is mostly on the imaging inputs, produced by the imaging inference model. On the other hand, the configuration of both intermediate fusion models is constituted by a single FCL, equivalent to the last layer of the imaging model, i.e., the classification layer, and it seems that this network was not able to fully capture the relationship between the clinical data and the features of the CT volumes, assuming its existence.

Limitations

When analyzing the results presented in the literature, existing methods can reach performance metrics above 0.90 [7,9–12]. However, the LIDC dataset is used for the development and evaluation of their proposed methodologies. The usage of this dataset results in these excellent metrics since the data do not represent a realistic view of the clinical context (they contain mostly easier cases) and do not translate the full heterogeneity of lung cancer patterns. Moreover, the LIDC dataset provides nodule contours as a result of the annotation process made by experts, and these nodules are labeled into five malignancy categories that can be further subclassified as benign, malignant and indeterminate. The indeterminate nodules, in some approaches, are discarded, which may lead to higher performance metrics [7]. On the other hand, this study made use of a dataset, the NLST dataset, different than what the vast majority of the proposed algorithms used. The NLST dataset presents cases with more complex lung cancer patterns (that are, therefore, more challenging) and, in addition, it does not provide nodule annotations. The regions of interest of the nodules used in this study were generated in a manual process susceptible to human errors, with some degree of uncertainty regarding the malignancy level. Ultimately, all these factors had an impact on the learning models, resulting in lower performance metrics. Considering the work that uses a mutual dataset [13], the NLST dataset, another two datasets were used by the authors of [13], with a total of 327 participants, whereas this study used a total of 1005 participants from the NLST dataset only, and it is not possible to ascertain if the same patients were used. Moreover, in this work, regions of interest of the nodules were manually generated, which adds another layer of divergence. As such, a comparison between the two works would not be fully equitable.

Additionally, the predictive capability of the clinical features seems to be very limited, which is corroborated by the clinical practice, in which physicians use these data in an initial phase of screening in order to discern patients that may have lung cancer. Afterwards, an initial diagnosis of this pathology is given to those patients through the visual assessment of medical images and subsequently confirmed with biopsy.

4. Conclusions

This work aimed at investigating the combination of more than one type of information for predicting lung cancer, specifically, extracted from CT nodules and clinical data. The study of each modality and the results obtained showed the utmost importance of the imaging data, essential for lung cancer diagnosis. The clinical features used, on the contrary, demonstrated poor predictive capability when used alone, which is understandable, as they are used as complementary information in the clinical context, serving as primary suspicion in the screening stage. The results obtained from the multimodality approaches showed the potential of fusing different data modalities. The future investigation could branch out from the described work, with the possibility of combining different strategies and architectures, such as implementing deep learning approaches for the extraction of features from the clinical data, with the goal of exploiting to its maximum potential the relationship shared between two distinct modalities.

Author Contributions: Conceptualization, J.V.S., P.M., F.S., T.P. and H.P.O.; methodology, J.V.S. and P.M.; data curation, P.M. and J.V.S.; writing—original draft preparation, J.V.S.; writing—review and editing, J.V.S., P.F. and T.P.; supervision, T.P. and H.P.O. All authors have read and agreed to the published version of the manuscript.

Funding: This work is financed by National Funds through the Portuguese funding agency, FCT-Foundation for Science and Technology Portugal, within project LA/P/0063/2020.

Institutional Review Board Statement: The databases used in the experiments of this work ensure, in the correspondent published description article, that the necessary ethical approvals regarding data access were obtained.

Informed Consent Statement: Participants read and signed a consent form that explained the National Lung Screening Trial in detail, including risks and benefits.

Data Availability Statement: The data was obtained from the dataset National Lung Screening Trial [16].

Acknowledgments: The authors thank the National Cancer Institute for access to the NCI data collected by the National Lung Screening Trial (NLST) from CDAS Project Number NLST-615. We thank Inês Neves for identifying the lung nodules on the CT scans of the NLST dataset. Additionally, we thank Component 5—Capitalization and Business Innovation, integrated in the Resilience Dimension of the Recovery and Resilience Plan within the scope of the Recovery and Resilience Mechanism (MRR) of the European Union (EU), framed in the Next Generation EU, for the period 2021–2026, within project HfPT, with reference 41, and National Funds through the Portuguese funding agency, FCT-Foundation for Science and Technology Portugal, a PhD Grant Number: 2021.05767.BD.

Conflicts of Interest: The authors declare no conflict of interest.

References

1. Sung, H.; Ferlay, J.; Siegel, R.L.; Laversanne, M.; Soerjomataram, I.; Jemal, A.; Bray, F. Global cancer statistics 2020: GLOBOCAN estimates of incidence and mortality worldwide for 36 cancers in 185 countries. *CA Cancer J. Clin.* **2021**, *71*, 209–249. [CrossRef] [PubMed]
2. Silva, F.; Pereira, T.; Neves, I.; Morgado, J.; Freitas, C.; Malafaia, M.; Sousa, J.; Fonseca, J.; Negrão, E.; Flor de Lima, B.; et al. Towards Machine Learning-Aided Lung Cancer Clinical Routines: Approaches and Open Challenges. *J. Pers. Med.* **2022**, *12*, 480. [CrossRef] [PubMed]
3. Malhotra, J.; Malvezzi, M.; Negri, E.; La Vecchia, C.; Boffetta, P. Risk factors for lung cancer worldwide. *Eur. Respir. J.* **2016**, *48*, 889–902. [CrossRef] [PubMed]
4. Srkalovic, G. Lung Cancer: Preventable Disease. *Acta Medica Acad.* **2018**, *47*, 39–49. [CrossRef] [PubMed]
5. Zhang, G.; Yang, Z.; Gong, L.; Jiang, S.; Wang, L.; Zhang, H. Classification of lung nodules based on CT images using squeeze-and-excitation network and aggregated residual transformations. *La Radiol. Medica* **2020**, *125*, 374–383. [CrossRef] [PubMed]
6. Huang, S.C.; Pareek, A.; Seyyedi, S.; Banerjee, I.; Lungren, M.P. Fusion of medical imaging and electronic health records using deep learning: A systematic review and implementation guidelines. *NPJ Digit. Med.* **2020**, *3*, 136. [CrossRef] [PubMed]
7. Lyu, J.; Bi, X.; Ling, S.H. Multi-Level Cross Residual Network for Lung Nodule Classification. *Sensors* **2020**, *20*, 2837. [CrossRef] [PubMed]
8. Armato III, S.G.; McLennan, G.; Bidaut, L.; McNitt-Gray, M.F.; Meyer, C.R.; Reeves, A.P.; Zhao, B.; Aberle, D.R.; Henschke, C.I.; Hoffman, E.A.; et al. The Lung Image Database Consortium (LIDC) and Image Database Resource Initiative (IDRI): A Completed Reference Database of Lung Nodules on CT Scans. *Med. Phys.* **2011**, *38*, 915–931. [CrossRef] [PubMed]
9. Calheiros, J.L.L.; de Amorim, L.B.V.; de Lima, L.L.; Filho, A.F.L.; Júnior, J.R.F.; de Oliveira, M.C. The Effects of Perinodular Features on Solid Lung Nodule Classification. *J. Digit. Imaging* **2021**, *34*, 798–810. [CrossRef] [PubMed]
10. Al-Shabi, M.; Shak, K.; Tan, M. 3D axial-attention for lung nodule classification. *Int. J. Comput. Assist. Radiol. Surg.* **2021**, *16*, 1319–1324. [CrossRef] [PubMed]
11. Kailasam, S.P.; Sathik, M.M. A Novel Hybrid Feature Extraction Model for Classification on Pulmonary Nodules. *Asian Pac. J. Cancer Prev.* **2019**, *20*, 457–468. [CrossRef] [PubMed]
12. Liu, D.; Liu, F.; Tie, Y.; Qi, L.; Wang, F. Res-trans networks for lung nodule classification. *Int. J. Comput. Assist. Radiol. Surg.* **2022**, *17*, 1059–1068. [CrossRef] [PubMed]
13. Uthoff, J.M.; Mott, S.L.; Larson, J.; Neslund-Dudas, C.M.; Schwartz, A.G.; Sieren, J.C.; Investigators, C. Computed Tomography Features of Lung Structure Have Utility for Differentiating Malignant and Benign Pulmonary Nodules. *Chronic Obstr. Pulm. Dis. J. COPD Found.* **2022**, *9*, 154–164. [CrossRef] [PubMed]

14. Regan, E.A.; Hokanson, J.E.; Murphy, J.R.; Make, B.; Lynch, D.A.; Beaty, T.H.; Curran-Everett, D.; Silverman, E.K.; Crapo, J.D. Genetic Epidemiology of COPD (COPDGene) Study Design. *COPD J. Chronic Obstr. Pulm. Dis.* **2011**, *7*, 32–43. [CrossRef] [PubMed]
15. Schwartz, A.G.; Lusk, C.M.; Wenzlaff, A.S.; Watza, D.; Pandolfi, S.; Mantha, L.; Cote, M.L.; Soubani, A.O.; Walworth, G.; Wozniak, A.; et al. Risk of Lung Cancer Associated with COPD Phenotype Based on Quantitative Image Analysis. *Cancer Epidemiol. Biomark. Prev.* **2016**, *25*, 1341–1347. [CrossRef] [PubMed]
16. National Lung Screening Trial Team; Aberle, D.R.; Berg, C.D.; Black, W.C.; Church, T.R.; Fagerstrom, R.M.; Galen, B.; Gareen, I.F.; Gatsonis, C.; Goldin, J.; et al. The National Lung Screening Trial: Overview and Study Design. *Radiology* **2011**, *258*, 243–253. [CrossRef] [PubMed]
17. Gong, J.; Liu, J.y.; Jiang, Y.j.; Sun, X.w.; Zheng, B.; Nie, S.d. Fusion of quantitative imaging features and serum biomarkers to improve performance of computer-aided diagnosis scheme for lung cancer: A preliminary study. *Med. Phys.* **2018**, *45*, 5472–5481. [CrossRef] [PubMed]
18. Stahlschmidt, S.R.; Ulfenborg, B.; Synnergren, J. Multimodal deep learning for biomedical data fusion: A review. *Briefings Bioinform.* **2022**, *23*, bbab569. [CrossRef] [PubMed]
19. Huang, J.; Ling, C.X. Using AUC and accuracy in evaluating learning algorithms. *IEEE Trans. Knowl. Data Eng.* **2005**, *17*, 299–310. [CrossRef]
20. Bayoudh, K.; Knani, R.; Hamdaoui, F.; Mtibaa, A. A survey on deep multimodal learning for computer vision: Advances, trends, applications, and datasets. *Vis. Comput.* **2022**, *38*, 2939–2970. [CrossRef] [PubMed]

Disclaimer/Publisher's Note: The statements, opinions and data contained in all publications are solely those of the individual author(s) and contributor(s) and not of MDPI and/or the editor(s). MDPI and/or the editor(s) disclaim responsibility for any injury to people or property resulting from any ideas, methods, instructions or products referred to in the content.

Communication

C2RL: Convolutional-Contrastive Learning for Reinforcement Learning Based on Self-Pretraining for Strong Augmentation

Sanghoon Park [1], Jihun Kim [1], Han-You Jeong [2], Tae-Kyoung Kim [3] and Jinwoo Yoo [4,*]

[1] Graduate School of Automotive Engineering, Kookmin University, Seoul 02707, Republic of Korea; ppp7326@naver.com (S.P.); wkdrns3847@gmail.com (J.K.)
[2] Department of Electrical Engineering, Pusan National University, Busan 46241, Republic of Korea; hyjeong@pusan.ac.kr
[3] Department of Electronic Engineering, Gachon University, Seongnam 13120, Republic of Korea; tkkim@gachon.ac.kr
[4] Department of Automobile and IT Convergence, Kookmin University, Seoul 02707, Republic of Korea
* Correspondence: jwyoo@kookmin.ac.kr

Abstract: Reinforcement learning agents that have not been seen during training must be robust in test environments. However, the generalization problem is challenging to solve in reinforcement learning using high-dimensional images as the input. The addition of a self-supervised learning framework with data augmentation in the reinforcement learning architecture can promote generalization to a certain extent. However, excessively large changes in the input images may disturb reinforcement learning. Therefore, we propose a contrastive learning method that can help manage the trade-off relationship between the performance of reinforcement learning and auxiliary tasks against the data augmentation strength. In this framework, strong augmentation does not disturb reinforcement learning and instead maximizes the auxiliary effect for generalization. Results of experiments on the DeepMind Control suite demonstrate that the proposed method effectively uses strong data augmentation and achieves a higher generalization than the existing methods.

Keywords: deep reinforcement learning; self-supervised learning; contrastive learning; generalization; data augmentation; network randomization

1. Introduction

Since the advent of AlphaGo, the potential of deep reinforcement learning has been demonstrated, and it has been applied in various fields, such as autonomous driving and automated robots. As Figure 1 shows, the combination of reinforcement learning and deep neural networks allows control tasks to be performed using high-dimensional observations, such as, images [1]. Notable successes include learning to play various games from raw images (board games [2] and video games [3,4]), controlling a car from a camera frame in the virtual environment [5], solving complicated problems from camera observations [6–8], and picking up objects in the real world [9].

However, the use of high dimensional observations, such as raw images, may lead to sample inefficiency [10,11]. In other words, learning the same number of steps shows a lower performance when using images rather than using a low-dimensional state vector. Among many studies, CURL increases the sample efficiency by learning the similarity between the input frames through contrastive learning, which is a self-supervised learning method that learns to extract richer representation from images while contrasting the query and key [12]. However, due to overfitting in the training environment, the reinforcement learning performance deteriorates even with minor background changes in the test environment that do not affect the action selection. In other words, in the unseen environment that is semantically similar to the seen environment, the improvement in the sample efficiency through contrastive learning is not guaranteed, and this is called a generalization problem in vision-based deep reinforcement learning [13,14].

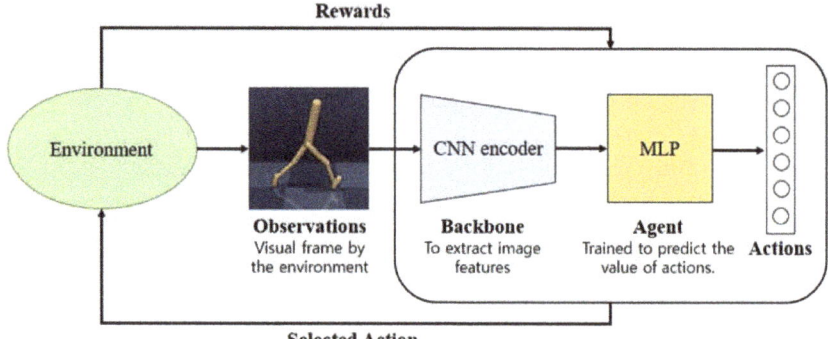

Figure 1. Vision-based reinforcement learning architecture.

Input image data are typically augmented to ensure a robust performance even in environments that the model has not observed [15]. Learning from various input distributions through augmentation can help prevent over-fitting in the training environment. In addition, data augmentation is essentially used for contrastive learning. Stronger data augmentation results in more effective contrastive learning, the auxiliary task of reinforcement learning, and generalization. However, the use of strong augmentation is limited because a large change in the input frame disturbs the downstream task (here, via reinforcement learning) [16]. By preventing the adverse effect of strong augmentation on reinforcement learning, the benefits of contrastive learning can be maximized, and generalization performance can be enhanced.

To improve the generalization of vision-based reinforcement learning, we propose a convolutional–contrastive learning for reinforcement learning (C2RL): a simple architecture that can be added to most reinforcement learning frameworks. Furthermore, we propose a self-pretraining method to overcome the trade-off associated with the augmentation strength and use strong augmentation for both reinforcement learning and contrastive learning without performance degradation. (i) Until the initial steps of the training stage, reinforcement learning and contrastive learning are performed without strong augmentation, such as random convolution. (ii) After training the encoder through self-pretraining, strong data augmentation, such as random convolution, is applied to the input frame and reinforcement, and contrastive learning is continued for the remaining training period. (iii) Although the input data significantly change due to strong augmentation (random convolution), robust feature extraction is possible, which does not significantly degrade the performance of reinforcement learning. (iv) Contrastive learning can induce a greater auxiliary effect on reinforcement learning due to strong augmentation.

One of the greatest contributions of this study is that strong augmentation is used more effectively in our method than when the same strong augmentation is applied consistently throughout training. Furthermore, our study introduces a new attempt on how to efficiently use image data in reinforcement learning. None of the existing studies have focused on contrastive learning using random convolution, despite its potential in achieving a stronger auxiliary effect. Experiments are performed in two modes of the DeepMind Control (DMControl) suite, as shown in Figure 2. The proposed approach significantly outperforms the existing generalization methods in both statically and dynamically changing test environments.

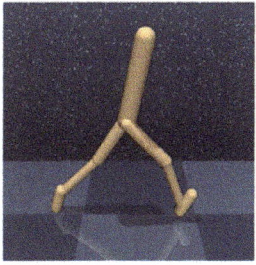
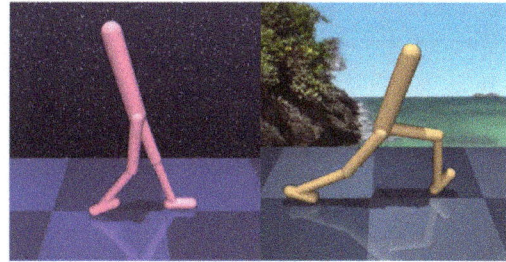

Figure 2. **Left:** Training environment (seen environment) of DMControl. **Right:** Test environments (unseen environments) of DMControl generalization benchmark (Color-hard and Video-easy mode) [17].

2. Related Work
2.1. Soft Actor Critic (SAC)

For continuous control from raw images, we use the SAC, which is a state-of-the-art, off-policy reinforcement learning algorithm that maximizes the expected sum of rewards [18]. The agent outputs action a_t from frame observations o_t, which are stored as transitions in the replay buffer D with reward r_t. The parameters of the SAC are ψ of the state value function V_ψ, θ of the soft Q-function Q_θ, and ϕ of policy π_ϕ. To learn a critic Q_θ, the critic parameters are trained by minimizing the Bellman error using transitions sampled from replay buffer D:

$$J_{Q_\theta} = E_{(o_t, a_t) \sim D} \left[\left(Q_\theta(o_t, a_t) - \left(r_t + \gamma V_\psi(o_{t+1}) \right) \right)^2 \right] \qquad (1)$$

The state value is estimated by sampling an action from the current policy π_ϕ, and $\overline{Q}_\theta$ denotes an exponential moving average of the critic network:

$$V_\psi(o_{t+1}) = E_{a' \sim \pi_\phi} \left[\left(\overline{Q}_\theta(o_{t+1}, a') - \alpha \log \pi_\phi(a'| o_{t+1}) \right] \qquad (2)$$

The policy parameter ϕ is trained by minimizing the divergence from the exponential of the soft-Q function, and α is a temperature parameter for the stochasticity of the optimal policy:

$$J_{\pi_\phi} = -E_{a_t \sim \pi_\phi} \left[\left(Q_\theta(o_t, a_t) - \alpha \log \pi_\phi(a_t|o_t) \right] \qquad (3)$$

2.2. Self-Supervised Learning

Self-supervised learning, an unsupervised learning strategy, is aimed at learning pretext tasks to improve the downstream task performance [19,20]. The trained model can extract rich representations from unlabeled data by learning appropriate pretext tasks that can facilitate downstream tasks, such as classification, object detection, or reinforcement learning, and can utilize them through transfer learning [21]. Recently, self-supervised learning models, such as MoCo [22], SimCLR [23], BYOL [24], and BERT [25], have made great advancements in natural language processing and computer vision tasks, and have also been actively applied to vision-based reinforcement learning.

Self-supervised learning can be divided into several types according to the pretext task. Among them, contrastive learning is a self-supervised learning method aimed at increasing the similarity between positive image pairs and decreasing the similarity between negative image pairs [26]. As shown in Figure 3, to define the positive and negative pairs, the input image is randomly augmented twice with each image acting as the query and key image. Based on the query, the key augmented from the same image is defined as the positive pair, and keys augmented from other images are defined as negative pairs. Contrastive learning allows a query encoder to extract rich representation vectors from unlabeled images, thereby improving the performance of downstream tasks such as reinforcement learning. In our study, InfoNCE is used as the loss function for contrastive learning. In

Equation (4), q is the query for contrast; k_+ and k_i are the positive and negative keys, respectively; and W is a matrix for bilinear products [27]. Through the log loss of a K-way softmax classifier with label k_+, the encoder can learn embeddings to determine the similarity between the query and keys.

$$L_{NCE} = \log \frac{exp(q^T W k_+)}{exp(q^T W k_+) + \sum_{i=0}^{K-1} exp(q^T W k_i)} \quad (4)$$

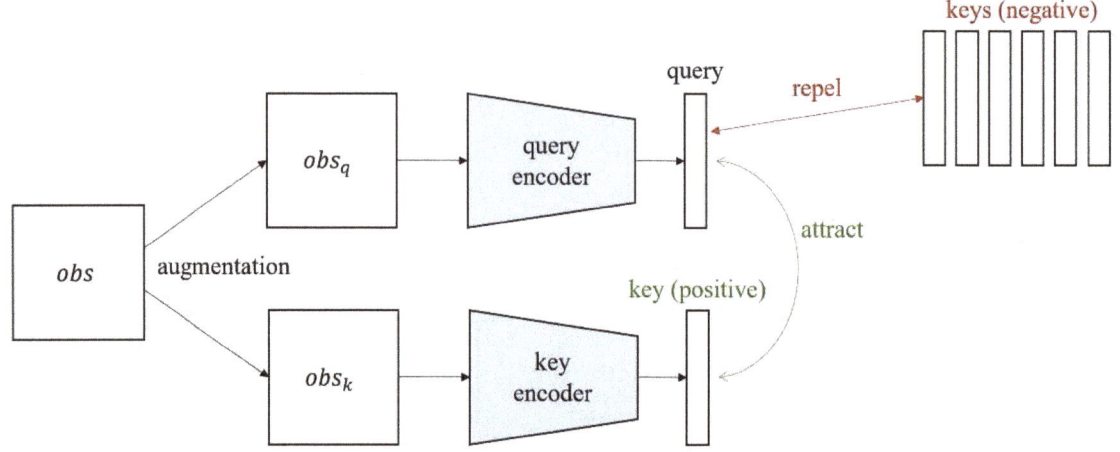

Figure 3. Conventional contrastive learning architecture.

2.3. Network Randomization

Random networks have been used to improve the various performance metrics associated with deep reinforcement learning. For example, researchers focusing on ensemble-based approaches used random networks to improve the uncertainty estimation and exploration of deep reinforcement learning [28]. Moreover, in unexplored state recognition tasks, randomly initialized neural networks were used to define intrinsic rewards for unexplored state visits [29]. In this study, we use a random network for improving the generalization in vision-based reinforcement learning. The input image is randomized by a single layer CNN with a kernel size of 3. Additionally, its output is padded in order to be in the same dimension as the input. For every training iteration, parameter ω is reinitialized with a prior distribution, such as Xavier normal distribution [30].

$$obs_{conv} = f_\omega(obs_{origin}) \quad (5)$$

When input images pass through a convolutional layer that is randomly initialized in every iteration of reinforcement learning, agents can be trained to be more invariant to the unseen environment. In other words, augmented images, as shown in Figure 4, can significantly improve the generalization of reinforcement learning as they vary the visual patterns of the input data and provide various perturbed low-level features, such as the color, shape, or texture [30]. Although strong data augmentation, such as random convolution, can improve the auxiliary effect on generalization, it cannot be applied independently because it significantly changes the distribution of images, resulting in instability and performance degradation of reinforcement learning.

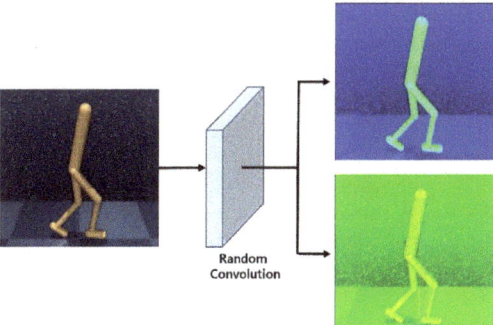

Figure 4. Example of a random convolution process.

3. Proposed Convolutional–Contrastive Learning for RL (C2RL)

This section describes C2RL, which is a simple, convolutional–contrastive learning architecture that can be attached to reinforcement learning frameworks. First, we describe convolutional–contrastive learning: a novel method to enhance the generalization of vision-based reinforcement learning. Subsequently, we introduce a training method that prevents strong augmentation from degrading the performance of reinforcement learning and maximizes the improvement in the generalization performance in unseen test environments.

3.1. Randomized Input Observation

The agent is trained using randomized input observations. To randomize the input observation, a single-layer convolutional neural network is added to the front of the feature extractor as a random network. In each iteration, the parameters of the random network are reinitialized along the Xavier normal distribution [31]. Through the use of the random network, the output has the same dimensions as the input, and various observations with different patterns are generated.

Image Blending

To prevent the loss of visual information due to excessive changes in the input image, we blend the image that passes through the random convolutional layer and the original image in a certain proportion, as shown in Figure 5. The image blending ratio is set through parameter α.

$$obs = \alpha \times obs_{origin} + (1 - \alpha) \times obs_{conv} \quad \dots \quad (0 \leq \alpha \leq 1) \qquad (6)$$

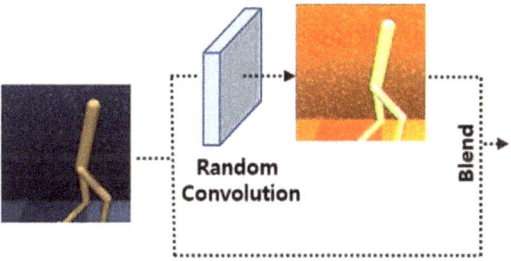

Figure 5. Principle of blending original and randomized images.

3.2. Strong Convolutional–Contrastive Learning

Equation (6) indicates that as α increases, the blending ratio of the original image increases, and convolutional–contrastive learning cannot achieve a sufficient auxiliary effect for the generalization performance. In contrast, when α is small, the large change in the input may confuse reinforcement learning. We introduce a learning method to overcome the trade-off associated with data augmentation strength and effectively exploit strong data augmentation. The training process is divided into two phases, as described in the following subsections.

3.2.1. Self-Pretraining for Strong Augmentation

In the initial stage of training, random convolution is not applied to the input image. Similar to CURL [12], the query and key representation vectors generated through the encoders are used for reinforcement learning and contrastive learning. As shown in Figure 6, no random convolutional layer is added, and the encoders are trained using only weak data augmentation for contrastive learning. After this self-pretraining process, the agent can use the strongly augmented image more efficiently. Unlike those in normal pretraining, data are self-generated in self-pretraining.

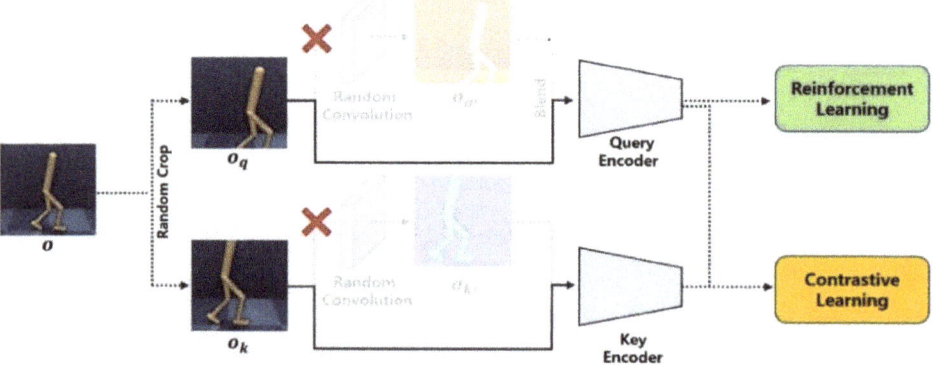

Figure 6. Reinforcement learning and contrastive learning without the random convolution.

3.2.2. Convolutional–Contrastive Learning Strategy for Reinforcement Learning

After self-pretraining in the early steps of training, a single, random, convolution layer is added to the front of the encoder to induce strong data augmentation as shown in Figure 7. Although strong augmentation is used only during the remaining time, the proposed approach outperforms the training methods that consistently use the same strong augmentation in all stages of training.

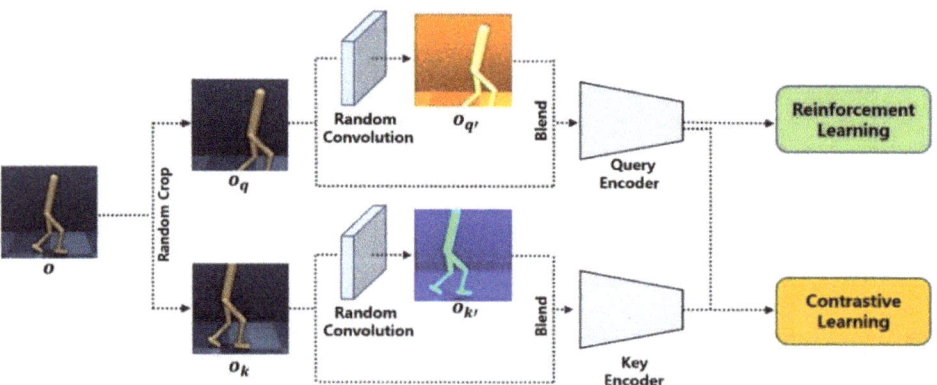

Figure 7. Reinforcement learning and contrastive learning with the random convolution.

4. Results

The objective of the proposed approach is to maximize the generalization effect through strong convolution–contrastive learning by preventing the performance degradation of reinforcement learning, owing to the strong augmentation. To evaluate the generalization performance, we compare the scores in various unseen test environments after training the agent via 500 k steps in DMControl [17]. Following the settings of PAD [32], we measure the generalization performance in the two types of test environments, i.e., those involving statically changing background (color-hard mode) and dynamically changing background (video-easy mode). We compare the test scores for the proposed augmentation methods of convolutional–contrastive learning and existing generalization methods. The test score is the average of episode returns obtained using 10 random seeds for each environment. Self-pretraining is performed for 200 k of the 500 k training steps.

4.1. Augmentation Methods for Convolutional–Contrastive Learning

We study the effect of various image blending parameters of our method(C2RL) on the generalization performance. Figure 8 shows the test scores for the color-hard mode of DMControl walker–walk environment. As shown in Figure 8a–d, a larger blending ratio of images passing through the random network corresponds to a smaller difference between the training score and test score, albeit with lower scores. In contrast, as shown in Figure 8e, the self-pretraining method proposed in Section 3.2 can help achieve higher scores in the test environment, even with considerable blending of the random images. Although the training and test scores are temporarily reduced when strong augmentation is applied after self-pretraining without random convolution, the proposed approach outperforms other methods that use the same augmentation throughout the training process.

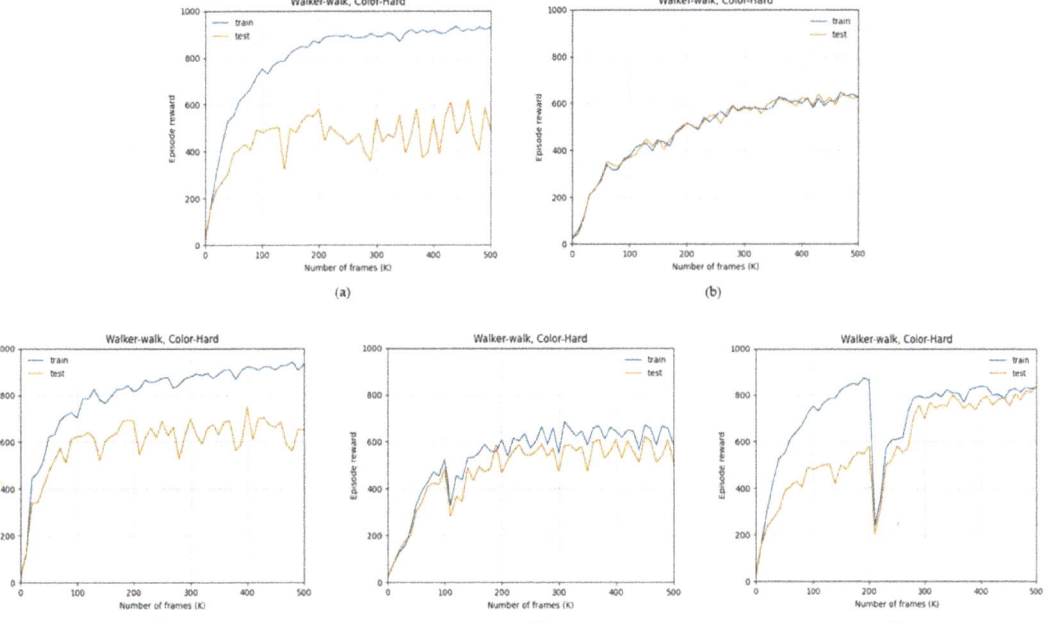

Figure 8. Learning curves on convolutional–contrastive learning. (**a**) uses only original image and (**b**) uses only random image. (**c**,**d**) use blended image with blending parameter α is 0.8 and 0.2 respectively. (**e**) uses blended image with blending parameter α (0.2) after self-pretraining.

Figure 9 shows the results according to the image blending ratio. After self-pretraining, we compare the results by setting the blending ratio α to 0.5, 0.2, and 0, and also shows the best performance at 0.2. If the blending ratio is 0.5, the generalization effect by random convolution is only half-used. However, we find that when the blending ratio is zero, a large change of the image makes reinforcement learning more difficult.

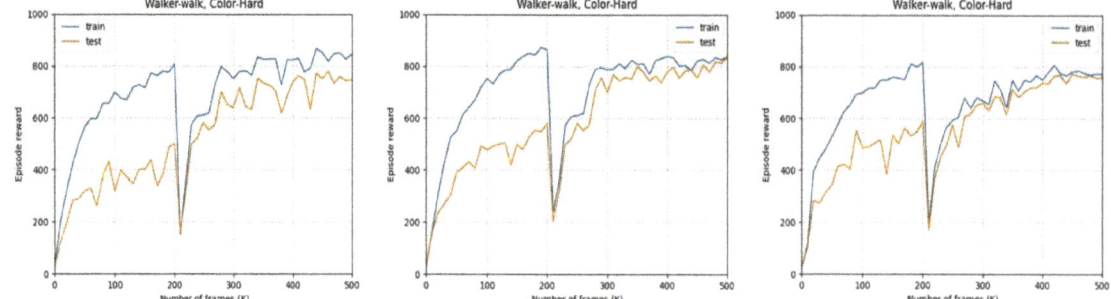

Figure 9. Results for the change in the multiple image blending ratio α after self-pretraining. From the left, 0.5, 0.2 and 0 are used as blending parameters α, respectively.

Moreover, we compare the test scores associated with different blending parameters of C2RL in various unseen environments of DMControl: normal **SAC**; **CURL**: using only weak augmentation(random crop) without random convolution, same as C2RL with α = 1; **C2RL(0.8)**: using a small ratio of random blending (α = 0.8) without self-pretraining; **C2RL(0.2)**: using a large ratio of random blending (α = 0.2) without self-pretraining; and **C2RL(+SP)**: C2RL(0.2) with self-pretraining. As shown in Tables 1 and 2, the highest score is obtained when self-pretraining is used in both modes of DMControl. In other words, self-pretraining allows strong data augmentation to be used efficiently for reinforcement learning and contrastive learning.

Table 1. Test scores for different augmentation methods in the DMControl color-hard mode.

Color-Hard	SAC	CURL	C2RL(0.8)	C2RL(0.2)	C2RL(+SP)
Walker, walk	414 ± 74	445 ± 99	707 ± 43	617 ± 46	899 ± 15
Walker, stand	719 ± 74	662 ± 54	874 ± 46	912 ± 27	954 ± 16
Cartpole, swingup	592 ± 50	454 ± 110	790 ± 59	375 ± 39	794 ± 20
Cartpole, balance	857 ± 60	782 ± 13	921 ± 15	970 ± 22	978 ± 12
Ball in cup, catch	411 ± 183	231 ± 92	713 ± 166	713 ± 93	893 ± 44
Finger, turn_easy	270 ± 43	202 ± 32	438 ± 95	454 ± 133	464 ± 111
Cheetah, run	154 ± 41	202 ± 22	251 ± 33	274 ± 13	292 ± 5
Reacher, easy	163 ± 45	325 ± 32	317 ± 67	212 ± 91	332 ± 61

Table 2. Test scores for different augmentation methods in the DMControl video-easy mode.

Video-Easy	SAC	CURL	C2RL(0.8)	C2RL(0.2)	C2RL(+SP)
Walker, walk	616 ± 80	556 ± 133	784 ± 34	689 ± 46	948 ± 15
Walker, stand	899 ± 53	852 ± 75	766 ± 47	891 ± 35	969 ± 23
Cartpole, swingup	375 ± 90	404 ± 67	589 ± 44	415 ± 38	600 ± 16
Cartpole, balance	693 ± 109	850 ± 91	926 ± 13	942 ± 18	948 ± 12
Ball in cup, catch	393 ± 175	316 ± 119	692 ± 85	643 ± 93	747 ± 79
Finger, turn_easy	355 ± 108	248 ± 56	461 ± 188	367 ± 154	421 ± 143
Cheetah, run	194 ± 30	154 ± 50	287 ± 21	234 ± 32	265 ± 24

4.2. Comparison with Existing Reinforcement Learning Networks

We compare the proposed approach with state-of-the-art methods of vision-based reinforcement learning; **CURL [12]**: a contrastive learning method using only weak augmentation (random crop) for reinforcement learning, same as C2RL with α = 1; **RAD [33]**: introduces two new data augmentations, i.e., random translate and random amplitude scale; **DrQ [34]**: uses value function regularization through data augmentation; **PAD [32]**: a self-supervised learning method for policy adaptation during the test. As shown in

Tables 3 and 4, in all environments of DMControl, the proposed method outperforms the state-of-the-art methods.

Table 3. Learning curves for various augmentation strategies (Color-hard).

Color-Hard	CURL	RAD	DrQ	PAD	C2RL + SP (Ours)
Walker, walk	445 ± 99	400 ± 61	520 ± 91	468 ± 47	899 ± 15
Walker, stand	662 ± 54	644 ± 88	770 ± 71	797 ± 46	954 ± 16
Cartpole, swingup	454 ± 110	590 ± 53	586 ± 52	630 ± 63	794 ± 20
Ball in cup, catch	231 ± 92	541 ± 29	365 ± 210	563 ± 50	893 ± 44

Table 4. Learning curves for various augmentation strategies (Video-easy).

Video-Easy	CURL	RAD	DrQ	PAD	C2RL + SP (Ours)
Walker, walk	556 ± 133	606 ± 63	682 ± 89	717 ± 79	948 ± 15
Walker, stand	852 ± 75	745 ± 146	873 ± 83	935 ± 20	969 ± 23
Cartpole, swingup	404 ± 67	373 ± 72	485 ± 105	521 ± 76	600 ± 16
Ball in cup, catch	316 ± 119	481 ± 26	318 ± 157	436 ± 55	747 ± 19

5. Conclusions

This paper proposes a novel, self-supervised learning method named C2RL, which allows the agent to use strong augmented images as the input. Self-pretraining without strong augmentation allows the agents to be trained by efficiently using strong data augmentation. Experimental results on the DMControl suite show that using part of the training process for self-pretraining, without strong augmentation, can promote the more efficient use of strong data augmentation, such as random convolution compared with that when the same strong data augmentation is used throughout the training. Moreover, the proposed method outperforms the state-of-the-art methods in extracting robust visual representations.

Author Contributions: Conceptualization and formal analysis, S.P.; investigation and validation, J.K.; methodology and software H.-Y.J.; software and writing, T.-K.K. and J.Y. All authors have read and agreed to the published version of the manuscript.

Funding: This work was supported by the National Research Foundation of Korea (NRF) grants funded by the Korean government (MSIT) (NRF-2021R1A5A1032937).

Institutional Review Board Statement: Not applicable.

Informed Consent Statement: Not applicable.

Data Availability Statement: Not applicable.

Conflicts of Interest: The authors declare no conflict of interest.

References

1. Mnih, V.; Kavukcuoglu, K.; Silver, D.; Graves, A.; Antonoglou, I.; Wierstra, D.; Riedmiller, M. Playing atari with deep reinforcement learning. *arXiv* **2013**, arXiv:1312.5602.
2. Silver, D.; Hubert, T.; Schrittwieser, J.; Antonoglou, I.; Lai, M.; Guez, A.; Lanctot, M.; Sifre, L.; Kumaran, D.; Graepel, T.; et al. A general reinforcement learning algorithm that masters chess, shogi, and Go through self-play. *Science* **2018**, *362*, 1140–1144. [CrossRef] [PubMed]
3. Mnih, V.; Kavukcuoglu, K.; Silver, D.; Rusu, A.A.; Veness, J.; Bellemare, M.G.; Graves, A.; Riedmiller, M.; Fidjeland, A.K.; Ostrovski, G.; et al. Human-level control through deep reinforcement learning. *Nature* **2015**, *518*, 529–533. [CrossRef]
4. Vinyals, O.; Ewalds, T.; Bartunov, S.; Georgiev, P.; Vezhnevets, A.S.; Yeo, M.; Makhzani, A.; Küttler, H.; Agapiou, J.; Schrittwieser, J.; et al. Starcraft ii: A new challenge for reinforcement learning. *arXiv* **2017**, arXiv:1708.04782.
5. Lillicrap, T.P.; Hunt, J.J.; Pritzel, A.; Heess, N.; Erez, T.; Tassa, Y.; Silver, D.; Wierstra, D. Continuous control with deep reinforcement learning. *arXiv* **2015**, arXiv:1509.02971.
6. Jaderberg, M.; Mnih, V.; Czarnecki, W.M.; Schaul, T.; Leibo, J.Z.; Silver, D.; Kavukcuoglu, K. Reinforcement learning with unsupervised auxiliary tasks. *arXiv* **2016**, arXiv:1611.05397.

7. Espeholt, L.; Soyer, H.; Munos, R.; Simonyan, K.; Mnih, V.; Ward, T.; Doron, Y.; Firoiu, V.; Harley, T.; Dunning, I.; et al. Impala: Scalable distributed deep-rl with importance weighted actor-learner architectures. In Proceedings of the International Conference on Machine Learning, PMLR, Stockholm, Sweden, 10–15 July 2018.
8. Jaderberg, M.; Czarnecki, W.M.; Dunning, I.; Marris, L.; Lever, G.; Castaneda, A.G.; Beattie, C.; Rabinowitz, N.C.; Morcos, A.S.; Ruderman, A.; et al. Human-level performance in 3D multiplayer games with population-based reinforcement learning. *Science* **2019**, *364*, 859–865. [CrossRef]
9. Kalashnikov, D.; Irpan, A.; Pastor, P.; Ibarz, J.; Herzog, A.; Jang, E.; Quillen, D.; Holly, E.; Kalakrishnan, M.; Vanhoucke, V.; et al. Scalable deep reinforcement learning for vision-based robotic manipulation. In Proceedings of the Conference on Robot Learning, PMLR, Zürich, Switzerland, 29–31 October 2018.
10. Lake, B.M.; Ullman, T.D.; Tenenbaum, J.B.; Gershman, S.J. Building machines that learn and think like people. *Behav. Brain Sci.* **2017**, *40*, e253. [CrossRef]
11. Kaiser, L.; Babaeizadeh, M.; Milos, P.; Osinski, B.; Campbell, R.H.; Czechowski, K.; Erhan, D.; Finn, C.; Kozakowski, P.; Levine, S.; et al. Model-based reinforcement learning for atari. *arXiv* **2019**, arXiv:1903.00374.
12. Laskin, M.; Srinivas, A.; Abbeel, P. Curl: Contrastive unsupervised representations for reinforcement learning. In Proceedings of the International Conference on Machine Learning, PMLR, Virtual, 13–18 July 2020.
13. Zhang, C.; Vinyals, O.; Munos, R.; Bengio, S. A study on overfitting in deep reinforcement learning. *arXiv* **2018**, arXiv:1804.06893.
14. Cobbe, K.; Klimov, O.; Hesse, C.; Kim, T.; Schulman, J. Quantifying generalization in reinforcement learning. In Proceedings of the International Conference on Machine Learning, PMLR, Long Beach, CA, USA, 9–15 June 2019.
15. Ma, G.; Wang, Z.; Yuan, Z.; Wang, X.; Yuan, B.; Tao, D. A comprehensive survey of data augmentation in visual reinforcement learning. *arXiv* **2022**, arXiv:2210.04561.
16. Hansen, N.; Wang, X. Generalization in reinforcement learning by soft data augmentation. In Proceedings of the 2021 IEEE International Conference on Robotics and Automation (ICRA), Xi'an, China, 30 May–5 June 2021; IEEE: Piscataway, NJ, USA, 2021.
17. Tassa, Y.; Doron, Y.; Muldal, A.; Erez, T.; Li, Y.; Casas, D.D.; Budden, D.; Abdolmaleki, A.; Merel, J.; Lefrancq, A.; et al. Deepmind control suite. *arXiv* **2018**, arXiv:1801.00690.
18. Haarnoja, T.; Zhou, A.; Abbeel, P.; Levine, S. Soft actor-critic: Off-policy maximum entropy deep reinforcement learning with a stochastic actor. In Proceedings of the International Conference on Machine Learning, PMLR, Stockholm, Sweden, 10–15 July 2018.
19. Doersch, C.; Gupta, A.; Efros, A.A. Unsupervised visual representation learning by context prediction. In Proceedings of the IEEE International Conference on Computer Vision, Santiago, Chile, 7–13 December 2015.
20. Zhang, R.; Yang, S.; Zhang, Q.; Xu, L.; He, Y.; Zhang, F. Graph-based few-shot learning with transformed feature propagation and optimal class allocation. *Neurocomputing* **2022**, *470*, 247–256. [CrossRef]
21. Ding, B.; Zhang, R.; Xu, L.; Liu, G.; Yang, S.; Liu, Y.; Zhang, Q. U^2D^2 Net: Unsupervised Unified Image Dehazing and Denoising Network for Single Hazy Image Enhancement. *IEEE Trans. Multimed.* **2023**, 1–16. [CrossRef]
22. He, K.; Fan, H.; Wu, Y.; Xie, S.; Girshick, R. Momentum contrast for unsupervised visual representation learning. In Proceedings of the IEEE/CVF Conference on Computer Vision and Pattern Recognition, Seattle, WA, USA, 13–19 June 2020.
23. Chen, T.; Kornblith, S.; Norouzi, M.; Hinton, G. A simple framework for contrastive learning of visual representations. In Proceedings of the International Conference on Machine Learning, PMLR, Virtual, 13–18 July 2020.
24. Grill, J.B.; Strub, F.; Altché, F.; Tallec, C.; Richemond, P.; Buchatskaya, E.; Doersch, C.; Avila Pires, B.; Guo, Z.; Gheshlaghi Azar, M.; et al. Bootstrap your own latent-a new approach to self-supervised learning. *Adv. Neural Inf. Process. Syst.* **2020**, *33*, 21271–21284.
25. Devlin, J.; Chang, M.W.; Lee, K.; Toutanova, K. Bert: Pre-training of deep bidirectional transformers for language understanding. *arXiv* **2018**, arXiv:1810.04805.
26. Wu, Z.; Xiong, Y.; Yu, S.X.; Lin, D. Unsupervised feature learning via non-parametric instance discrimination. In Proceedings of the IEEE Conference on Computer Vision and Pattern Recognition, Salt Lake City, UT, USA, 18–23 June 2018.
27. Oord, A.V.; Li, Y.; Vinyals, O. Representation learning with contrastive predictive coding. *arXiv* **2018**, arXiv:1807.03748.
28. Osband, I.; Aslanides, J.; Cassirer, A. Randomized prior functions for deep reinforcement learning. *Adv. Neural Inf. Process. Syst.* **2018**, *31*, 8626–8638.
29. Burda, Y.; Edwards, H.; Storkey, A.; Klimov, O. Exploration by random network distillation. *arXiv* **2018**, arXiv:1810.12894.
30. Lee, K.; Lee, K.; Shin, J.; Lee, H. Network randomization: A simple technique for generalization in deep reinforcement learning. *arXiv* **2019**, arXiv:1910.05396.
31. Glorot, X.; Bengio, Y. Understanding the difficulty of training deep feedforward neural networks. In Proceedings of the Thirteenth International Conference on Artificial Intelligence and Statistics. JMLR Workshop and Conference Proceedings, Sardinia, Italy, 13–15 May 2010.
32. Hansen, N.; Jangir, R.; Sun, Y.; Alenyà, G.; Abbeel, P.; Efros, A.A.; Pinto, L.; Wang, X. Self-supervised policy adaptation during deployment. *arXiv* **2020**, arXiv:2007.04309.

63. Laskin, M.; Lee, K.; Stooke, A.; Pinto, L.; Abbeel, P.; Srinivas, A. Reinforcement learning with augmented data. *Adv. Neural Inf. Process. Syst.* **2020**, *33*, 19884–19895.
64. Kostrikov, I.; Yarats, D.; Fergus, R. Image augmentation is all you need: Regularizing deep reinforcement learning from pixels. *arXiv* **2020**, arXiv:2004.13649.

Disclaimer/Publisher's Note: The statements, opinions and data contained in all publications are solely those of the individual author(s) and contributor(s) and not of MDPI and/or the editor(s). MDPI and/or the editor(s) disclaim responsibility for any injury to people or property resulting from any ideas, methods, instructions or products referred to in the content.

Article

TransNet: Transformer-Based Point Cloud Sampling Network

Hookyung Lee [1], Jaeseung Jeon [1], Seokjin Hong [1], Jeesu Kim [2] and Jinwoo Yoo [3,*]

[1] Graduate School of Automotive Engineering, Kookmin University, Seoul 02707, Republic of Korea; gnrud099@gmail.com (H.L.); sing5386@naver.com (J.J.); cheongsu030536@gmail.com (S.H.)
[2] Departments of Cogno-Mechatronics Engineering and Optics and Mechatronics Engineering, Pusan National University, Busan 46241, Republic of Korea; jeesukim@pusan.ac.kr
[3] Department of Automobile and IT Convergence, Kookmin University, Seoul 02707, Republic of Korea
* Correspondence: jwyoo@kookmin.ac.kr

Abstract: As interest in point cloud processing has gradually increased in the industry, point cloud sampling techniques have been researched to improve deep learning networks. As many conventional models use point clouds directly, the consideration of computational complexity has become critical for practicality. One of the representative ways to decrease computations is downsampling, which also affects the performance in terms of precision. Existing classic sampling methods have adopted a standardized way regardless of the task-model property in learning. However, this limits the improvement of the point cloud sampling network's performance. That is, the performance of such task-agnostic methods is too low when the sampling ratio is high. Therefore, this paper proposes a novel downsampling model based on the transformer-based point cloud sampling network (TransNet) to efficiently perform downsampling tasks. The proposed TransNet utilizes self-attention and fully connected layers to extract meaningful features from input sequences and perform downsampling. By introducing attention techniques into downsampling, the proposed network can learn about the relationships between point clouds and generate a task-oriented sampling methodology. The proposed TransNet outperforms several state-of-the-art models in terms of accuracy. It has a particular advantage in generating points from sparse data when the sampling ratio is high. We expect that our approach can provide a promising solution for downsampling tasks in various point cloud applications.

Keywords: deep learning; transformer; self-attention; multi-head attention; point cloud; down sampling; classification; network

1. Introduction

The technology for creating point clouds using 3D sensings, such as RGB-D cameras and LiDAR, is advancing rapidly, and increases in computing speeds and interest in the 3D point cloud field are drawing attention as well [1–3]. This has raised the importance of point clouds in various fields. Point clouds provide a detailed and accurate representation of real-world objects and environments, allowing for the precise measurements, analysis, and manipulation of 3D data. They have numerous applications in fields such as robotics, autonomous vehicles, virtual reality, architecture, and cultural heritage preservation. As the technology for creating point clouds continues to improve, we can anticipate an even greater reliance on these data structures for a wide range of applications.

Because the form of 3D point cloud data differs from that of a typical image or natural language processing (NLP) data, when point cloud research first began, new methods of point cloud generation were needed because point clouds were contained in irregular spaces with varying densities.

Initially, a method was proposed to convert 3D point cloud data into 2D images for processing. This method converts the points of 3D point cloud data into pixels of an image and treats them as images. While it has been successfully applied in the field of image processing, it does not fully reflect the complexity and diversity of 3D point cloud data. Thus, other metrics are needed to process 3D point cloud data.

Initially, projection-based [4,5] and volumetric convolution-based methods [6–8] were proposed to convert each point into a grid to handle a 3D point cloud and perform feature extraction using convolutional layers in the same way as conventional 2D images on the grid. Because these methods convert irregular points in 3D space into a grid format, the number of points in the grid cell is uneven, resulting in the loss of information or wasted calculations in certain cells. To overcome the problems of grid transformation, direct point-based strategies have emerged. Some methods independently model each point using multiple shared multi-layer perceptrons (MLPs) [9–11]. Depending on the type of convolution kernel, 3D convolution methods have emerged [12–16].

Point clouds are being applied to various fields, such as classification, semantic segmentation [17,18], and registration [19], instance segmentation [20,21]. These methods use point cloud data as input and aggregate local features in the last step. While they maintain accurate location information, computational costs increase linearly with the number of points, and processing high-capacity, dense 3D point cloud data remains challenging. Accordingly, a method of sampling data is proposed to reduce the amount of data in the 3D point cloud and improve processing efficiency. Previously, heuristic-based sampling methods, such as task-agnostic random sampling, fast point sampling, and grid voxel sampling, have been used. However, these methods can degrade performance because they lose information or select meaningless data from downstream tasks. Recently, a task-oriented sampling network [22–24] was proposed, allowing the generation of sampling that optimized the performance of downstream tasks. S-Net and SampleNet performed well for specific tasks with sampling strategies using deep learning. In addition, APSNet used the attention-based method to focus on relationships among the points. Still, these models do not fully consider the relationship information between point clouds.

In this paper, we propose a methodology that leverages the complete information from the input sequence to effectively interact with the task model for task-oriented sampling. TransNet is a novel transformer-based model that handles an entire sequence in parallel, capturing a long range of point cloud information and point-to-point interaction information more effectively. Feature extraction is performed by adding the embedding layer of the input and positional encoding. After generating the query, key, and value, it proceeds through the transformer [25] layer with the self-attention mechanism to effectively capture the complex interrelationships among points within the input point cloud data. By focusing on the most informative points, our method can selectively sample only the most relevant areas of the point cloud, thereby improving the efficiency of the network and the accuracy of the output. Through this approach, we can gain a more comprehensive understanding of point cloud data and easily extract meaningful features that are essential for downstream tasks (see Figure 1). Our proposed model has achieved state-of-the-art performance improvements in the field of point cloud classification. In particular, the effect is remarkable for sparse points due to the high sampling ratio. To summarize, our main contributions are threefold:

1. We propose TransNet, a novel self-attention-based point cloud sampling network, as a task-oriented objective.
2. Our approach demonstrates enhanced performance on point cloud tasks, outperforming both task-agnostic and task-oriented methods.
3. This approach effectively addresses the long-range dependency issues that are commonly encountered in point clouds. Thus, it has a notable impact on the sparsely sampled point clouds, where a high sampling ratio is required to effectively capture the underlying geometric structures.

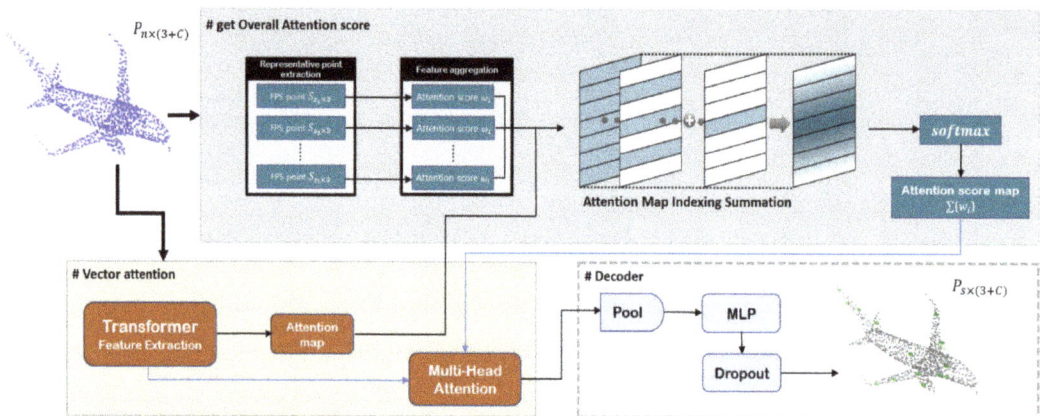

Figure 1. Overview of TransNet. TransNet divides the initial input into two parts. It calculates a comprehensive attention score map for each part (**top**) and vector attention (**lower left**). Then, multi-head attention is performed to understand the relationships between the points, and this process is repeated for training. In the decoder part (**lower right**), pooling and MLP are used to extract the final features. Detailed implementation methods are provided in Sections 3.1 and 3.2.

2. Related Work

Deep learning on point clouds: Deep learning has been applied to various point cloud-related tasks, such as object detection, segmentation, and classification. For example, Qi et al. [9,10] proposed point cloud classification and segmentation and achieved state-of-the-art performance on several benchmark datasets. In addition, a deep learning model was proposed for object detection in point clouds [26,27]. Other models have been proposed for point cloud processing utilizing local features [16] and adaptive convolution operations [28]. Additionally, some studies have explored the use of generative models [29,30] for point cloud generation and reconstruction tasks. One study [31] used a local self-attention mechanism, unlike the global attention scheme used in previous studies. Furthermore, it demonstrated that vector attention methods outperformed scalar attention methods and introduced position encoding methods to properly process location information in point clouds. Although the application was different in this paper, the self-attention technique was encoded by applying the point cloud technique similar to the point transformer [31]. To preserve the location information, positional encoding was utilized, and a decoder was constructed without undergoing multiple stages.

Point cloud sampling: Task-agnostic algorithms, such as random sampling, uniform sampling, farthest point sampling (FPS), and grid sampling, have been widely used in the past. Among them, FPS remains a popular choice in many recent studies [31,32]. While FPS has been widely used, it may not fully consider the downstream tasks for which the sampled points are used, leading to potential performance degradation. Thus, alternative downsampling methods have recently been proposed [22–24]. According to Dovrat et al. [22], the efficiency and accuracy of sampling could be improved through a learnable point cloud sampling method. Lang et al. [23] introduced a novel differentiable relaxation for point cloud sampling. The authors of [24] proposed sampling attention mechanisms to enhance the relationships among points by assigning importance weights, allowing for a more effective sampling process. Our TransNet is a task-oriented sampling method that mitigates long-range dependency while viewing the relationships between points globally and locally.

Transformer and self-attention: Transformer and self-attention models have revolutionized machine translation and NLP [25,33]. Considering this, such methods have been increasingly used in the field of 2D image recognition [34,35]. Inspired by these findings, researchers have also attempted to apply self-attention networks to point cloud data. However, previous studies have utilized global attention on the entire point cloud, which limits their applicability to understanding large-scale 3D scenes due to high computational costs. Recently, Hengshuang et al. [31] developed a highly accurate and scalable self-attention network, specifically for large 3D scenes, using vector attention applied locally. In contrast to prior approaches, we applied a transformer locally to handle the input's point cloud sampling, which has been shown to be highly effective.

Nearest neighbor selection: In recent studies, nearest neighbor (NN) methods have been widely used for information fusion. However, in the context of neural networks, the main drawback is that the selection rule is not differentiable. To address this, Goldberger et al., proposed the probabilistic relaxation of NN rules by defining categorical distributions over a set of candidate neighbors. In our study, we applied self-attention to point clouds using k NNs, allowing for a better grasp of the relationships between the points. Additionally, we incorporated skip connections to consider information from the global area.

Positional encoding: In the domain of deep learning models, positional encoding has been commonly utilized to encode the positional information of input data. With respect to point clouds, previous research has employed basic encoding techniques, such as Cartesian, spherical, and polar coordinates, to incorporate the position information of the points. However, these methods have limitations in terms of information loss and insufficient expressiveness. To address this issue, some studies have employed learned positional encoding techniques to incorporate more informative position information into the model. These techniques usually involve learning a continuous function to represent the position information, which can capture complex spatial relationships and patterns in point clouds.

3. Proposed TransNet

Here, we briefly explain the transformer and self-attention concepts. Transformer and self-attention networks are innovative and have shown impressive results in NLP [25,33] and 2D image analysis [34,35]. Recently, networks have also been applied to 3D point cloud scenes [31,32]. Self-attention can be classified into two types: dot-product attention [25] and vector attention [36]. The standard formula for dot-product attention is as follows:

$$y_i = \sum_{x_j \in X} \tau\left(\alpha(x_i)^T \beta(x_j) + \delta\right) \gamma(x_j) \quad (1)$$

where $x_i \in X$ is a set of feature vectors, x_i and y_i are the input feature and output feature, respectively, α, β, and γ are pointwise feature transformations (e.g., MLP, linear layer), and τ and δ are normalization functions (a *softmax* and a positional encoding function, respectively).

Unlike dot product attention, vector attention measures show similarity by calculating the distance between the input vector and the weight vector:

$$y_i = \sum_{x_j \in X} \tau(\varepsilon(\mu(\alpha(x_i), \beta(x_j))) + \delta) \odot \gamma(x_j) \quad (2)$$

where μ is a relation function (e.g., subtraction, multiplication) and ε is a mapping function (e.g., MLP) that produces attention vectors for feature aggregation.

3.1. Transformer-Based Sampling Layer

Traditional task-oriented sampling methods, such as S-Net [22], SampleNet [23], and APSNet [24], use PointNet models that employ convolution networks to perform feature extraction. Moreover, S-Net and SampleNet generate m points at a time, and APSNet proceeds through the sequential generation process. In this study, we introduce a novel deep-learning sampling model based on self-attention. We processed inputs by defining the query, key, and value without using the convolution network. We used vector attention and the subtraction operation between the query and key. Our vector attention process was as follows:

$$y_i = \sum_{x_j \in X_{knn}} \tau(\varepsilon(\mu(\alpha(x_i), \beta(x_j))) + \delta) \odot \gamma(x_j) \qquad (3)$$

Moreover, taking inspiration from [31], we performed self-attention within a local neighborhood to avoid the high computational costs that arise from global self-attention.

Here, $P \in R^{n \times 3}$ denotes a point cloud that contains a given point cloud n, which is the number of point clouds. We applied feature transformation to the local region selected by KNN to generate value v. We created an attention map of the same size as the value and used the indexing sum to create an attention score map W based on the relationship between representative points (a detailed explanation is given in Section 3.2). Finally, we obtained the attention value by using multi-head attention on the attention score map W and value:

$$\text{Attention value} = \sum_{i=1}^{n} w_i \odot v \qquad (4)$$

3.2. Attention Score Map

We defined FPS as an algorithm for extracting representative points and performed multiple rounds of self-attention on the points extracted by FPS. With this result, we obtained the query and key for each sampled point, proceeded through a subtraction relationship, and then created a similarity in addition to Positional encoding. A description of the figure is shown in Figure 2. This process was repeated for the number of farthest point sampling performed. This resulted in the generation of an attention map that effectively encompassed all the generated values called the scatter sum. We describe the process in detail below.

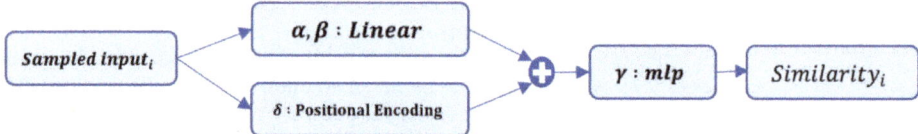

Figure 2. Similarity calculation. Prior to performing value v and multi-head attention, we generated attention scores for the sampled points via FPS and created an attention score map for multiple similarities. Additionally, we incorporated positional encoding to retain positional information.

First, we performed the KNN algorithm to find local features for the initial input and defined the value for local vectors. Then, we created an attention map for the empty space corresponding to the shape of the vector to calculate the distance between the input vector and the weight vector using vector attention. For the points S obtained through the FPS algorithm as $S \in P$, we generated query and key vectors for the representative S points among the N points generated through the FPS algorithm and examined their similarity through the subtraction relationship of the two generated vectors, as described in Equation (5).

$$w_i = \sum_{x_j \in X_{knn}} \alpha(x_i) - \beta(x_j) \qquad (5)$$

We added this to the index corresponding to the attention score map W. We repeated this process for all set ratios R and applied a normalization activation function to the resulting attention score map:

$$W = idx(w_1) + idx(w_2) + \ldots + idx(w_i) \qquad (6)$$

followed by performing multi-head attention with the initially obtained value v. Details are described in Figure 3.

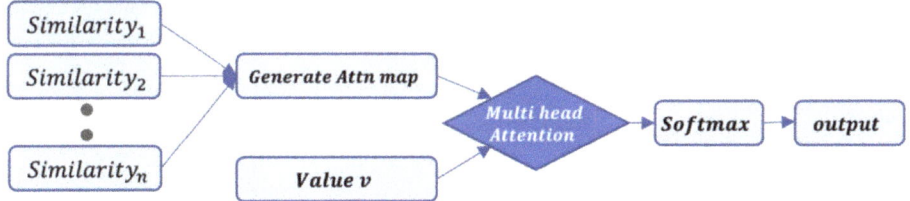

Figure 3. Transformer sampling layer.

To summarize the process before entering the Decoder, the initial point cloud was divided into two steps. In the first step (yellow area in Figure 1), the value v of the input was obtained, and an empty attention map of the same size was created. In the second step (gray area in Figure 1), a representative point was drawn through the FPS, and the query and key were obtained for each of the selected $S_{s_i \times 3}$; the similarity was obtained through this. Subsequently, we added weights to each index of the attention map generated in the first step to create a single attention map, which was then used to perform vector attention with the value v.

By identifying the interrelationships between the points in the input sequence, self-attention facilitated a better understanding of the relationships between each point, resulting in superior performance in handling long-term dependencies. Compared to traditional models that use LSTM [37], our proposed model enables the parallel processing of input sequences, resulting in superior performance, particularly in scenarios with high sampling ratios.

3.3. Decoder

In the decoding stage, a max-pooling operation was conducted to collect the extracted features. To generate the final output, a fully connected network (FCN) was utilized. To mitigate the loss of positional information that was caused by employing a linear MLP and positional encoding was further incorporated. Additionally, dropout was employed as a regularization technique to prevent overfitting. The result contained a number of s points for the task. With this result, we performed multi-task learning. By adopting a multi-task learning approach, all tasks could be efficiently processed within a single model. The concurrent training of these two models enabled us to effectively leverage the shared data distribution and consider more inter-task correlations, thus further improving the model's performance. Further details regarding the model's architecture are illustrated in Figure 1 (lower right).

3.4. Loss

We applied supervised learning and used two types of loss: task loss L_{task} and sampling loss L_{sample}, to train TransNet, where Total loss is the sum of the weights added to these two losses. The sampling loss L_{sample} aimed to minimize the distance between the points sampled from S, and the corresponding points in P, while also ensuring that the sampled points were spread out as much as possible across the original point cloud P.

L_{task} can be defined as the cross-entropy loss for classification. Additionally, the formula for this is as follows:

$$L(\hat{y}, y) = -\sum_{i=1}^{C} y_i \log(\hat{y}_i) \qquad (7)$$

L_{sample} is the sum of two things: average neighbor loss and maximum neighbor loss. Given two-point sets Q_1 and Q_2, the average nearest neighbor loss can be denoted as:

$$L_a(Q_1, Q_2) = \frac{1}{|Q_1|} \sum_{q_1 \in Q_1} \min_{q_2 \in Q_2} ||q_1 - q_2||_2^2 \tag{8}$$

Additionally, maximal nearest neighbor loss can be given as:

$$L_m(Q_1, Q_2) = \max_{q_1 \in Q_1} \min_{q_2 \in Q_2} ||q_1 - q_2||_2 \tag{9}$$

The sampling loss is then given by:

$$L_{sample}(Q, P) = L_a(Q, P) + \beta L_m(Q, P) + (\gamma + \delta |Q|) L_a(P, Q) \tag{10}$$

where β, γ and δ are the hyperparameter that adjusts the size between losses. In conclusion, the total loss is then as follows:

$$L_{total} = L_{task} + \lambda L_{sample}(Q, P) \tag{11}$$

where λ is a hyperparameter value that adjusts the value between the task loss and the sample loss.

4. Experimental Results and Discussion

In this section, we demonstrate that the performance of our TransNet is superior to that of existing sampling methods in various fields. We conducted experiments in the classification and registration domains and proved that the performance was particularly good in areas with high point cloud sampling ratios. We experimented with two variations of TransNet (W.O indexing summation) and TransNet. The former is a model that applies the transformer architecture. This is a method of applying the Transformer method by generating queries, keys, and values based on existing information without using FPS points. The latter is a model that incorporates an attention map into the transformer architecture. This last model performed better, and here we experimented by comparing the latter model with those of other papers.

We implemented TransNet in PyTorch, setting the batch size, SGD optimizer with momentum, and weight decay to 128, 0.9, and 0.0001, respectively. In addition, we conducted 400 epochs of training in all the experiments. For classification, we used the ModelNet 40 dataset [38]. We performed experiments on 1024 points that were uniformly sampled. To train and evaluate our models, we used the train-test split dataset provided on the official website. We used instance-wise accuracy as a metric to evaluate the classification results of each sample in multi-class classification problems. Each sample belonged to a single class, and if the predicted class by the classification model matched the actual class, the sample was considered "correctly classified". Therefore, instance-wise accuracy represents the proportion of samples that were correctly classified among all the samples. Furthermore, we focused our experiments on sparse points with sampling ratios of 16, 32, 64, and 128, which had previously shown an inferior performance in all papers. In this study, we conducted an experimental analysis using PointNet, which led us to assume the outcomes of Table 1.

The sampled experiment would not surpass the accuracy achieved by the original PointNet prior to sampling. Therefore, we contended that using other state-of-the-art models can also lead to higher sampling accuracy. For example, in the case of classification, since PointNet was defined as the underlying model, it was assumed that the performance of the unsampled PointNet would not be exceeded by any sampled model.

As Figure 4 shows, we have created a model that exceeds the performance of existing models for the experimental results of sampling 8, 16, 32, 64 for 1024 points each. In particular, we demonstrated that the sparse points (the results of sampling 8 or 16) resulted in greater deviations from other models and that our model was more robust in sparse data. In Figure 5, we present a sampling comparison experiment of our model and comparison model on the same object, from which we can clearly see the difference. Our model samples objected much more evenly and reasonably than an APSNet model's sampling result, demonstrating its superior performance in classification results. As a result, our TransNet had a better grasp of corners and characteristic parts than existing methods. However, all deep learning models exhibited a tendency to mislead in certain areas (table legs, flower in a pot, etc.), with many weak characteristics in common, and this problem remains to be solved.

Table 1. Classification accuracies of five sampling methods on ModelNet40. All experiments were conducted in the same environment.

Sampling Ratio	128	64	32	16
RS	8.7	24.87	54.53	79.26
FPS	24.31	55.12	76.92	87.53
SampleNet [23]	80.71	85.32	86.38	87.10
APSNet [24]	82.72	84.89	86.66	**88.00**
TransNet	**87.47**	**88.16**	**88.49**	87.88

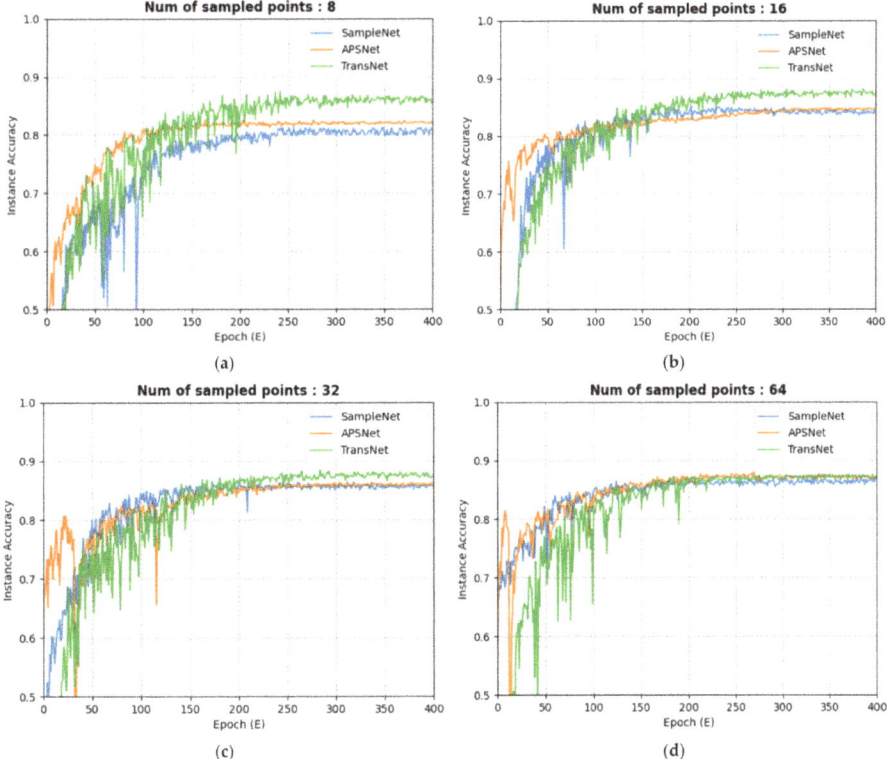

Figure 4. Instance accuracy. This is the result of training three sampling models (SampleNet, APSNet, TransNet) on the ModelNet40 dataset after uniformly sampling 1024 examples. The sampling ratios used were 128, 64, 32, and 16. For example, if 8 examples were sampled from 1024, the sampling ratio would be 128. TransNet achieved much better performance in sparse areas such as (**a**,**b**) and showed superior performance in other areas. Less sparse (**c**,**d**) can also see similar or higher levels of results than existing papers. The classification accuracies are shown in Table 1.

Figure 5. Visualized Sampled Points. Results for 8, 16, and 32 sampling points (green) on ModelNet40. The results on the left are from TransNet (**a**), and those on the right are from APSNet (**b**). Moreover, from top to bottom, the objects are a laptop, nightstand, chair, and airplane. The gray points represent the original ground truth, and the green points were generated. From the overall shape, the result value of our TransNet (**a**) szx more evenly expressed than the result value of APSNet (**b**). Overall, for square objects (laptop, desk), the method tended to have a better grasp of the ends, while for objects such as chairs, it tended to have a better grasp of the legs. Detailed class-specific results are shown in Table 2.

Table 2. Class-specific accuracies of two sampling methods on ModelNet40. All experiments were conducted in the same environment.

Sampling Ratio		Laptop	Chair	Nightstand	Airplane
128	SampleNet	81.32	91.38	38.18	97.51
	APSNet	83.33	95.56	42.18	98.52
	TransNet	**85.32**	**96.49**	**51.88**	**99.1**
64	SampleNet	85.23	94.38	59.38	98.11
	APSNet	88.88	95.52	64.67	99.49
	TransNet	**91.42**	**96.52**	**65.92**	**99.55**
32	SampleNet	89.29	86.38	67.21	98.89
	APSNet	90.90	94.68	67.39	**99.50**
	TransNet	**90.91**	**97.05**	**72.50**	99.00

5. Conclusions

Transformer algorithms have been expanded beyond the natural language process into various fields. We applied this algorithm to the field of point cloud sampling and achieved successful results. In this paper, we proposed a novel transformer-based point cloud sampling network to achieve precision performance. While S-Net and SampleNet generated the sampling process using MLP-based methods, and APSNet used an attention-based model, TransNet employed a multi-head self-attention technique in the down-sampling process. In addition, assuming that FPS is a representative point for viewing general-purpose information, we created an attention map that collected information after proceeding through several FPS algorithms and indexed it for each location while applying multi-head attention. This strategy improved the relationship between each point in the learning procedure while it also learned simultaneously with task models and reasonably understood relationships alongside alleviating long-range dependencies. The proposed TransNet demonstrated a better performance in terms of precision, especially on sparse data. Moreover, the proposed sampling method could be applied to various kinds of point cloud deep learning networks. Thus, its usefulness would be meaningful in many practical scenarios.

6. Ablation Study

6.1. K-Nearest

After extracting the representative points with FPS, we applied the K-nearest neighbor algorithm to find the neighbor points for the representative points and identify the association between the points. The neighborhood size k is the number of neighbors in the representative point P of a point $q \in Q$. We evaluated the impact of the hyperparameter, k, by training multiple progressive TransNet for classification with different values of k. TransNet was applied as k = 16, and experiments were conducted on $k \in \{4, 8, 16, 32\}$, respectively. The results of the experiment are shown in Table 3. If the neighbor was smaller (k = 4 or k = 8), there might not have been enough context for the model to make predictions. If the neighbor was larger (k = 32), each self-attention layer had a large number of data points, many of which could be further away and less relevant. This could result in excessive noise during processing with the potential to degrade the accuracy of the model.

Table 3. The accuracy of TransNet according to K. All experiments were conducted in the same environment, and we adopted K = 16.

K-Size	TransNet-4			TransNet-8			TransNet-16			TransNet-32		
Sampling ratio	128	64	32	128	64	32	128	64	32	128	64	32
Instance Accuracy	85.36	86.24	85.32	85.95	86.68	85.93	**87.47**	**88.16**	**88.49**	86.21	87.67	87.58

6.2. Additional Experiments

As mentioned briefly earlier, when applying the Task model without the sampling model, higher accuracy was achieved compared to the application of the sampling model. When tested with PointNet, an accuracy of 90.08 was obtained. To be as close to this target as possible, we made several attempts, such as adding dropout, skipping connections, or experimenting with added positional encoding to add positional information in different parts of the model. The activation function also conducted many experiments, such as ReLu, Leaky ReLu, and ELU. As a result, we set the probability of the dropout to 0.1 and added positional encoding to the encoder and decoder portions of the model, respectively. The activation function performed best when using ReLu.

Author Contributions: Conceptualization and formal analysis, H.L.; investigation and validation, J.J.; methodology and software, S.H.; software and writing, J.K. and J.Y. All authors have read and agreed to the published version of the manuscript.

Funding: This work was supported by the National Research Foundation of Korea (NRF) grants funded by the Korean government (MSIT) (NRF-2021R1A5A1032937).

Institutional Review Board Statement: Not applicable.

Informed Consent Statement: Not applicable.

Data Availability Statement: Not applicable.

Conflicts of Interest: The authors declare no conflict of interest.

References

1. Liang, Z.; Guo, Y.; Feng, Y.; Chen, W.; Qiao, L.; Zhou, L.; Zhang, J.; Liu, H. Stereo matching using multi-level cost volume and multi-scale feature constancy. *IEEE Trans. Pattern Anal. Mach. Intell.* **2019**, *43*, 300–315. [CrossRef] [PubMed]
2. Guo, Y.; Sohel, F.; Bennamoun, M.; Lu, M.; Wan, J. Rotational projection statistics for 3D local surface description and object recognition. *Int. J. Comput. Vis.* **2013**, *105*, 63–86. [CrossRef]
3. Guo, Y.; Bennamoun, M.; Sohel, F.; Lu, M.; Wan, J. 3D object recognition in cluttered scenes with local surface features: A survey. *IEEE Trans. Pattern Anal. Mach. Intell.* **2014**, *36*, 2270–2287. [CrossRef] [PubMed]
4. Lawin, F.J.; Danelljan, M.; Tosteberg, P.; Bhat, G.; Khan, F.S.; Felsberg, M. Deep Projective 3D Semantic Segmentation. In Proceedings of the International Conference on Computer Analysis of Images and Patterns, Ystad, Sweden, 22–24 August 2017; pp. 95–107.
5. Boulch, A.; Le Saux, B.; Audebert, N. Unstructured Point Cloud Semantic Labeling Using Deep Segmentation Networks. In Proceedings of the Workshop 3D Object Retrieval, Lyon, France, 23–24 April 2017; pp. 17–24.
6. Maturana, D.; Scherer, S. Voxnet: A 3d Convolutional Neural Network for Real-Time Object Recognition. In Proceedings of the 2015 IEEE/RSJ International Conference on Intelligent Robots and Systems (IROS), Hamburg, Germany, 28 September–2 October 2015.
7. Riegler, G.; Ulusoy, A.O.; Geiger, A. Octnet: Learning Deep 3d Representations at High Resolutions. In Proceedings of the IEEE Conference on Computer Vision and Pattern Recognition, Honolulu, HI, USA, 21–26 July 2017.
8. Wang, P.-S.; Liu, Y.; Guo, Y.X.; Sun, C.Y.; Tong, X. O-cnn: Octree-based convolutional neural networks for 3d shape analysis. *ACM Trans. Graph. (TOG)* **2017**, *36*, 1–11. [CrossRef]
9. Qi, C.R.; Su, H.; Mo, K.; Guibas, L.J. Pointnet: Deep Learning on Point Sets for 3d Classification and Segmentation. In Proceedings of the IEEE Conference on Computer Vision and Pattern Recognition, Honolulu, HI, USA, 21–26 July 2017.
10. Qi, C.R.; Yi, L.; Su, H.; Guibas, L.J. Pointnet++: Deep hierarchical feature learning on point sets in a metric space. In Proceedings of the 2017 Conference on Neural Information Processing Systems (NIPS 2017), Long Beach, CA, USA, 4–9 December 2017; Volume 30.
11. Zaheer, M.; Kottur, S.; Ravanbakhsh, S.; Poczos, B.; Salakhutdinov, R.R.; Smola, A.J. Deep sets. In Proceedings of the 2017 Conference on Neural Information Processing Systems (NIPS 2017), Long Beach, CA, USA, 4–9 December 2017; Volume 30.
12. Song, W.; Liu, Z.; Tian, Y.; Fong, S. Pointwise CNN for 3d object classification on point cloud. *J. Inf. Process. Syst.* **2021**, *17*, 787–800.
13. Thomas, N.; Smidt, T.; Kearnes, S.; Yang, L.; Li, L.; Kohlhoff, K.; Riley, P. Tensor field networks: Rotation-and translation-equivariant neural networks for 3d point clouds. *arXiv* **2018**, arXiv:1802.08219.
14. Groh, F.; Wieschollek, P.; Hendrik; Lensch, P.A. Flex-Convolution: Million-Scale Point-Cloud Learning Beyond Grid-Worlds. In *Proceedings of the Computer Vision–ACCV 2018: 14th Asian Conference on Computer Vision, Perth, Australia, 2–6 December 2018*; Revised Selected Papers, Part I 14; Springer International Publishing: Berlin/Heidelberg, Germany, 2019.
15. Wu, W.; Qi, Z.; Fuxin, L. Pointconv: Deep Convolutional Networks on 3d Point Clouds. In Proceedings of the IEEE/CVF Conference on Computer Vision and Pattern Recognition, Long Beach, CA, USA, 15–20 June 2019.

6. Thomas, H.; Qi, C.R.; Deschaud, J.E.; Marcotegui, B.; Goulette, F.; Guibas, L.J. Kpconv: Flexible and Deformable Convolution for Point Clouds. In Proceedings of the IEEE/CVF International Conference on Computer Vision, Seoul, Republic of Korea, 27 October–2 November 2019.
7. Li, J.; Chen, B.M.; Lee, G.H. So-net: Self-Organizing Network for Point Cloud Analysis. In Proceedings of the IEEE Conference on Computer Vision and Pattern Recognition, Salt Lake City, UT, USA, 18–22 June 2018.
8. Wang, Y.; Sun, Y.; Liu, Z.; Sarma, S.E.; Bronstein, M.M.; Solomon, J.M. Dynamic graph cnn for learning on point clouds. *ACM Trans. Graph. (TOG)* **2019**, *38*, 1–12. [CrossRef]
9. Aoki, Y.; Goforth, H.; Srivatsan, R.A.; Lucey, S. Pointnetlk: Robust & Efficient Point Cloud Registration Using Pointnet. In Proceedings of the IEEE/CVF Conference on Computer Vision and Pattern Recognition, Long Beach, CA, USA, 15–20 June 2019.
10. Wang, W.; Yu, R.; Huang, Q.; Neumann, U. SGPN: Similarity Group Proposal Network for 3d Point Cloud Instance Segmentation. In Proceedings of the IEEE Conference on Computer Vision and Pattern Recognition, Salt Lake City, UT, USA, 18–22 June 2018.
11. Wang, X.; Liu, S.; Shen, X.; Shen, C.; Jia, J. Associatively Segmenting Instances and Semantics in Point Clouds. In Proceedings of the IEEE/CVF Conference on Computer Vision and Pattern Recognition, Long Beach, CA, USA, 15–20 June 2019.
12. Dovrat, O.; Lang, I.; Avidan, S. Learning to Sample. In Proceedings of the IEEE/CVF Conference on Computer Vision and Pattern Recognition, Long Beach, CA, USA, 15–20 June 2019.
13. Lang, I.; Manor, A.; Avidan, S. SampleNet: Differentiable Point Cloud Sampling. In Proceedings of the IEEE/CVF Conference on Computer Vision and Pattern Recognition, Seattle, WA, USA, 13–19 June 2020.
14. Ye, Y.; Yang, X.; Ji, S. APSNet: Attention Based Point Cloud Sampling. *arXiv* **2022**, arXiv:2210.05638.
15. Vaswani, A.; Shazeer, N.; Parmar, N.; Uszkoreit, J.; Jones, L.; Gomez, A.N.; Kaiser, L.; Polosukhin, I. Attention is all you need. In Proceedings of the 2017 Conference on Neural Information Processing Systems (NIPS 2017), Long Beach, CA, USA, 4–9 December 2017; Volume 30.
16. Shi, S.; Wang, X.; Li, H. Pointrcnn: 3D Object Proposal Generation and Detection from Point Cloud. In Proceedings of the IEEE/CVF Conference on Computer Vision and Pattern Recognition, Long Beach, CA, USA, 15–20 June 2019.
17. Shi, S.; Guo, C.; Jiang, L.; Wang, Z.; Shi, J.; Wang, X.; Li, H. Pv-rcnn: Point-Voxel Feature Set Abstraction for 3d Object Detection. In Proceedings of the IEEE/CVF Conference on Computer Vision and Pattern Recognition, Seattle, WA, USA, 13–19 June 2020.
18. Jiang, M.; Wu, Y.; Zhao, T.; Zhao, Z.; Lu, C. Pointsift: A sift-like network module for 3d point cloud semantic segmentation. *arXiv* **2018**, arXiv:1807.00652.
19. Yang, G.; Huang, X.; Hao, Z.; Liu, M.Y.; Belongie, S.; Hariharan, B. Pointflow: 3d Point Cloud Generation with Continuous Normalizing Flows. In Proceedings of the IEEE/CVF International Conference on Computer Vision, Seoul, Republic of Korea, 27 October–2 November 2019.
20. Vakalopoulou, M.; Chassagnon, G.; Bus, N.; Marini, R.; Zacharaki, E.I.; Revel, M.P.; Paragios, N. Atlasnet: Multi-Atlas Non-Linear Deep Networks for Medical Image Segmentation. In *Medical Image Computing and Computer Assisted Intervention–MICCAI 2018: 21st International Conference, Granada, Spain, September 16–20, 2018, Part IV 11*; Springer International Publishing: Berlin/Heidelberg, Germany, 2018.
21. Zhao, H.; Jiang, L.; Jia, J.; Torr, P.H.; Koltun, V. Point Transformer. In Proceedings of the IEEE/CVF International Conference on Computer Vision, Montreal, BC, Canada, 11–17 October 2021.
22. Lai, X.; Liu, J.; Jiang, L.; Wang, L.; Zhao, H.; Liu, S.; Qi, X.; Jia, J. Stratified Transformer for 3d Point Cloud Segmentation. In Proceedings of the IEEE/CVF Conference on Computer Vision and Pattern Recognition, New Orleans, LA, USA, 18–24 June 2022.
23. Wu, F.; Fan, A.; Baevski, A.; Dauphin, Y.N.; Auli, M. Pay less attention with lightweight and dynamic convolutions. *arXiv* **2019**, arXiv:1901.10430.
24. Hu, H.; Zhang, Z.; Xie, Z.; Lin, S. Local Relation Networks for Image Recognition. In Proceedings of the IEEE/CVF International Conference on Computer Vision, Seoul, Republic of Korea, 27 October–2 November 2019.
25. Liu, Z.; Lin, Y.; Cao, Y.; Hu, H.; Wei, Y.; Zhang, Z.; Lin, S.; Guo, B. Swin Transformer: Hierarchical Vision Transformer using Shifted Windows. In Proceedings of the IEEE/CVF International Conference on Computer Vision, Montreal, BC, Canada, 11–17 October 2021.
26. Zhao, H.; Jia, J.; Koltun, V. Exploring Self-Attention for Image Recognition. In Proceedings of the IEEE/CVF Conference on Computer Vision and Pattern Recognition, Seattle, WA, USA, 13–19 June 2020.
27. Hochreiter, S.; Schmidhuber, J. Long short-term memory. *Neural Comput.* **1997**, *9*, 1735–1780. [CrossRef] [PubMed]
28. Wu, Z.; Song, S.; Khosla, A.; Yu, F.; Zhang, L.; Tang, X.; Xiao, J. 3d Shapenets: A Deep Representation for Volumetric Shapes. In Proceedings of the IEEE Conference on Computer Vision and Pattern Recognition, Boston, MA, USA, 7–12 June 2015.

Disclaimer/Publisher's Note: The statements, opinions and data contained in all publications are solely those of the individual author(s) and contributor(s) and not of MDPI and/or the editor(s). MDPI and/or the editor(s) disclaim responsibility for any injury to people or property resulting from any ideas, methods, instructions or products referred to in the content.

Article

DCFF-MTAD: A Multivariate Time-Series Anomaly Detection Model Based on Dual-Channel Feature Fusion

Zheng Xu [1,2], Yumeng Yang [1,2], Xinwen Gao [1,3,*] and Min Hu [1,2]

1. SHU-SUCG Research Centre of Building Information, Shanghai University, Shanghai 201400, China
2. SILC Business School, Shanghai University, Shanghai 201800, China
3. School of Mechatronic Engineering and Automation, Shanghai University, Shanghai 200444, China
* Correspondence: gxw@shu.edu.cn

Abstract: The detection of anomalies in multivariate time-series data is becoming increasingly important in the automated and continuous monitoring of complex systems and devices due to the rapid increase in data volume and dimension. To address this challenge, we present a multivariate time-series anomaly detection model based on a dual-channel feature extraction module. The module focuses on the spatial and time features of the multivariate data using spatial short-time Fourier transform (STFT) and a graph attention network, respectively. The two features are then fused to significantly improve the model's anomaly detection performance. In addition, the model incorporates the Huber loss function to enhance its robustness. A comparative study of the proposed model with existing state-of-the-art ones was presented to prove the effectiveness of the proposed model on three public datasets. Furthermore, by using in shield tunneling applications, we verify the effectiveness and practicality of the model.

Keywords: multivariate time-series; anomaly detection; short-time Fourier transform

1. Introduction

In real-world applications, complex systems and devices, such as smart grids, water treatment systems, shield machines, and self-driving cars, typically contain multiple sensors. During operation, these sensors generate large quantities of time-series data that are often interrelated. Abnormal changes in the data from one sensor can affect the data from other sensors, making it increasingly difficult to detect anomalies in this growing volume and dimension of data. The development of efficient and accurate multivariate time-series anomaly detection algorithms is crucial for the continuous monitoring of key indicators or parameters in these systems and devices, ultimately increasing their level of automation.

In recent years, deep learning has become a widely adopted tool for time-series data analysis [1], with the ability to effectively extract both temporal and spatial features. At present, multivariate time series anomaly detection is mainly divided into the following three methods: (1) Temporal correlation-based method, where each variable is impacted by historical data values. Recurrent neural networks (RNNs) [2–6] and temporal convolutional networks (TCNs) [7] have gained popularity for their ability to extract temporal features. RNNs store past information in time and automatically extract advanced features from historical data, while TCNs provide greater flexibility in changing the size of the receptive field, enabling better control over the memory length of the model. (2) Spatial correlation-based method, which describes the correlation between different variables. In terms of extracting spatial features, graph neural networks (GNNs) and their variants [8–10] have played a critical role. GNNs view each variable in multivariate data as a node and the relationships between variables as edges. This allows the GNN to learn relationships between variables and extract potential spatial features. However, with an increasing number of sensors, the use of GNNs can lead to higher space complexity and greater memory requirements when learning these relationships. (3) The method for spatial-temporal correlation fusion, of which Transformer is a typical model [11–13]. Transformer's self-attention mechanism can capture the potential correlation between sequences, and its position encoding and upsampling algorithm can capture multi-scale temporal information. However, it is worth noting that the training process of the transformer-based model requires powerful computer hardware and the complexity of the model may hinder its deployment in practical engineering projects.

Recurrent neural networks are prone to gradient disappearance or gradient explosion when processing large amounts of data. The greater the number of sensors, the greater the memory occupied by the graph neural network in extracting spatial features from multivariate data. The model based on transformer has high requirements on equipment hardware in the training stage, coupled with the harsh site environment of shield engineering projects, so it is difficult to apply to shield engineering. Therefore, we propose a stable and practical model for multivariate time-series anomaly detection. The model is based on a dual-channel feature extraction module. Unlike previous studies in the literature, our model combines spatial STFT to extract spatial features and the time-based graph attention layer to extract temporal features. These features are then fused and fed into the subsequent network structure. As a result of the fusion of different feature information, the proposed method can be more accurate and robust in its detection of anomalies.

The contributions of the paper are as follows:

1. A multivariate time-series anomaly detection model based on dual-channel feature fusion (DCFF-MTAD) is proposed.
2. A spatial short-time Fourier transform module is presented for fully extracting spatial features from multivariate data.
3. In order to improve the robustness of the anomaly detection model, the Huber loss is introduced.
4. Our network shows good performance on three publicly available datasets.
5. Extensive ablation experiments are conducted to investigate the key factors improving anomaly detection performance.

The structure of this paper is organized as follows. In Section 2, we present the literature related to deep learning methods and short-time Fourier transform in anomaly detection of multivariate time-series. Section 3 shows the dual channel feature extraction module, feature fusion module, and anomaly detection module. In Section 4, we conducted the comparative experiment with existing state-of-the-art models and the ablation study of our model. In Section 5, we verify the effectiveness and practicability of our model in shield tunneling engineering. In Section 6, we present the conclusion and the prospect of future research work.

2. Related Work

In this section, we first review deep learning method in multivariate time-series anomaly detection, and since our model relies on the short-time Fourier transform, we also summarize related research works on the topic.

2.1. Deep Learning Method

Since data in complex systems or devices often lack anomaly labels, the problem of anomaly detection is usually regarded as an unsupervised learning problem. During the past few years, researchers have proposed many effective methods for unsupervised anomaly detection.

2.1.1. Temporal Correlation-Based Method

Zong et al. [14] proposed a deep autoencoder Gaussian mixture model (DAGMM) for unsupervised anomaly detection, which utilizes a deep autoencoder to generate a low-dimensional representation and reconstruction error for each input data point, and further fed it into a Gaussian mixture model. DAGMM jointly optimizes the parameters of the deep autoencoder and the hybrid model in an end-to-end manner and utilizes a separate estimation network to facilitate the parameter learning of the hybrid model. Hundman et al. [15] used LSTM for anomaly detection in spacecraft telemetry systems and proposed a new non-parametric dynamic threshold (NDT) method that does not rely on scarce labels or spurious parametric assumptions. Li et al. [16] proposed an unsupervised multivariate anomaly detection method based on generative adversarial network (GAN), which uses long short-term memory recurrent neural network (LSTM-RNN) as the basic model in the GAN framework (i.e., generator and discriminator) to capture the temporal correlation of the time-series distribution. Instead of processing each data stream independently, the method considers the entire set of variables simultaneously to capture potential interactions between variables. Audibert et al. [17] proposed USAD, an autoencoder-based approach for unsupervised anomaly detection in multivariate time-series, and conducted adversarial training inspired by generative adversarial networks. Its autoencoder structure makes it an unsupervised method and enables it to show great stability in adversarial training. Su et al. [18] proposed OmniAnomaly, which uses key technologies such as random variable connection and plane normalization flow to learn a robust representation of multivariate time-series, obtain its normal pattern, reconstruct input data according to these representations, and detect anomalies using reconstruction probability. Abdulaal et al. [19] extracted the priors of multivariate signals through spectrum analysis of latent spatial representation to synchronize the representation of the original sequence, then input the random subset of the synchronous multivariate into an automatic encoder array that learns the loss of minimum quantile reconstruction, and finally infer and locate anomalies through majority voting. Liang et al. [20] proposed the multi-time-scale deep convolutional generative adversarial network (MTS-DCGAN) framework, which used cross-correlation calculation based on multi-time-scale sliding Windows to transform multivariate time-series into multi-channel feature matrix. By inputting the feature matrix into DCGAN, multi-layer CNN can capture the nonlinear interrelated features hidden in the original multivariate time-series without prior knowledge.

2.1.2. Spatial Correlation-Based Method

Deng and Hooi [21] proposed an attention-based graph neural network method—GDN—which can learn the dependency graph between sensors, and identify and explain the deviation of these relationships, helping users infer the cause of anomalies. Chen, X. et al. [22] proposed GraphAD, a new multivariate time series anomaly detection model based on graph neural networks. They extracted patterns of key indicators from attribute, entity, and temporal perspectives via graph neural networks. Wang, Y. et al. [23] proposed a simple but effective graph self-supervised learning scheme called Deep Cluster Infomax for node representation learning, which captures intrinsic graph attributes in a more concentrated feature space by clustering the whole graph into multiple parts. Yang, J., and Yue, Z. [24] proposed a hierarchical spatial-temporal graph representation that constructs discriminative decision boundaries by learning hierarchical normality closed hyperspheres on the generated graph structural representation, without requiring a predefined topological prior. Razaque, A. et al. [25] proposed an anomaly detection paradigm called novel matrix profile to address the full-pair similarity search problem for time-series data in healthcare. A novel matrix profile can be used on large multivariate datasets and produces high-quality approximate solutions in a reasonable amount of time.

2.1.3. The Method for Spatial-Temporal Correlation Fusion

Li, Y. et al. [26] proposed an inflated convolutional transformer-based GAN to improve the accuracy and generalization of the model. They utilized multiple generators and a single discriminator to mitigate the pattern collapse problem. Each generator consists of an inflated convolutional neural network and a transform module to obtain both fine-grained and coarse-grained information of the time series. Qin, S. et al. [27] proposed a novel method for time series anomaly detection based on transformer and signal decomposition.

They provide a multi-view embedding method to capture temporal and correlation features of the signal. To make full use of temporal patterns, a frequency attention module is designed to extract periodic oscillation features. Kim, J. et al. [28] proposed an unsupervised prediction-based time series anomaly detection method using the transformer, which learns the dynamic patterns of sequential data through a self-attentive mechanism. The output representation of each transformer layer is accumulated in the encoder to obtain a representation with multiple levels and rich information. The decoder fuses this representation by means of a one-dimensional convolution operation. Therefore, the model can forecast while considering both the global trend and local variability of the input time series.

2.2. Short-Time Fourier Transform

In recent years, short-time Fourier transform has become a successful method for fault diagnosis or anomaly detection. In the existing research, most scholars have utilized the short-time Fourier transform to convert the signal into a spectrogram, then extract features from the spectrogram by using a convolutional neural network, and finally detect or classify abnormal signals. Gultekin et al. [29] proposed a data fusion method based on convolutional neural network, which used short-time Fourier transform to detect and identify operational faults in automatic transfer vehicles (ATVs). Li and Boulanger [30] combined the short-time Fourier transform spectra of the ECG signal with hand-made features to detect more complex cardiac abnormalities, including 16 distinct rhythm abnormalities and 13 heartbeat abnormalities. Zhou et al. [31] proposed a radio anomaly detection algorithm based on an improved GAN, which uses short-time Fourier transform to obtain the spectral graph image from the received signal, then reconstructs the spectral graph by combining the encoder network in the original GAN, and detects the anomaly according to the reconstruction error and discriminator loss. Chong et al. [32] studied the feasibility of detecting adverse substructure conditions by using bullet train load through finite element numerical simulation. All the synthesized signals obtained from the numerical simulation are analyzed using fast Fourier transform (FFT) in the frequency domain and short-time Fourier transform in the time-frequency domain. The three-dimensional Fourier scattering transform proposed by Kavalerov et al. [33] is a fusion of time-frequency representation and neural network architecture, taking advantage of short-time Fourier transform and the numerical computational efficiency of a deep learning network structure. Khan et al. [34] proposed a new network intrusion detection system (NIDS) framework based on deep convolutional neural networks, which utilized the network spectrum image generated by short-time Fourier transform to improve the accuracy of intrusion detection. Haleem et al. [35] used short-time Fourier transform to convert ECG beats into 2D images to automatically distinguish normal ECG from cardiac adverse events such as arrhythmia and congestive heart failure. Sanakkayala et al. [36] used short-time Fourier transform to convert bearing vibration signals into spectral graphs and then used the convolutional neural network VGG16 to extract features and classify health conditions.

As mentioned above, both deep learning techniques and short-time Fourier transform in multivariate time-series anomaly detection have shown superiority in some specific cases. In our paper, we combine the advantages of these two techniques to research multivariate time-series anomaly detection.

3. Methods

In this paper, we describe our multivariate time-series anomaly detection model DCFF-MTAD in detail. Figure 1 shows the general framework of DCFF-MTAD, which includes the following parts:

(1) Preprocessing module: multivariate time-series data collected from multiple sensors are normalized and used as input to the model; (2) dual-channel feature extraction module composed of time-based graph attention layer and short-term Fourier transform; (3) feature fusion module based on gated recurrent unit (GRU); (4) anomaly detection mod-

ule composed of prediction model, reconstruction model, and anomaly score calculation method. Below is a detailed description of each of these modules.

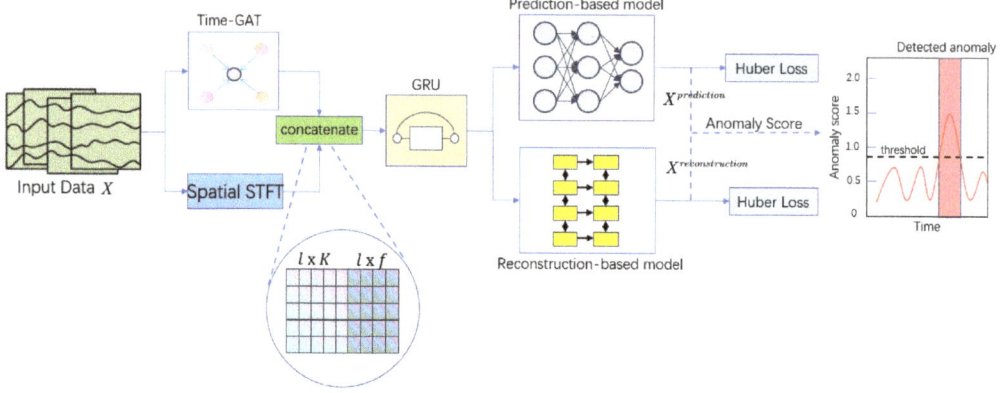

Figure 1. The overall architecture of our proposed DCFF-MTAD.

3.1. Preprocessing Module

3.1.1. Data Format

A univariate time-series is generated by a single sensor and is strictly arranged by timestamp. Multiple univariate time-series from the same entity form a multivariate time-series. In this paper, $t_1, t_2, \cdots, t_N$ is used to represent the timestamp; there are K sensors in total, and the matrix X is used to represent the multivariate time-series as follows:

$$X = \begin{bmatrix} x_{11} & x_{12} & \cdots & x_{1K} \\ x_{21} & x_{22} & \cdots & x_{2K} \\ \vdots & \vdots & & \vdots \\ x_{N1} & x_{N2} & \cdots & x_{NK} \end{bmatrix} \quad (1)$$

where x_{nk} ($1 \leq n \leq N, 1 \leq k \leq K$) represents the value of the kth sensor at the timestamp t_n. In this paper, we use the sliding window of size $l \times K$ ($1 < l < N$) to obtain data.

3.1.2. Data Normalization

The input parameters of our model are multivariate data, which contain multiple variables with different dimensional units. In the training process, this can lead to low prediction accuracy of multivariate data, low reconstruction accuracy of multivariate data, and slow gradient descent of the optimal solution. To resolve these issues, the data normalization method is used to map the data values of multiple variables to the same scale. We use the maximum and minimum values of each sensor data for normalization:

$$\tilde{x}_{nk} = \frac{x_{nk} - x_k^{\min}}{x_k^{\max} - x_k^{\min}} \quad (2)$$

where $x_k^{\max}$ and $x_k^{\min}$ respectively represent the maximum and minimum values of the kth sensor data, and $\tilde{x}_{nk}$ represents the result of x_{nk} after normalization processing.

3.2. Spatial Short-Time Fourier Transform

The curve forms of multivariate data are complex and varied, and most of the curves are non-stationary signals. Since time-domain methods cannot obtain frequency information, and frequency-domain methods cannot obtain instantaneous features, time-frequency transform methods are often used in non-stationary signal analysis to diagnose industrial machinery faults [37]. STFT is a conventional and classic linear time-frequency analysis method, which overcomes the shortcomings of the traditional Fourier transform, which cannot reveal the local characteristics of the signal. Unlike STFT, the Wigner distribution is a nonlinear distribution. Since the Wigner distribution does not use a window function similar to the STFT definition, it has no loss of resolution when analyzing signals. Wigner distribution is a traditional quadratic time-frequency distribution, and suffers from the cross-terms presence, which will generate redundant interference information when processing multi-component signals. The cross-terms-free Wigner distribution (frequently named the S-method) unifies the desirable properties of STFT (cross-terms-free nature) and Wigner distribution (optimal auto-terms presentation, high concentration, resolution, and selectivity). Besides, the S-method provides both the noise influence reduction in comparison to the conventional time-frequency tools (the spectrogram and the Wigner distribution) [38] and the best performances in estimation of the instantaneous frequency [39]. Compared with STFT, the operation of S-method is more complex, and it needs more computation when dealing with multivariate data. Although the S-method has better time-frequency analysis performance, its more complex computation is not conducive to practical engineering applications. Considering the advantages and disadvantages of various methods, STFT is used to extract features from a large number of multivariate time series data.

Unlike the previous approach of applying STFT in the time dimension, we apply STFT in the spatial dimension of multivariate data. Figure 2 illustrates how we perform the short time Fourier transform in the spatial dimension.

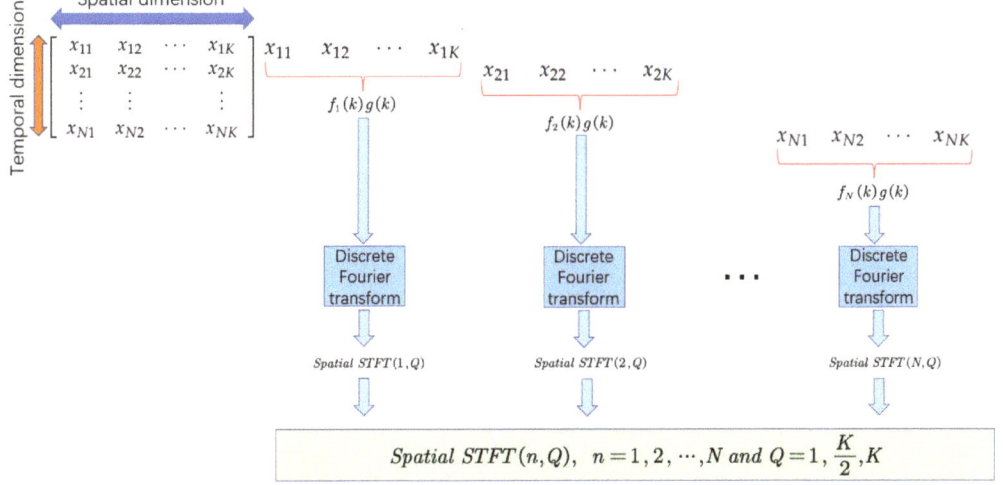

Figure 2. The calculation process of spatial STFT.

The spatial short-time Fourier transform is defined as follows:

$$\text{Spatial STFT}(n, Q) = \text{DFT}\{f_n(k)g(k)\}, n = 1, 2, \cdots N \qquad (3)$$

where DFT denotes discrete Fourier transform. $f_n(k)$ denotes a function formed by multiple variables in the spatial dimension and $g(k)$ denotes a window function. Q denotes the location of the computed Fourier transform.

The spatial short-time Fourier transform is carried out on the multivariate time-series data whose tensor dimension is (l, K). Set the window size (nperseg) of STFT to the dimension of the dataset and the number of window function overlap to the default value of 50%. After the transformation, the three-dimensional tensor with dimension

(l, f, t) is obtained. f denotes the dimension of the extracted spatial features and t denotes the dimension of the centroid of the window function.

3.3. Time-Based Graph Attention Layer (Time-GAT)

The graph attention network (GAT) uses attention weights to replace connections between nodes that are either 0 or 1 and is able to automatically learn and optimize connections between nodes by using the attention mechanism. The relationship between nodes can be optimized into continuous values so that the expression of information is more abundant, and the correlation between nodes can be better integrated into the model. The graph attention network does not use the Laplace matrix for complex calculation, and the calculation of its attention value is carried out in parallel between nodes, so the calculation process of the graph attention network is very efficient. In this paper, we use the same method as Zhao et al. [40] to construct the graph attention layer based on time.

As can be seen in Figure 3, all the data in the sliding window is considered to be a complete graph. Specifically, the feature vector at a timestamp is represented by a node, and the relationship between two timestamps is represented by an edge. In this way, the time dependence in the time-series data can be captured.

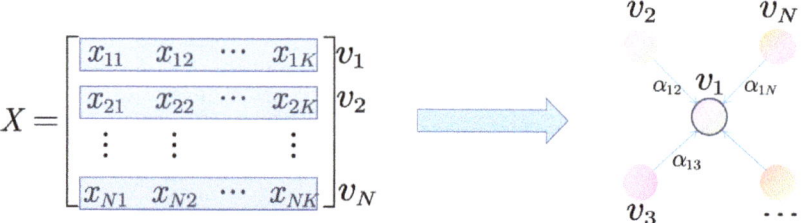

Figure 3. Schematic diagram of constructing a time-based graph attention layer.

Generally, given a graph with N nodes, i.e., $\{v_1, v_2, \cdots, v_N\}$, where v_i represents the feature vector of each node, the calculated output of each node is as follows:

$$h_i = \sigma\left(\sum_{j=1}^{L} \alpha_{ij} v_j\right) \quad (4)$$

where σ represents the sigmoid activation function. α_{ij} represents the attention score that measures the contribution of node j to node i, and node j is one of the adjacent nodes of node i. L represents the total number of adjacent nodes of node i.

The calculation formula of the attention score α_{ij} is as follows:

$$e_{ij} = LeakyReLU\left(\omega^T \cdot (v_i \oplus v_j)\right) \quad (5)$$

$$\alpha_{ij} = \frac{\exp(e_{ij})}{\sum_{l=1}^{L} \exp(e_{il})} \quad (6)$$

where $\oplus$ represents the concatenation of two node representations, $\omega \in R^{2m}$ is a column vector of learnable parameters, and m represents the dimension of the feature vector of each node.

3.4. Feature Fusion Module Based on GRU

The dimension of the output data of the time-based graph attention layer is (l, K), and the dimension of the output data after spatial short-time Fourier transform is (l, f, t). Since the data after the short-time Fourier transform is a complex number, we conduct modulo operation on it. To match the dimensions of the output data of the two channels, we conduct average operation on the third dimension t of the three-dimensional tensor (l, f, t). Finally, we get the two-dimensional tensor (l, f).

The GRU is a variant of the traditional RNN, which can effectively capture the correlation between time-series data and alleviate the phenomenon of gradient disappearance or gradient explosion in traditional neural networks. Compared with LSTM network, GRU has a simpler structure and fewer parameters and requires less time in the training phase. Therefore, GRU is chosen as the model of the feature fusion module.

The time and spatial features of the multivariate time-series are respectively extracted through the time-based graph attention layer and the spatial short-time Fourier transform. In this paper, the output data of the two channels are concatenated in the way shown in Figure 4 to obtain a tensor with dimension $(l, K + f)$, which is then sent to the GRU and mapped into a tensor with dimension (l, K) through the GRU.

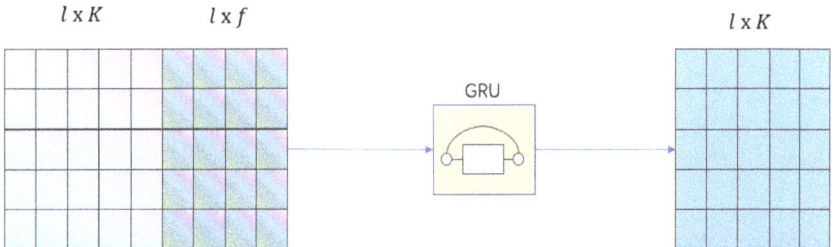

Figure 4. Feature fusion map.

3.5. Loss Function

In multivariate time-series anomaly detection, loss function plays a very important role. The loss function is a function used to measure the gap between the predicted data and the actual data. For the same neural network, the selection of loss function will affect the quality of model training to a certain extent.

Huber Loss is a piece-wise loss function for regression problems, which combines the advantages of both mean absolute error (MAE) and mean square error (MSE). A model using MSE as a loss function is likely to forcibly fit outliers to reduce the value of the loss function, thereby affecting the output of the model. Compared with MSE, Huber Loss has better robustness to outliers. Huber Loss is selected for its reliability and validity, which is defined as follows:

$$L_\delta(y, \hat{y}) = \begin{cases} \frac{1}{2}(y - \hat{y})^2, & |y - \hat{y}| \leq \delta \\ \delta|y - \hat{y}| - \frac{1}{2}\delta^2, & |y - \hat{y}| > \delta \end{cases} \quad (7)$$

where $\hat{y}$ represents the predicted value of y, and δ is a boundary. When the absolute value of the difference between the actual value and the predicted value is less than or equal to δ, the square error is used; When the absolute value of the difference between the actual value and the predicted value is greater than δ, Loss is reduced and a linear function is used. This approach can reduce the weight of outliers in the calculation of Loss and prevent overfitting of the model.

A prediction-based model is used to predict the value of the next timestamp. We use the same approach as Zhao et al. [40], stacking three fully-connected layers after GRU as the prediction-based model. A reconstruction-based model is used to capture the data distribution across the time-series. We use a GRU-based decoder as the reconstruction-based model.

During training, the parameters of the prediction-based and reconstruction-based models are updated simultaneously. Both models use Huber Loss as the loss function, and the loss function of the entire model is defined as the sum of the loss functions of the two models, as shown in the following formula.

$$Loss = Loss_{prediction} + Loss_{reconstruction} \qquad (8)$$

where $Loss_{prediction}$ represents the loss function of the prediction-based model and $Loss_{reconstruction}$ represents the loss function of the reconstruction-based model.

3.6. Anomaly Determination

There are two results for each timestamp: one is the predicted value based on the prediction model and the other is the reconstructed value based on the reconstruction model. We calculate the anomaly score for each variable and take the average of the anomaly scores for all variables as the final anomaly score. The final anomaly score is calculated as follows:

$$\text{Anomaly Score} = \frac{1}{K}\sum_{k=1}^{K}\left(\left|x_k^{prediction} - x_k\right| + \gamma\left|x_k^{reconstruction} - x_k\right|\right) \qquad (9)$$

where $\left|x_k^{prediction} - x_k\right|$ represents the deviation degree between the predicted value and the actual value of variable k and $\left|x_k^{reconstruction} - x_k\right|$ represents the deviation degree between the reconstructed value and the actual value of variable k. K is the total number of variables. γ is a hyperparameter that balances the error based on the prediction model with the error based on the reconstruction model.

The threshold is set to give the best boundary between normal and abnormal data. Hundman et al. [15] proposed a non-parametric dynamic threshold method, which is an unsupervised method and does not depend on labeled data and statistical assumptions about errors. The non-parametric dynamic threshold method is selected for its low computational cost and high performance. We calculate the threshold for each variable and take the average of the thresholds for all variables as the final threshold. If the anomaly score on a timestamp is greater than the final threshold, the timestamp is marked as an anomaly.

4. Experiment

We have trained and evaluated DCFF-MTAD on three publicly available datasets, compared it with existing state-of-the-art models for detecting anomalies in multivariate time-series, and analyzed the experimental results. The datasets, evaluation metrics, implementation details, comparative experiments, ablation study, and sensitivity analysis are presented in this section.

4.1. Datasets

We selected three publicly available datasets, including soil moisture active passive (SMAP) [41], server machine dataset (SMD) [18], and mars science laboratory (MSL) [41], which are widely used anomaly detection datasets.

SMAP is a publicly available dataset of the amount of water in the earth's topsoil collected by NASA. The observation mission uses both active and passive sensors. The active sensor is the L-band radar, and the passive sensor is the L-band microwave radiometer.

SMD is a 5-week public dataset collected from 28 computers of a large Internet company.

MSL is a publicly available dataset from NASA that contains telemetry data from spacecraft monitoring systems for unexpected event anomaly reports.

Table 1 shows the detailed statistics of the three datasets, including dataset name, number of entities, dimensionality, size of training set and test set, and the ratio of anomalies in the test set.

Table 1. Dataset statistics.

Dataset Name	Number of Entities	Number of Dimensions	Training Set Size	Testing Set Size	Anomaly Ratio (%)
SMAP	55	25	135,183	427,617	13.13
SMD	28	38	708,405	708,420	4.16
MSL	27	55	58,317	73,729	10.72

4.2. Evaluation Metrics

We evaluate the anomaly detection performance of all models using precision, recall, and F1-score on the test dataset. Precision indicates the percentage of correctly detected anomalies among all detected anomalies, recall indicates the percentage of correctly detected anomalies among all anomalies, and the F1 value is the harmonic mean of precision and recall.

$$\text{Precision} = \frac{TP}{TP + FP} \tag{10}$$

$$\text{Recall} = \frac{TP}{TP + FN} \tag{11}$$

$$F1 = \frac{2 \times \text{Precision} \times \text{Recall}}{\text{Precision} + \text{Recall}} \tag{12}$$

where TP represents the number of samples correctly detected as abnormal in abnormal samples; TN represents the number of samples correctly detected as normal in normal samples; FP represents the number of samples correctly detected as abnormal in normal samples; FN represents the number of samples correctly detected as normal in abnormal samples.

4.3. Comparative Experiments

4.3.1. Implementation Details

We implemented our model in Pytorch 1.10.0 and CUDA 10.2. The model was fully trained on a server equipped with Intel(R) Xeon(R) Silver 4110 CPU @2.10GHz and an NVIDIA Tesla P100 GPU (16G memory). To be fair, we set the learning rate for all models to 0.001, using the Adam optimizer for 10 training sessions and the non-parametric dynamic threshold method. In our model, we set γ to 1, the batch size to 256, and the hidden layer dimensions of the GRU-based feature fusion module, the prediction-based model and the reconstruction-based model are all 150. In the feature fusion module based on GRU, the prediction-based model, and the reconstruction-based model, the dropout mechanism [42] is used to prevent the overfitting problem of the complex model. The key idea is to drop some neurons (set output to zero) randomly during the training process with some probability, which helps to prevent complex co-adaptations on training data. The dropout ratio adopted in this paper is 0.3.

4.3.2. Experimental Results

We compared DCFF-MTAD with existing state-of-the-art models for multivariate time-series anomaly detection, including LSTM [15], OmniAnomaly [18], USAD [17], MAD-GAN [16], DAGMM [14], and GDN [21]. Table 2 shows the F1, precision (P), and recall (R) of all models on the three public datasets. In terms of the F1 value, compared with existing state-of-the-art models, our network structure performs well in multivariate time-series anomaly detection, especially on the SMAP and SMD datasets. On the MSL dataset, OmniAnomaly shows the best anomaly detection performance, but our model is also second only to OmniAnomaly and outperforms the remaining models.

Table 2. Anomaly detection performance metrics on three datasets. The best results are highlighted in bold.

Models	SMAP			SMD			MSL		
	F1	P	R	F1	P	R	F1	P	R
LSTM	0.5800	0.9809	0.4118	0 *	0 *	0 *	0.8322	0.7126	0.9999
OmniAnomaly	0.5789	0.9747	0.4118	0.8709	0.7713	0.9999	**0.9509**	0.9064	0.9999
USAD	0.8360	0.9734	0.7326	0.8704	0.7713	0.9989	0.8959	0.8115	0.9999
MAD-GAN	0.5725	0.9390	0.4118	0 *	0 *	0 *	0.8367	0.7193	0.9999
DAGMM	0.5752	0.9536	0.4118	0.7892	0.6518	0.9999	0.9351	0.8782	0.9999
GDN	0.5773	0.9655	0.4118	0.7486	0.5991	0.9974	0.9051	0.8267	0.9999
Our Model	**0.8955**	0.9767	0.8268	**0.9140**	0.8416	0.9999	0.9366	0.9257	0.9478

* The place of 0 in this table is because the threshold selected by the non-parametric dynamic threshold method is too high, resulting in TP and FP being 0. Therefore, the three calculation indicators are all 0.

As can be seen from Table 2, LSTM's performance is suboptimal because it only captures time information, as indicated by its F1 values of 0.5800 and 0.8322 on the SMAP and MSL datasets, respectively. MAD-GAN's performance is unstable, with F1 values of 0.5725 and 0.8367 on the SMAP and MSL datasets, respectively, due to the difficulty of training GAN-based network, which may suffer from issues such as mode collapse and non-convergence. USAD performs well on the SMAP, SMD, and MSL datasets, with F1 values of 0.8360, 0.8704, and 0.8959, respectively. USAD's encoder-decoder architecture combines the advantages of autoencoders and adversarial training, and is able to learn how to amplify the reconstruction error of inputs containing anomalies. USAD is more stable and robust than methods based on GAN architectures. The F1 value of DAGMM on MSL dataset is 0.9351, which is lower than that of OmniAnomaly and Our Model. DAGMM saves the key information of input samples in low-dimensional space, including the features discovered through dimensionality reduction and reconstruction errors, so it shows good performance on MSL dataset with higher dimensions. F1 values of GDN on SMAP, SMD, and MSL datasets are 0.5773, 0.7486, and 0.9051, respectively. The GDN captures the unique characteristics of each sensor, and its graph structure can learn the relationships between high-dimensional sensors to detect deviations in those relationships. The higher the dimension of the dataset, the better the GDN can play its advantages, and the higher its F1 value. The F1 values of OmniAnomaly on the SMD and MSL datasets are 0.8709 and 0.9509, respectively. OmniAnomaly uses stochastic recurrent neural networks to model the explicit time dependence between random variables, and planar normalization streams to better capture the complex distribution of input data. Therefore, OmniAnomaly shows good performance on the more complex high-dimensional MSL dataset. LSTM focuses on the temporal features of multivariate data, and GDN focuses on the spatial features of multivariate data. Our model extracts the spatial and temporal features of multivariate data through spatial STFT and time-based graph attention layer, and these two features are further fused. Therefore, our model contains richer feature information, which is an important reason why our model is superior to other models. Our model improves F1 values by 7.12% and 4.95% on SMAP and SMD datasets, respectively. In addition, Huber Loss is used to improve the robustness of the model, which makes the performance of our model on the three datasets very stable. Although our model performs well on F1, the complexity of our model is high. As shown in Table 3, parameters, FLOPs, and runtime of our model are larger than other models.

Table 3. Model complexity.

Model	Parameters	FLOPs	Runtime
LSTM	3962	20,120	1.8 ms
OmniAnomaly	18,490	37,760	3.8 ms
USAD	10,609	33,760	3.7 ms
MAD_GAN	9903	25,920	3.1 ms
DAGMM	10,516	21,234	3.4 ms
GDN	4160	61,956	13.9 ms
Our Model	325,934	3,207,080	51 ms

4.4. Ablation Study

In this section, we conducted a series of experiments to verify the effectiveness of the different components in the proposed DCFF-MTAD.

4.4.1. Effectiveness of Dual-Channel Feature Extraction Module

We propose a dual-channel module to extract features from multivariate time-series data, which makes the feature information richer and improves anomaly detection accuracy. To further demonstrate the effectiveness of our proposed dual-channel feature extraction module, we conducted a series of experiments, including single-channel, dual-channel, and triple-channel, and the experimental results are shown in Table 4. The feature-based graph attention layer (Feat-GAT) in the ablation experiment was constructed in the same way as in the study of Zhao et al. [40].

Table 4. The effects of dual-channel feature extraction module on three datasets. The best results are highlighted in bold. ✓ indicates that the module is used for experiments.

STFT	Time-GAT	Feat-GAT	SMAP			SMD			MSL		
			F1	P	R	F1	P	R	F1	P	R
✓			0.8086	0.9242	0.7187	0.9019	0.8213	0.9999	0.9196	0.9048	0.9348
	✓		0.8060	0.9759	0.6864	0.8335	0.7146	0.9999	0.8952	0.9037	0.8869
✓	✓		**0.8955**	0.9767	0.8268	**0.9140**	0.8416	0.9999	0.9366	0.9257	0.9478
✓		✓	0.6730	0.9866	0.5106	0.8844	0.7928	0.9999	**0.9542**	0.9662	0.9426
✓	✓	✓	0.8938	0.9588	0.8372	0.8744	0.7768	0.9999	0.9258	0.9607	0.8934

On the SMAP dataset, our model shows better detection performance than the network composed of spatial STFT and Feat-GAT, with 33.06% improvement of the F1 value. Although the F1 value of our model is only 0.19% higher than that of the triple-channel network, our model has fewer parameters, less computation, and shorter training time.

On the SMD dataset, our model performs better than the network composed of spatial STFT and Feat-GAT, and the F1 value is improved by 3.35%. Our model performs better than the triple-channel network, and the F1 value increases by 4.53%.

On the MSL dataset, our model performs weaker than the network composed of spatial STFT and Feat-GAT, because the MSL dataset has a higher dimension than SMAP and SMD. A major advantage of Feat-GAT is that it can capture the relationship between higher-dimensional data. Therefore, Feat-GAT has given full play to its advantage on the MSL dataset. Compared with the triple-channel network, our model still performs well, and the F1 value has increased by 1.17%.

The experimental results of the above three public datasets show that our dual-channel feature extraction module can extract richer data information, both in the time dimension and in the spatial dimension. Therefore, our dual-channel feature extraction module achieves better detection results than single-channel networks. In addition, our dual-channel feature extraction module has fewer parameters than the triple-channel network, which shortens the training time of the model.

4.4.2. Effectiveness of Huber Loss

In this section, we investigate the effect of Huber Loss on model performance. To verify the effectiveness of Huber Loss, we choose ordinary MSE Loss as the baseline. Above the baseline, MSE Loss is replaced by Huber Loss, and other modules in the model remain unchanged. We conduct this ablation experiment on three public datasets, and Table 5 shows the results of the ablation experiments.

Table 5. The effects of Huber Loss. The best results are highlighted in bold.

Loss	SMAP			SMD			MSL		
	F1	P	R	F1	P	R	F1	P	R
MSE	0.8481	0.9458	0.7686	0.8913	0.8039	0.9999	0.9287	0.9103	0.9478
Huber	**0.8955**	**0.9767**	**0.8268**	**0.9140**	**0.8416**	0.9999	**0.9366**	**0.9257**	0.9478

When Huber Loss is used as the loss function of the model, the detection index is significantly improved. On the SMAP dataset, F1 value, precision, and recall are increased by 5.59%, 3.27%, and 7.57%, respectively. On the SMD dataset, the F1 value and precision are increased by 2.55% and 4.69%, respectively. On the MSL dataset, the F1 value and precision are improved by 0.85% and 1.69%, respectively. The experimental results show that Huber Loss as a loss function can improve the detection performance of the model.

4.4.3. Effectiveness of Threshold Calculation Method

A study [43] showed that determining an appropriate threshold is as important as the algorithm itself. The experiments in the part will further prove the point. In multivariate time-series anomaly detection, there are two common threshold calculation methods: non-parametric dynamic threshold (NDT) and peak over threshold (POT) [44]. We used the two threshold calculation methods to conduct experiments on three public datasets, and Table 6 shows the experimental results.

Table 6. The effects of the threshold calculation method. The best results are highlighted in bold.

Method	SMAP				SMD				MSL			
	F1	P	R	Threshold	F1	P	R	Threshold	F1	P	R	Threshold
POT	0 *	0 *	0 *	1.0967	0.7137	0.5549	0.9999	0.1424	0.7467	0.9865	0.6007	0.9086
NDT	**0.8955**	**0.9767**	**0.8268**	0.6390	**0.9140**	**0.8416**	0.9999	0.2111	**0.9366**	0.9257	0.9478	0.5845

* On the SMAP dataset, the threshold obtained by POT is too high, resulting in TP and FP being 0. Therefore, the three calculation indicators are all 0.

When NDT is used as threshold calculation method, the detection metrics are significantly improved. On the SMD dataset, the F1 value and precision increased by 28.07% and 51.67%, respectively. On the MSL dataset, the F1 value and recall increased by 25.43% and 57.78%, respectively. The experimental results show that NDT can significantly improve the detection performance of the model.

4.5. Sensitivity Analysis

The model proposed in this paper involves many parameters, such as sliding window size, learning rate, γ, batch size, and the hidden layer dimensions of the GRU-based feature fusion module, the prediction-based model, and the reconstruction-based model. Sliding window size and γ are more important than other parameters. Therefore, we focus on analyzing the sensitivity of the sliding window size and γ.

4.5.1. Sensitivity to the Window Size.

The sliding window size is an important parameter that affects model performance and training time. We conducted several sets of experiments with different window sizes on three public datasets, and the experimental results are shown in Figures 5 and 6. Compared with SMAP, F1 values fluctuate less on SMD and MSL. When we use smaller windows, our model takes less time to train. If the window is too small, our model cannot capture local contextual information well. If the window is too large, the training time of the model will increase. The window size of 100 used in the experiments balances the F1 score and training time.

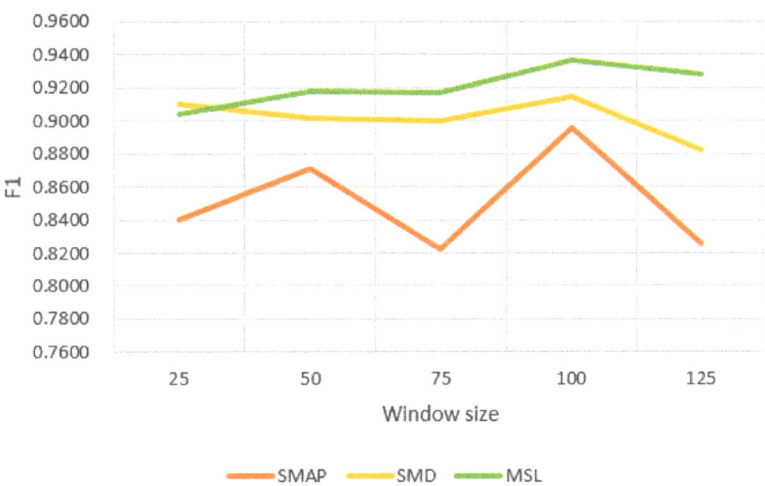

Figure 5. F1 score with window size.

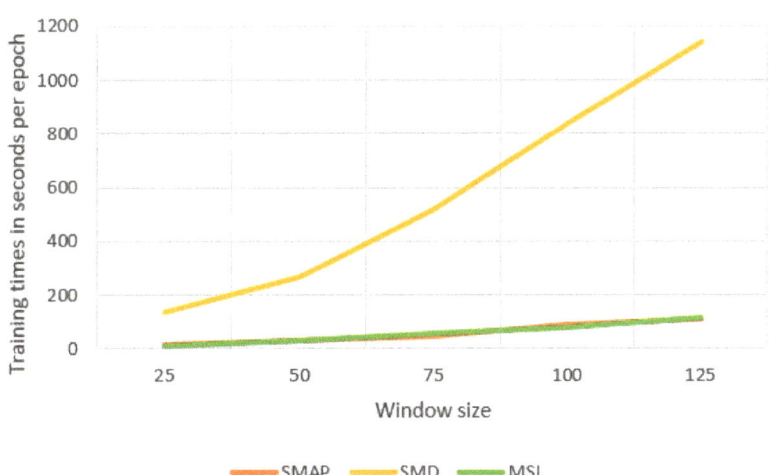

Figure 6. Training times with window size.

4.5.2. Sensitivity to γ

γ is an important parameter used to balance the error of the prediction model and the error of the reconstruction model. We conducted several experiments with different γ values on three public datasets, and the experimental results are shown in Figure 7. Compared with MSL, the F1 values fluctuate more on SMAP and SMD. The γ of 1 we use in the experiments balances prediction-based and reconstruction-based models well.

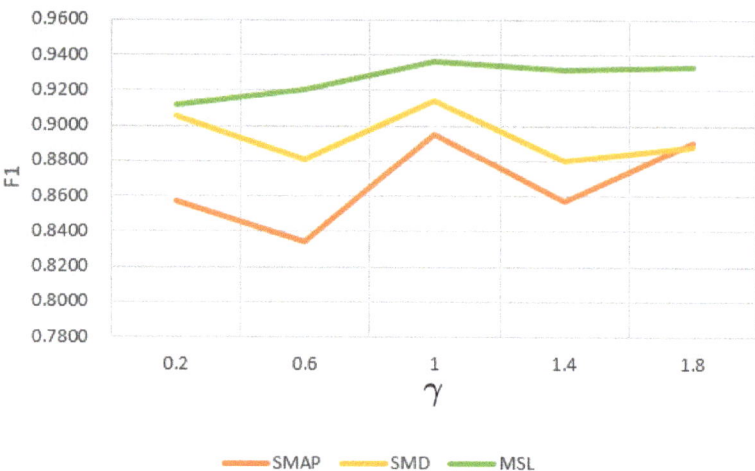

Figure 7. F1 score with γ.

5. Case Study

In this section, we verify the effectiveness and practicality of DCFF-MTAD on our dataset, which focuses on the shield machine—a critical piece of engineering equipment used for tunnel excavation in projects such as subways and highway tunnels. Due to the complexity and uncertainty of the underground environment, engineering safety and quality problems occur from time to time. For our study, we consider a tunnel project section with a total length of 943.5 m, where the shield tunneling method is employed for construction. Specifically, we analyze an abnormal event report which indicates that the shield machine passed through a karst cave between 13:23 and 21:32 on 2 November 2020. The geological survey report shows that the cave fillings consist of soft plastic silty clay, gravel sand, gravel, and other materials, which pose significant challenges for anomaly detection due to their inhomogeneity and complexity.

We used 31,333 pieces of normal data from 13:35 on 23 October 2020 to 14:08 on 29 October 2020 as training data, and 6475 pieces of data from 06:01 on 2 November 2020 to 21:56 on 2 November 2020 as testing data. We selected the total thrust, advance speed, cutter head torque, and screw speed recorded in the abnormal event report as the input parameters of our model. Firstly, the input parameters were normalized, and then the processed data were input into the spatial STFT module and the time-based graph attention layer in parallel to extract the spatial and temporal features of the input parameters, respectively. The output representations of the spatial STFT module and the time-based graph attention layer were concatenated and fed into the GRU to fuse spatial and temporal features. The output of the GRU was fed into the prediction-based and reconstruction-based models in parallel to obtain the predicted and reconstructed values of the input parameters.

The experimental results on the testing data are shown in Figure 8. The green curve represents the actual value, the orange curve represents the predicted value obtained by the prediction-based model, and the dark blue curve represents the reconstructed value obtained by the reconstruction-based model, all of which are normalized values. Our dual-channel feature extraction module fully extracts the time and spatial features of the construction parameters, and the prediction and reconstruction models use these two features to predict and reconstruct the data. We used the anomaly score to measure the difference between the value obtained by the model and the actual value, and an appropriate threshold was calculated using the non-parametric dynamic threshold method. When the anomaly score consistently exceeds the threshold, we consider it an abnormal phenomenon, which is indicated by the light red box.

As shown in Figure 8, the time of the purple line is 13:11 on 2 November 2020. From 06:01 to 13:11, total thrust, advance speed, cutter head torque, and screw speed of the shield machine fluctuated around 0.75, 0.1, 0.5, and 0.1, respectively, and the anomaly score was always below the threshold. From 13:11 to 13:23, the shield machine gradually approached the cave area, total thrust showed a downward trend, advance speed and screw speed showed an upward trend, cutter head torque showed obvious fluctuations, and the anomaly score gradually increased but did not exceed the threshold. From 13:23 to 21:32, when the shield machine tunneled forward in the cave area, the total thrust and cutter head torque decreased, the advance speed and screw speed increased, and the anomaly score exceeds the threshold—this period is marked as abnormal. After 21:32, the shield machine left the cave area and the four construction parameters gradually returned to the normal range, with the anomaly score gradually falling below the threshold.

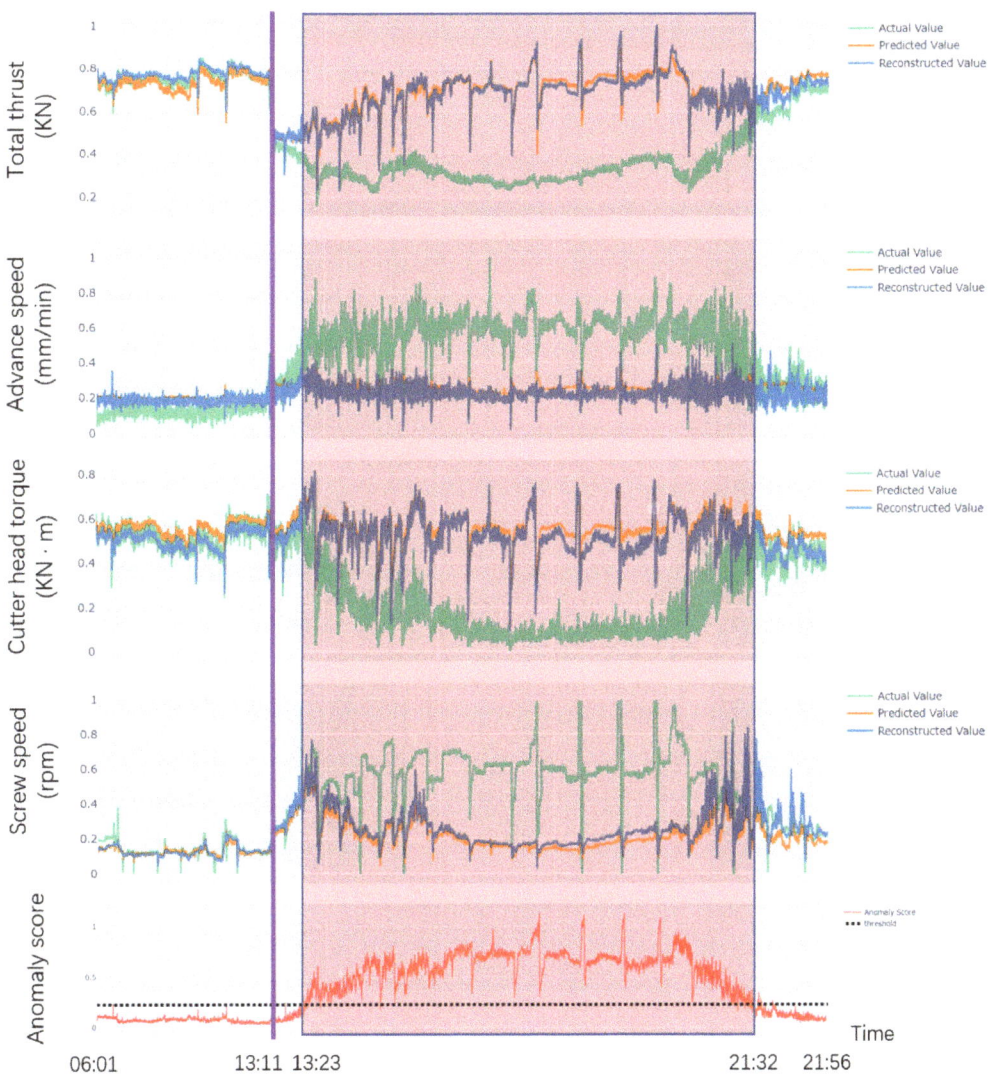

Figure 8. Anomaly detection result graph.

Our model extracts both spatial and temporal features of the construction parameters through spatial STFT and time-based graph attention layer. The two features are fused. This is because our model extracts the rich information hidden in the data, so the model successfully can learn the normal behavior pattern of construction parameters. When the construction parameter is abnormal, our model can detect the anomaly quickly. From this, it can be seen that DCFF-MTAD is able to monitor whether the shield machine is operating normally or not without much prior knowledge of the detection of abnormal behavior. This would be of great significance for ensuring the safe operation of the shield machine and improving its reliability by combining the proposed method with the automatic intelligent control technology of the shield machine.

6. Conclusions and Future Work

In this paper, we propose DCFF-MTAD, a stable and practical multivariate time series anomaly detection model, which employs dual-channel feature fusion. The performance of the model is mainly guaranteed by (1) the temporal features of multivariate data extracted by time-based graph attention layer; (2) the spatial features of multivariate data extracted by spatial STFT; (3) the feature-level fusion of temporal features and spatial features enriches the feature representation of data; and (4) the introduction of Huber Loss makes the anomaly detection performance of the model more robust. Extensive experiments are conducted to tune and validate the hyperparameters of the model to achieve the best performance. Experimental results show that compared with the state-of-the-art approaches, our method has the best anomaly detection performance on the SMAP and SMD public datasets, with 7.12% and 4.95% improvement in F1 values, respectively. The F1 score of our method is second only to OmniAnomaly on the MSL public dataset. In a practical application scenario, our method can continuously monitor important parameters during shield tunneling. Anomalies in construction parameters can be well detected when the parameters deviate from normal behavior.

In future research, we will improve our method so that it can be widely used in multivariate time series anomaly detection. The following suggestions can further improve the performance of the method and increase the availability of the method in anomaly detection: (1) The fully connected neural network used in the prediction-based model and the GRU used in the reconstruction-based model are more commonly used methods. If lower model complexity is required, the computational efficiency and performance of the entire model could be improved by replacing them with faster and better modules. (2) The choice of threshold is very important for the anomaly detection performance of the model. If the selected threshold is too high, only a few anomalies can be detected; if the selected threshold is too low, normal samples will be misjudged as abnormal. To further improve the anomaly detection performance and universality of the model, two existing threshold calculation methods (peak over threshold and non-parametric dynamic threshold) can be optimized or a new one can be developed. (3) In the case study, our model learned features from the normal behavior of construction parameters. On small-scale datasets, unsupervised learning may be extended to supervised learning, which means that both normal behavior and abnormal behavior can be used for learning data features.

Author Contributions: Conceptualization, X.G. and M.H.; Data curation, Z.X. and Y.Y.; Formal analysis, Z.X. and Y.Y.; Funding acquisition, X.G. and M.H.; Investigation, Z.X., Y.Y. and X.G.; Methodology, Z.X., Y.Y. and X.G.; Project administration, X.G. and M.H.; Resources, X.G. and M.H.; Software, Z.X. and Y.Y.; Supervision, X.G. and M.H.; Validation, Z.X., Y.Y., X.G., and M.H.; Visualization, Z.X. and Y.Y.; Writing—original draft, Z.X. and Y.Y.; Writing—review and editing, X.G. All authors have read and agreed to the published version of the manuscript.

Funding: This research was funded by the project of Shanghai Science and Technology Commission: research on digital twin scenario construction and dynamic inference technology for accurate maintenance of transportation infrastructure (No. 21ZR1423800) and funded by the project of Shanghai

Municipal Transportation Commission: the key technology and application research of the whole life cycle operation management of urban infrastructure (cross-river bridge) (No. JT2021-KY-013).

Institutional Review Board Statement: Not applicable.

Informed Consent Statement: Not applicable.

Data Availability Statement: This research employed publicly available datasets for its experimental studies. The data in the case study are not publicly available due to the confidentiality requirement of the project.

Acknowledgments: The authors would like to thank Li Zhou for her help. The authors would like to thank the editor and the anonymous reviewers for their valuable comments and suggestions on earlier versions of the manuscript.

Conflicts of Interest: The authors declare no conflict of interest.

Abbreviations

The following abbreviations are used in this manuscript:

STFT	Short-time Fourier Transform
RNN	Recurrent Neural Network
TCN	Temporal Convolutional Network
GNN	Graph Neural Network
DAGMM	Deep Autoencoder Gaussian Mixture Model
LSTM	Long Short-Term Memory
NDT	Non-parametric Dynamic Threshold
GAN	Generative Adversarial Network
GDN	Graph Deviation Network
USAD	UnSupervised Anomaly Detection
MTS-DCGAN	Multi-Time-Scale Deep Convolutional Generative Adversarial Network
CNN	Convolutional Neural Network
ATV	Automatic Transfer Vehicle
ECG	electrocardiogram
FFT	Fast Fourier Transform
NIDS	Network Intrusion Detection System
VGG	Visual Geometry Group
GRU	Gated Recurrent Unit
GAT	Graph Attention Network
MAE	Mean Absolute Error
MSE	Mean Square Error
SMAP	Soil Moisture Active Passive
SMD	Server Machine Dataset
MSL	Mars Science Laboratory
NASA	National Aeronautics and Space Administration
MAD	Multivariate Anomaly Detection
FLOPs	floating point operations
POT	Peak Over Threshold

References

1. Ismail Fawaz, H.; Forestier, G.; Weber, J.; Idoumghar, L.; Muller, P.A. Deep learning for time series classification: A review. *Data Min. Knowl. Discov.* **2019**, *33*, 917–963. [CrossRef]
2. Chauhan, S.; Vig, L. Anomaly detection in ECG time signals via deep long short-term memory networks. In Proceedings of the IEEE International Conference on Data Science and Advanced Analytics (DSAA), Paris, France, 19–21 October 2015; pp. 1–7.
3. Goh, J.; Adepu, S.; Tan, M.; Lee, Z.S. Anomaly detection in cyber physical systems using recurrent neural networks. In Proceedings of the 2017 IEEE 18th International Symposium on High Assurance Systems Engineering (HASE), Singapore, 12–14 January 2017; pp. 140–145.
4. Ding, N.; Ma, H.; Gao, H.; Ma, Y.; Tan, G. Real-time anomaly detection based on long short-Term memory and Gaussian Mixture Model. *Comput. Electr. Eng.* **2019**, *79*, 106458. [CrossRef]

5. Wu, W.; He, L.; Lin, W.; Su, Y.; Cui, Y.; Maple, C.; Jarvis, S. Developing an unsupervised real-time anomaly detection scheme for time series with multi-seasonality. *IEEE Trans. Knowl. Data Eng.* **2020**, *34*, 4147–4160. [CrossRef]
6. Shen, L.; Li, Z.; Kwok, J. Timeseries anomaly detection using temporal hierarchical one-class network. *Adv. Neural Inf. Process. Syst.* **2020**, *33*, 13016–13026.
7. He, Y.; Zhao, J. Temporal convolutional networks for anomaly detection in time series. *J. Phys. Conf. Ser.* **2019**, *1213*, 042050. [CrossRef]
8. Chen, Z.; Chen, D.; Zhang, X.; Yuan, Z.; Cheng, X. Learning graph structures with transformer for multivariate time-series anomaly detection in IoT. *IEEE Internet Things J.* **2021**, *9*, 9179–9189. [CrossRef]
9. Dai, E.; Chen, J. Graph-augmented normalizing flows for anomaly detection of multiple time series. *arXiv* **2022**, arXiv:2202.07857.
10. Han, S.; Woo, S.S. Learning Sparse Latent Graph Representations for Anomaly Detection in Multivariate Time Series. In Proceedings of the 28th ACM SIGKDD Conference on Knowledge Discovery and Data Mining, Washington, DC, USA, 14–18 August 2022; pp. 2977–2986.
11. Xu, J.; Wu, H.; Wang, J.; Long, M. Anomaly transformer: Time series anomaly detection with association discrepancy. *arXiv* **2021**, arXiv:2110.02642.
12. Tuli, S.; Casale, G.; Jennings, N.R. Tranad: Deep transformer networks for anomaly detection in multivariate time series data. *arXiv* **2022**, arXiv:2201.07284.
13. Wang, X.; Pi, D.; Zhang, X.; Liu, H.; Guo, C. Variational transformer-based anomaly detection approach for multivariate time series. *Measurement* **2022**, *191*, 110791. [CrossRef]
14. Zong, B.; Song, Q.; Min, M.R.; Cheng, W.; Lumezanu, C.; Cho, D.; Chen, H. Deep autoencoding gaussian mixture model for unsupervised anomaly detection. In Proceedings of the International Conference on Learning Representations, Vancouver, BC, Canada, 30 April–3 May 2018.
15. Hundman, K.; Constantinou, V.; Laporte, C.; Colwell, I.; Soderstrom, T. Detecting spacecraft anomalies using lstms and nonparametric dynamic thresholding. In Proceedings of the 24th ACM SIGKDD International Conference on Knowledge Discovery & Data Mining, London, UK, 19–23 August 2018; pp. 387–395.
16. Li, D.; Chen, D.; Jin, B.; Shi, L.; Goh, J.; Ng, S.K. MAD-GAN: Multivariate anomaly detection for time series data with generative adversarial networks. In Proceedings of the Artificial Neural Networks and Machine Learning–ICANN 2019: Text and Time Series: 28th International Conference on Artificial Neural Networks, Munich, Germany, 17–19 September 2019; Proceedings Part IV; Springer: Berlin/Heidelberg, Germany, 2019; pp. 703–716.
17. Audibert, J.; Michiardi, P.; Guyard, F.; Marti, S.; Zuluaga, M.A. Usad: Unsupervised anomaly detection on multivariate time series. In Proceedings of the 26th ACM SIGKDD International Conference on Knowledge Discovery & Data Mining, Online, 6–10 July 2020; pp. 3395–3404.
18. Su, Y.; Zhao, Y.; Niu, C.; Liu, R.; Sun, W.; Pei, D. Robust anomaly detection for multivariate time series through stochastic recurrent neural network. In Proceedings of the 25th ACM SIGKDD International Conference on Knowledge Discovery & Data Mining, Anchorage, AK, USA, 4–8 August 2019; pp. 2828–2837.
19. Abdulaal, A.; Liu, Z.; Lancewicki, T. Practical approach to asynchronous multivariate time series anomaly detection and localization. In Proceedings of the 27th ACM SIGKDD Conference on Knowledge Discovery & Data Mining, Virtual, 14–18 August 2021; pp. 2485–2494.
20. Liang, H.; Song, L.; Wang, J.; Guo, L.; Li, X.; Liang, J. Robust unsupervised anomaly detection via multi-time scale DCGANs with forgetting mechanism for industrial multivariate time series. *Neurocomputing* **2021**, *423*, 444–462. [CrossRef]
21. Deng, A.; Hooi, B. Graph neural network-based anomaly detection in multivariate time series. In Proceedings of the AAAI Conference on Artificial Intelligence, Virtual, 2–9 February 2021; Volume 35, pp. 4027–4035.
22. Chen, X.; Qiu, Q.; Li, C.; Xie, K. GraphAD: A Graph Neural Network for Entity-Wise Multivariate Time-Series Anomaly Detection. In Proceedings of the 45th International ACM SIGIR Conference on Research and Development in Information Retrieval, Madrid, Spain, 11–15 July 2022; pp. 2297–2302.
23. Wang, Y.; Zhang, J.; Guo, S.; Yin, H.; Li, C.; Chen, H. Decoupling representation learning and classification for gnn-based anomaly detection. In Proceedings of the 44th International ACM SIGIR Conference on Research and Development in Information Retrieval, Madrid, Spain, 11–15 July 2021; pp. 1239–1248.
24. Yang, J.; Yue, Z. Learning Hierarchical Spatial-Temporal Graph Representations for Robust Multivariate Industrial Anomaly Detection. *IEEE Trans. Ind. Inform.* **2022**. [CrossRef]
25. Razaque, A.; Abenova, M.; Alotaibi, M.; Alotaibi, B.; Alshammari, H.; Hariri, S.; Alotaibi, A. Anomaly detection paradigm for multivariate time series data mining for healthcare. *Appl. Sci.* **2022**, *12*, 8902. [CrossRef]
26. Li, Y.; Peng, X.; Zhang, J.; Li, Z.; Wen, M. DCT-GAN: Dilated convolutional transformer-based gan for time series anomaly detection. *IEEE Trans. Knowl. Data Eng.* **2021**, *35*, 3632–3644. [CrossRef]
27. Qin, S.; Zhu, J.; Wang, D.; Ou, L.; Gui, H.; Tao, G. Decomposed Transformer with Frequency Attention for Multivariate Time Series Anomaly Detection. In Proceedings of the 2022 IEEE International Conference on Big Data (Big Data), Osaka, Japan, 17–20 December 2022; pp. 1090–1098.
28. Kim, J.; Kang, H.; Kang, P. Time-series anomaly detection with stacked Transformer representations and 1D convolutional network. *Eng. Appl. Artif. Intell.* **2023**, *120*, 105964. [CrossRef]

19. Gültekin, Ö.; Cinar, E.; Özkan, K.; Yazıcı, A. Multisensory data fusion-based deep learning approach for fault diagnosis of an industrial autonomous transfer vehicle. *Expert Syst. Appl.* **2022**, *200*, 117055. [CrossRef]
20. Li, H.; Boulanger, P. Structural Anomalies Detection from Electrocardiogram (ECG) with Spectrogram and Handcrafted Features. *Sensors* **2022**, *22*, 2467. [CrossRef] [PubMed]
21. Zhou, X.; Xiong, J.; Zhang, X.; Liu, X.; Wei, J. A radio anomaly detection algorithm based on modified generative adversarial network. *IEEE Wirel. Commun. Lett.* **2021**, *10*, 1552–1556. [CrossRef]
22. Chong, S.H.; Cho, G.C.; Hong, E.S.; Lee, S.W. Numerical study of anomaly detection under rail track using a time-variant moving train load. *Geomech. Eng.* **2017**, *13*, 161–171.
23. Kavalerov, I.; Li, W.; Czaja, W.; Chellappa, R. 3-D Fourier scattering transform and classification of hyperspectral images. *IEEE Trans. Geosci. Remote Sens.* **2020**, *59*, 10312–10327. [CrossRef]
24. Khan, A.S.; Ahmad, Z.; Abdullah, J.; Ahmad, F. A spectrogram image-based network anomaly detection system using deep convolutional neural network. *IEEE Access* **2021**, *9*, 87079–87093. [CrossRef]
25. Haleem, M.S.; Castaldo, R.; Pagliara, S.M.; Petretta, M.; Salvatore, M.; Franzese, M.; Pecchia, L. Time adaptive ECG driven cardiovascular disease detector. *Biomed. Signal Process. Control.* **2021**, *70*, 102968. [CrossRef]
26. Sanakkayala, D.C.; Varadarajan, V.; Kumar, N.; Soni, G.; Kamat, P.; Kumar, S.; Patil, S.; Kotecha, K. Explainable AI for Bearing Fault Prognosis Using Deep Learning Techniques. *Micromachines* **2022**, *13*, 1471. [CrossRef] [PubMed]
27. Liu, C.; Cichon, A.; Królczyk, G.; Li, Z. Technology development and commercial applications of industrial fault diagnosis system: A review. *Int. J. Adv. Manuf. Technol.* **2021**, *118*, 3497–3529. [CrossRef]
28. Stankovic, L.; Ivanovic, V.; Petrovic, Z. Unified approach to the noise analysis in the spectrogram and Wigner distribution. *Ann. Telecommun.* **1996**, *51*, 585–594. [CrossRef]
29. Friedlander, B.; Scharf, L. On the structure of time-frequency spectrum estimators. *IEEE Trans. Signal Process.* **2023**, *in press.*
30. Zhao, H.; Wang, Y.; Duan, J.; Huang, C.; Cao, D.; Tong, Y.; Xu, B.; Bai, J.; Tong, J.; Zhang, Q. Multivariate time-series anomaly detection via graph attention network. In Proceedings of the 2020 IEEE International Conference on Data Mining (ICDM), Sorrento, Italy, 17–20 November 2020; pp. 841–850.
31. Entekhabi, D.; Njoku, E.G.; O'Neill, P.E.; Kellogg, K.H.; Crow, W.T.; Edelstein, W.N.; Entin, J.K.; Goodman, S.D.; Jackson, T.J.; Johnson, J.; et al. The soil moisture active passive (SMAP) mission. *Proc. IEEE* **2010**, *98*, 704–716. [CrossRef]
32. Srivastava, N.; Hinton, G.; Krizhevsky, A.; Sutskever, I.; Salakhutdinov, R. Dropout: A simple way to prevent neural networks from overfitting. *J. Mach. Learn. Res.* **2014**, *15*, 1929–1958.
33. Goldstein, M.; Uchida, S. A comparative evaluation of unsupervised anomaly detection algorithms for multivariate data. *PLoS ONE* **2016**, *11*, e0152173. [CrossRef]
34. Siffer, A.; Fouque, P.A.; Termier, A.; Largouet, C. Anomaly detection in streams with extreme value theory. In Proceedings of the 23rd ACM SIGKDD International Conference on Knowledge Discovery and Data Mining, Halifax, NS, Canada, 13–17 August 2017; pp. 1067–1075.

Disclaimer/Publisher's Note: The statements, opinions and data contained in all publications are solely those of the individual author(s) and contributor(s) and not of MDPI and/or the editor(s). MDPI and/or the editor(s) disclaim responsibility for any injury to people or property resulting from any ideas, methods, instructions or products referred to in the content.

Article

MTGEA: A Multimodal Two-Stream GNN Framework for Efficient Point Cloud and Skeleton Data Alignment

Gawon Lee and Jihie Kim *

Department of Artificial Intelligence, Dongguk University, 30 Pildong-ro 1 Gil, Seoul 04620, Republic of Korea
* Correspondence: jihie.kim@dgu.edu

Abstract: Because of societal changes, human activity recognition, part of home care systems, has become increasingly important. Camera-based recognition is mainstream but has privacy concerns and is less accurate under dim lighting. In contrast, radar sensors do not record sensitive information, avoid the invasion of privacy, and work in poor lighting. However, the collected data are often sparse. To address this issue, we propose a novel Multimodal Two-stream GNN Framework for Efficient Point Cloud and Skeleton Data Alignment (MTGEA), which improves recognition accuracy through accurate skeletal features from Kinect models. We first collected two datasets using the mmWave radar and Kinect v4 sensors. Then, we used zero-padding, Gaussian Noise (GN), and Agglomerative Hierarchical Clustering (AHC) to increase the number of collected point clouds to 25 per frame to match the skeleton data. Second, we used Spatial Temporal Graph Convolutional Network (ST-GCN) architecture to acquire multimodal representations in the spatio-temporal domain focusing on skeletal features. Finally, we implemented an attention mechanism aligning the two multimodal features to capture the correlation between point clouds and skeleton data. The resulting model was evaluated empirically on human activity data and shown to improve human activity recognition with radar data only. All datasets and codes are available in our GitHub.

Keywords: human activity recognition; mmWave radar; Kinect V4 sensor; point clouds; skeleton data; multimodal; two stream; attention mechanism

Citation: Lee, G.; Kim, J. MTGEA: A Multimodal Two-Stream GNN Framework for Efficient Point Cloud and Skeleton Data Alignment. *Sensors* **2023**, *23*, 2787. https://doi.org/10.3390/s23052787

Academic Editor: Yi Qin

Received: 29 January 2023
Revised: 28 February 2023
Accepted: 1 March 2023
Published: 3 March 2023

Copyright: © 2023 by the authors. Licensee MDPI, Basel, Switzerland. This article is an open access article distributed under the terms and conditions of the Creative Commons Attribution (CC BY) license (https://creativecommons.org/licenses/by/4.0/).

1. Introduction

As the world population ages, older persons are a growing group in society. According to World Population Prospects 2019 (United Nations, 2019), by 2050, the number of persons aged 65 years or over globally will surpass those aged 15–24. In addition to this, single-person households have increased tremendously in the last few years due to societal changes. With these population changes, home care systems have emerged as a promising venue of intelligent technologies for senior and single-person households. In addition, the recent COVID-19 pandemic has further increased the importance of developing home care systems. The current mainstream home care systems are based on cameras [1]; however, people can feel uncomfortable being recorded by cameras and hence might refuse to be monitored by camera-based techniques. The biggest problem is the invasion of privacy. If the personal data recorded by the camera is leaked, it may have devastating consequences. There is also a problem with the accuracy of the camera being affected by the lighting and its placement. Consequently, alternative approaches to home care are needed.

With the advances in Frequency-Modulated Continuous Wave (FMCW) mmWave technology, human activity recognition by mmWave radar sensors has recently attracted significant attention. A radar sensor can collect 3D coordinates called point clouds while emitting and absorbing radio waves to and from objects. Moreover, depending on the hardware or data collection tool type, other data (e.g., range and velocity) can be captured simultaneously. A radar sensor also does not require a strict environment setting. In other words, it works correctly even in poor lighting and with poor camera placement. Because a radar sensor does not record personal information as an image or video, the issue of invasion of privacy is significantly reduced. However, radar produces sparse point clouds due to the radar sensor's radio wavelength and inherent noise. Many researchers have devoted effort to processing sparse radar data [2–5] and have thus devised voxelization. Voxelization is a method that converts point clouds into voxels with constant dimensions, which researchers decide empirically. Singh et al. [6] voxelized point clouds with dimensions 60 × 10 × 32 × 32 (depth = 10) and then fed them into a set of classifiers. Although voxelization is a well-known pre-processing method, it is inefficient, as researchers must decide the dimensions empirically. Using upsampling techniques to deal with the sparsity of the point clouds is another popular method. Palipana et al. [7] resampled the number of points to achieve a fixed number. They used Agglomerative Hierarchical Clustering (AHC) for upsampling. The AHC algorithm adds a cluster's centroid as a new point after clustering the point clouds.

Another popular sensor is the Microsoft Kinect [8,9], which provides various data such as RGB videos, depth sequences, and skeleton information. In recent years, many studies have taken advantage of skeleton data because of their robustness to human appearance change as well as illumination. Hence, plenty of related skeleton data (e.g., NTURGB+D [10] and NTU-RGB+D 120 [11]) has been collected and used. Rao et al. [12] proposed learning the pattern invariance of actions using a momentum Long Short-Term Memory (LSTM) after seven augmentation strategies to boost action recognition accuracy via 3D skeleton data. To overcome the sparsity of point clouds, we propose exploiting this skeleton data in radar-based recognition, and we designed a multimodal framework that can effectively combine point clouds with useful skeleton information.

Depth video recordings gathered using Kinect were also utilized for human activity recognition. In the [13], the authors pre-processed the dataset recorded by depth cameras. To avoid misleading context, separating poses and removing context were needed. However, the opportunities for learning more from the background rather than a real person's data remain, and recorded videos have privacy issues.

In the case of wearable sensors, Wozniak et al. [14] identify the user's body position using wearable sensor data from various body parts, such as the ankle, wrist, waist, and chest. They have decided only two sensors are enough to obtain up to 100% accuracy in a thorough examination. Although proposed models in [14] achieved 99.89% accuracy rates, wearable devices which touch body parts, such as the chest, during data collection, can be quite cumbersome in actual use, especially for children or elderly people.

Various multimodal frameworks that take advantage of data from multiple sources have already been studied. As such, fusion strategies for combining multimodal features have been devised. These include concatenation [15], attention mechanisms [16], and a simple weight-sum manner [17].

Based on these results, this paper proposes a novel Multimodal Two-stream GNN Framework for Efficient Point Cloud and Skeleton Data Alignment (MTGEA) to improve human activity recognition with radar data. The proposed framework utilizes spatial temporal graph convolutional networks (ST-GCNs) as graph neural networks (GNNs), which can effectively capture both temporal and spatial features. Three upsampling techniques were used to address the sparsity of point clouds. In addition, unlike previous work, which uses the single-modal framework, we constructed a multimodal framework with skeletal data so that reliable features could be obtained. While strict one-to-one mapping is difficult due to the different types of environmental settings, in the proposed model, the point clouds and skeleton data can be used together as 3D coordinates. Based on the embedded representations generated from applying ST-GCN to both data, we incorporated an attention mechanism in aligning the point clouds and skeleton data and attained structural similarity and accurate key features from the two datasets. Then, the aligned features and embedded features of point clouds were concatenated to form the final classification decision. For the reasoning of human activity recognition, we used the radar data only, with the Kinect part frozen. We evaluated MTGEA empirically with seven human activity data, including falling. All data were collected by mmWave radar and Kinect v4 sensors simultaneously. In summary, our main contributions are as follows:

- We propose a novel MTGEA. Our major contribution is presenting a new approach for incorporating accurate Kinect skeletal features into the radar recognition model, enabling human activity recognition using sparse point clouds alone without having to use the Kinect stream during reasoning;
- We propose skeleton data with an attention mechanism as a tool for generating reliable features for the multimodal alignment of point clouds. We also utilize three upsampling techniques to address the sparsity of radar point clouds;
- We provide a new point cloud and skeleton dataset for human activity recognition. All data simultaneously collected by mmWave radar and Kinect v4 sensors are open source, along with the entire code and pre-trained classifiers.

2. Related Works

Early research on detecting human actions usually used images. Ogundokun et al. [18] proposed a deep convolutional neural network (DCNN) framework for human posture classification. They chose DCNN for deriving abstract feature maps from input data. However, the pixels of images and image sequences have various backgrounds, so features should be carefully extracted due to the risk of privacy invasion.

So, in the case of radar sensors, most researchers focused on pre-processing sparse point clouds. One of the popular methods was voxelization. Sengupta et al. [19] presented mmPose-NLP, an mmWave radar-based skeletal keypoint inspired by natural language processing (NLP). In their study, point clouds were first pre-processed through voxelization. Authors regarded this method as a process similar to the tokenization of NLP. The mmPose-NLP architecture was applied to predict the voxel indexes, corresponding to 25 skeleton key points. To measure the accuracy of the proposed system, the authors used the Mean Absolute Error (MAE) metric. However, voxelization pre-processing methods, which usually require a fixed shape, are augmented sequences. In the case of point clouds, Palipana et al. [7] proposed an upsampling method to expand sparse point clouds. They used AHC for upsampling until they achieved a fixed number of point clouds. In the AHC algorithm, all point clouds formed clusters first, and each cluster's centroid was added to the point clouds as a new point. We provide more detailed information regarding the AHC algorithm in Section 3.2.

In [20], a pre-trained model based on two consecutive convolution neural networks (CNNs) was used to extract reliable features in skeleton form from sparse radar data. Then, the GNN-based model was applied for classification. It achieved above 90% accuracy on the MMActivity dataset [6]. However, two-phase flow models such as this can be inefficient.

In this paper, we utilized the two-stream multimodal framework and alignment method to exploit an accurate skeleton dataset from Kinect. Many previous researchers have devised various alignment methods for proper feature fusion. Yang et al. [17] built a shallow graph convolutional network with a two-stream structure for bone and joint skeleton data and proposed a weight-sum manner to obtain the final prediction. This method requires a lower computational cost and is relatively simple. Concatenation is one of the popular methods for feature fusion. Pan et al. [21] proposed a Variational Relational Point Completion Network (VRCNet) to construct complete shapes for partial point clouds. VRCNet had two consecutive encoder–decoder sub-networks named probabilistic modeling (PMNet) and relational enhancement (RENet). In the PMNet, the concatenation of coarse complete point clouds and incomplete point clouds occurred, which led to the generation of the overall skeletons. Weiyao et al. [15] proposed a multimodal action recognition model based on RGB-D and adopted skeleton data as the multimodal data. The proposed network consisted of GCN and CNN. The GCN network took the skeletal sequence, and R (2+1)D based on the CNN network architecture took the RGB video. Then, the outer product of two compressed features was obtained to make the final classification decision. Zheng et al. [16] designed a Multimodal Relation Extract Neural Network with Efficient

Graph Alignment (MEGA). To identify textual relations using visual clues, MEGA utilized visual objects in an image and textual entities in a sentence as multimodal data. The authors conducted experiments using the MNRE dataset, demonstrating that the alignment of visual and textual relations by attention could improve the relation extraction performance. In this paper, we created a skeleton and point cloud dataset and used these sensor data as multimodal data. Then, we utilized an attention mechanism to integrate these two features to assist in generating more reliable features.

3. Methodology

3.1. Subsection Experimental Environments and Dataset

Training and test data were collected following a study protocol approved by the Institutional Review Board of Dongguk University (Approval number: DUIRB-202104-04). We recruited 19 subjects to collect the new dataset, the DGUHA (Dongguk University Human Activity) dataset, which includes both point cloud and skeleton data. All subjects were in their twenties (the average age was 23 years). In the environment shown in Figure 1a, each subject performed seven movements: running, jumping, sitting down and standing up, both upper limb extension, falling forward, right limb extension, and left limb extension, as illustrated in Figure 2 (This figure was captured from the authors and thus did not require approval from IRB). All of the subjects performed each activity for about 20 s. Including break time, data collection was performed for 1 h, and all activities were repeated approximately 5–6 times during this time. We utilized an mmWave radar sensor and Microsoft Kinect v4 sensor to collect the data.

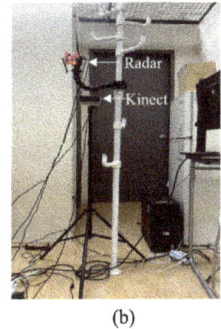

(a) (b)

Figure 1. Experimental environments for the DGUHA dataset. (**a**) Data collection environments, and (**b**) Data collection setup.

In the case of the mmWave radar sensor, TI's IWR1443BOOST radar (Texas Instruments, city and country: Dallas, TX, USA), which includes four receivers and three transmitters, was used. It is based on FMCW, of which a chirp signal is a fundamental component. After transmitters emit an FMCW signal, receivers detect objects in a 3D plane by measuring the delay time according to the distance to the target as a frequency difference. The sensor was mounted parallel to the ground at a height of 1.2 m, as shown in Figure 1b. The sampling rate of the radar was 20 fps, and we collected the data using a robot operating system [22]. We stored five primary data modalities: 3D coordinates (x, y, and z in m), range, velocity, bearing angle (degrees), and intensity. The 3D coordinates are usually called point clouds.

The Microsoft Kinect v4 sensor was also mounted parallel to the ground at a height of 1 m, as shown in Figure 1b. A total of 25 skeleton data represented the 3D locations of 25 major body parts: spine, chest, neck, left shoulder, left elbow, left wrist, left hand, left hand tip, left thumb, right shoulder, right elbow, right wrist, right hand, right hand tip, right thumb, left hip, left knee, left ankle, left foot, right hip, right knee, right ankle, right foot, and head. It captured skeleton data at a sampling rate of 20 fps. We collected the two datasets on Ubuntu 18.04 system simultaneously, and they were saved as a text file, as illustrated in Figure 3.

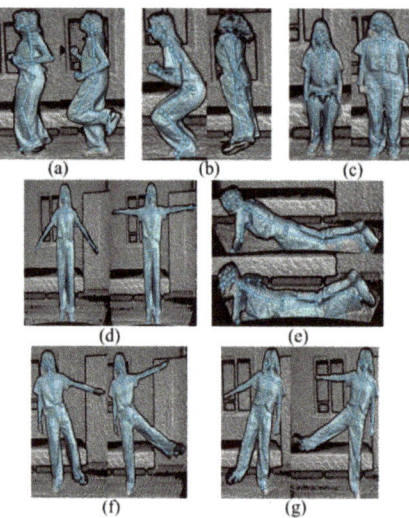

Figure 2. The DGUHA dataset collected in our experiments. (**a**) Running, (**b**) Jumping, (**c**) Sitting down and standing up, (**d**) Both upper limb extension, (**e**) Falling forward, (**f**) Right limb extension, and (**g**) Left limb extension.

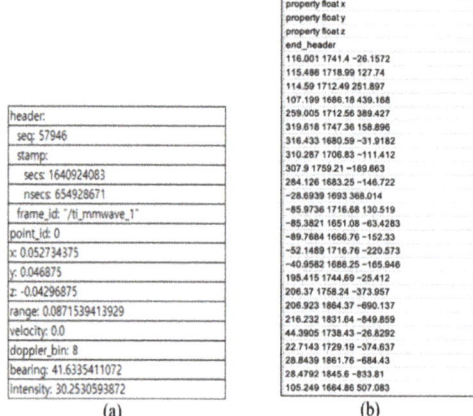

Figure 3. The DGUHA dataset format. (**a**) mmWave radar data format in DGUHA, and (**b**) Kinect data format in DGUHA.

3.2. Data Augmentation

The sampling rates of both sensors were the same, and each activity was performed for 20 s, as mentioned in Section 3.1. Although exact one-to-one mapping was difficult due to the different types of hardware and data collection tools, the two datasets were stored at 400 frames per activity. If there were fewer than 400 frames, we replaced missing frames with the last ones. In contrast, extra frames were removed to maintain 400 frames. We randomly picked data files from each activity to check the average, median, and mode of the number of point clouds. As shown in Table 1, the point clouds were sparse. This sparsity is because of the radar sensor's radio wavelength and inherent noise. To address the above challenge, we applied three upsampling techniques introduced in [7,12] to the point clouds.

Table 1. Descriptive statistics for data samples.

Activity	Mean	Median	Mode	Min	Max
Running	12.79	13.0	13	3	24
Jumping	9.65	10.0	10	3	3
Sitting down and standing up	5.86	6.0	5	2	13
Both upper limb extension	6.18	5.0	3	2	17
Falling forward	3.78	4.0	3	2	8
Right limb extension	6.11	6.0	5	2	13
Left limb extension	5.22	5.0	5	2	13

To use the skeleton data collected from Kinect simultaneously with those from the radar sensor as multimodal data, our upsampling techniques aimed to augment the number of point clouds to 25 per frame to match the number of joints in the collected skeleton data. To augment the number of point clouds, we used the following techniques for upsampling:

(1) Zero-Padding (ZP): ZP is the simplest and most efficient of the many data augmentation methods. We padded the remaining points with zeros to obtain 25-point clouds;

(2) Gaussian Noise (GN): The GNs were generated based on the standard derivations (SDs) of the original datasets. After ZP, we added Gaussian noise $N(0, 0.05)$ over point clouds according to the following formula:

$$N\left(x|\mu,\sigma^2\right) = \frac{1}{(2\pi\sigma^2)^{\frac{1}{2}}} \exp\left\{-\frac{1}{2\sigma^2}(x-\mu)^2\right\} \tag{1}$$

(3) Agglomerative Hierarchical Clustering (AHC): This algorithm is a bottom-up and iterative clustering approach. It consists of three steps. First, the dissimilarity between all data is calculated. Generally, Euclidean distance or Manhattan distance can be calculated. Second, the two closest data are clustered to create a class. Finally, the dissimilarity between the cluster and other data or between clusters is calculated. These three steps are repeated until all data become one cluster. Maximum, minimum, and mean can be calculated to measure the dissimilarity of the two clusters.

3.3. Feature Extraction Using ST-GCNs

We obtained 25 point clouds through upsampling to match the skeleton data. We then used the ST-GCN architecture to acquire multimodal representation, as illustrated in Figure 4. The GNN used in the proposed MTGEA is the ST-GCN. ST-GCN achieved promising performance by utilizing a graph representation of the skeleton data [23]. In the skeleton structure, human joints can be considered a vertex or node of a graph, and connections between them can be regarded as an edge or relation of the graph. In addition to a spatial graph based on human joints, there are temporal edges connecting joints between the previous and next steps within a movement. If a spatio-temporal graph for a movement is denoted as $\mathcal{G} = (V, E)$, V denotes the set of the joints, and E denotes both spatial and temporal edges. The authors [23] adopted a propagation rule similar to that of GCNs [24], which is defined as follows:

$$f_{out} = \widehat{A}^{-\frac{1}{2}}(A+I)\widehat{A}^{-\frac{1}{2}} f_{in} W, \tag{2}$$

where $\widehat{A}^{ii} = \sum_j (A^{ij} + I^{ij})$ and W is the weight matrix. The authors also used partitioning strategies such as distance partitioning, spatial configuration partitioning, and dismantled adjacency matrix into multiple matrixes A_j, where $A + I = \sum_j A_j$. Therefore, Equation (2) is transformed into:

$$f_{out} = \sum_j \widehat{A}_j^{-\frac{1}{2}} A_j \widehat{A}_j^{-\frac{1}{2}} f_{in} W_j, \tag{3}$$

where $\widehat{A}_j^{ii} = \sum_k(A_j^{ik}) + \varepsilon$ and $\varepsilon = 0.001$ is used to avoid empty rows in A_j. Then, the element-wise product is conducted between A_j and M to implement the learnable edge importance weighting. M is a learnable weight matrix and is initialized as an all-one matrix. Consequently, Equation (3) is substituted with:

$$f_{out} = \sum_j \widehat{A}_j^{-\frac{1}{2}}(A_j \otimes M)\widehat{A}_j^{-\frac{1}{2}} f_{in} W_j, \quad (4)$$

where $\otimes$ denotes the element-wise product. In our model, the three channels, which made up the 3D coordinates, were the input. As illustrated in Figure 4, two consecutive ST-GCN layers had the same 128 channels, and the final output of the ST-GCN contained 32 channels.

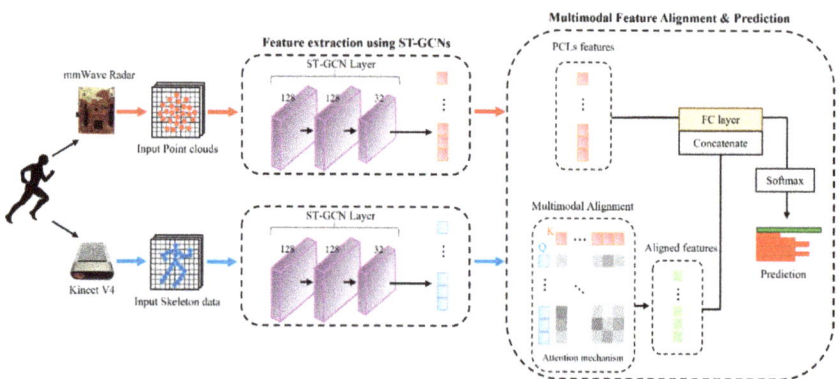

Figure 4. Illustration of the MTGEA. Three ST-GCN layers with the same channels were used to extract features in both point cloud and skeleton data. After passing the ST-GCN layers, features were extracted in the spatio-temporal domain from 3D coordinate data. Their features were then transformed into the matrixes Q, K, and V by three learnable matrixes, and the attention function was calculated, after which an aligned feature was obtained. The aligned and point cloud features were then concatenated and sent to the fully connected layer to form a final classification decision.

3.4. Multimodal Feature Alignment by Attention

In the field of NLP, an attention mechanism was first introduced in [25]. This mechanism allows a decoder to find parts to pay attention to from the source sentence. We implemented an attention mechanism to align point clouds and skeleton data. Unlike previous feature fusion methods [26–29], which operate by concatenating the features or simply calculating a weight-sum, an attention mechanism can find the structural similarity and accurate key features between two features, resulting in the generation of reliable features. These reliable features can help our model address sparse point clouds and recognize human activities more accurately. The input of the attention function, (scaled dot-product attention) [30], consists of a query, a key of the dimension d_k and values of the dimension d_v. We set d_k and d_v to the same number d_t, as proposed in [16], for simplicity. Queries, keys, and values were packed into matrixes Q, K, and V, respectively, and the matrix of outputs was calculated as:

$$Attention(Q, K, V) = softmax\left(\frac{QK^T}{\sqrt{d_t}}\right)V, \quad (5)$$

where the dot products of the query with all keys are scaled down by d_t. In practice, we projected each point cloud and skeleton data into a common t-dimensional space using an ST-GCN, achieving point cloud representation $X \in \mathbb{R}^{N \times d_t}$ and skeleton representation $Y \in \mathbb{R}^{N \times d_t}$. Then, we used three learnable matrixes $W_q \in \mathbb{R}^{d_t \times d_t}$, $W_k \in \mathbb{R}^{d_t \times d_t}$ and $W_v \in \mathbb{R}^{d_t \times d_t}$ empirically to generate the matrixes Q, K, and V as:

$$Q = W_q Y + bias_q, \tag{6}$$

$$K = W_k X + bias_k, \tag{7}$$

$$V = W_v X + bias_v, \tag{8}$$

where $bias_q$, $bias_k$ and $bias_v$ are the learnable biases. After generating the matrixes Q, K, and V, we computed the attention function and obtained the aligned feature $Z \in \mathbb{R}^{N \times d_t}$, as illustrated in Figure 4.

3.5. Feature Concatenation & Prediction

As shown in the rightmost box of Figure 4, we concatenated the aligned and point cloud features and sent them to the fully connected layer to obtain the final classification decision. Finally, the classification decision was normalized by the softmax function.

4. Results

In this section, we demonstrate the effectiveness of the proposed MTGEA components with the training and test sets of the DGUHA dataset. We performed all experiments on a machine with an Intel Xeon-Gold 6226 CPU, 192GB RAM (Intel Corporation, Santa Clara, CA, USA), and RTX 2080 Ti (Gigabyte, New Taipei City, Taipei) graphic card. We report the accuracy and weighted F1 score value as the evaluation metrics. The weighted F1 score is one of the metrics that take imbalanced data into account. Originally, the F1 score was calculated as follows:

$$F_1 \ score = \frac{2 \cdot Recall \cdot Precision}{Recall + Precision}, \tag{9}$$

where $Recall$ is True Positive/True Positive + False Negative, and $Precision$ is True Positive/True Positive + False Positive. We considered the weighted F1 score so that the ratio of the classes was balanced. (Approximately, running: 0.1432, jumping: 0.1419, sitting down and standing up: 0.1419, both upper limb extension: 0.1432, falling forward: 0.1432, right limb extension: 0.1432, and left limb extension: 0.1432.)

Three MTGEA models were trained using the three augmented types of data. We trained each model with a batch size of 13 for 300 epochs and used stochastic gradient descent with a learning rate of 0.01. Then, we froze the weights of the Kinect stream to verify the possibility of human activity recognition using radar data only. Therefore, only the test dataset of the point cloud was fed into the network during the test process, and the results are shown in Table 2.

Table 2. Test Accuracy on the DGUHA dataset.

Model	Accuracy (%)	Weighted F$_1$ Score (%)
MTGEA (ZP + Skeleton)	85.09	79.35
MTGEA (GN + Skeleton)	95.03	95.13
MTGEA (AHC + Skeleton)	98.14	98.14

Among the three augmented point cloud datasets, the MTGEA model that used the ZP augmentation strategy for sparse point clouds performed poorly in terms of prediction since the missing points were replaced by zeros only. However, the other models using multiple different augmentation strategies achieved higher accuracies of around 90%. In our evaluation, the best-performing MTGEA model, which was the one that used the AHC augmentation strategy, achieved a test accuracy of 98.14% and a weighted F1 score of 98.14%. This was 13.05% higher than the accuracy of the MTGEA model that used the ZP augmentation strategy and 3.11% higher than that using the GN augmentation strategy. This result indicates that the AHC algorithm can augment sparse point clouds more effectively. The confusion matrixes for the visualization of classification performance for our DGUHA dataset are illustrated in Figure 5, and the a–g labels denote the seven types of activity shown in Figure 2. According to the confusion matrix in Figure 5c, the MTGEA model that used the AHC augmentation strategy classified (a) running, (c) sitting down and standing up, (f) right limb extension, and (g) left limb extension 100% correctly. However, a few activities were confused with other activities; these were (b) jumping, (d) both upper limb extension, and (e) falling forward. However, these activities still achieved a high accuracy of over 95%.

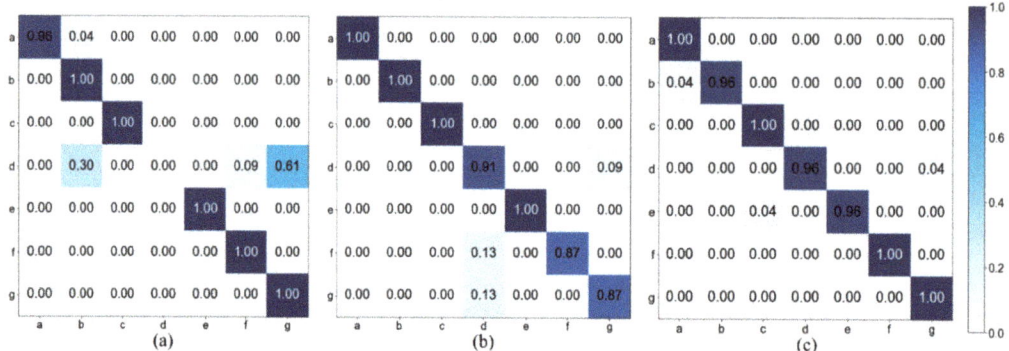

Figure 5. Confusion matrixes of three MTGEA models with different augmented data. (**a**) MTGEA (ZP + Skeleton), (**b**) MTGEA (GN + Skeleton), and (**c**) MTGEA (AHC + Skeleton).

According to the confusion matrix in Figure 5b, the MTGEA model that used the GN augmentation strategy achieved an accuracy under 95% for three out of seven activities. The three activities, (d) both upper limb extension, (f) right limb extension, and (g) left limb extension, are somewhat similar, as the arms or arms and legs moved away from the body and then moved back toward the body.

The MTGEA model that used the ZP augmentation strategy achieved 0% accuracy for (d) both upper limb extension activity, as this activity was somewhat confused with (b) jumping, (f) right limb extension, and (g) left limb extension, as shown in Figure 5a.

From these observations, we found that simple movements in which the body remains still and only the arms or legs move are generally harder to recognize than complex movements requiring the whole body, such as moving from left to right or running. Finally, the MTGEA model that used the AHC augmentation strategy achieved 95% accuracy for all activities, indicating the robustness of the model for simple activities that do not have complex movements distinct from other activities.

In addition, ablation studies were performed to demonstrate the necessity of the multimodal framework and attention mechanism in the proposed model.

5. Ablation Studies

5.1. Ablation Study for the Multimodal Framework

Ablation experiments were performed to justify the multimodal design of the proposed model. Single-modal models were created using a one-stream ST-GCN, and the ST-GCN architecture was the same as that of the MTGEA. The accuracy and weighted F1 score of the single-modal models are shown in Table 3. Compared to the multimodal models with the same augmented data, the single-modal models generally showed lower performance.

In the case of point clouds, the single-modal model used augmented point clouds with ZP and achieved 81.99% accuracy and a weighted F1 score of 81.51%. This was 3.1% lower in accuracy than the MTGEA model that used ZP. Notably, however, the single-modal model achieved a 2.16% higher weighted F1 score, as it classified (d) both upper limb extension activities 57% correctly. However, it classified the remaining activities incorrectly more often than the MTGEA model.

The second single-modal model that used augmented point clouds with GN achieved 92.55% accuracy and a weighted F1 score of 92.45%. These were 2.48% and 2.68% lower, respectively, than those of the MTGEA model that used the GN. The third single modal model used augmented point clouds with AHC and achieved 93.79% accuracy and a weighted F1 score of 93.80%, and both values were over 4% lower than those of the MTGEA.

Table 3. Performance comparison of single-modal models on the DGUHA dataset.

Model	Accuracy (%)	Weighted F_1 Score (%)
Augmented point clouds using ZP	81.99	81.51
Augmented point clouds using GN	92.55	92.45
Augmented point clouds using AHC	93.79	93.80
Skeleton data	97.52	97.51

The single-modal model that used skeleton data showed the best performance in this ablation experiment. It achieved an accuracy of 97.52% and a weighted F1 score of 97.51%, which were only 0.62% and 0.63% lower, respectively, than those of the MTGEA model that used the AHC augmentation strategy. These results seem to imply that since two useful datasets could be exploited by a multimodal framework, the multimodal models' performance was generally better than that of the single-modal models'.

5.2. Ablation Study for the Attention Mechanism

Ablation experiments without an attention mechanism were conducted. Many feature fusion strategies have been studied to combine features effectively, and concatenation is one of the most popular methods. In this experiment, we concatenated two feature representations extracted by the ST-GCN before sending them to the fully connected layer instead of the attention mechanism, as illustrated in Figure 6. Then, we fed them to a softmax classifier to form a prediction.

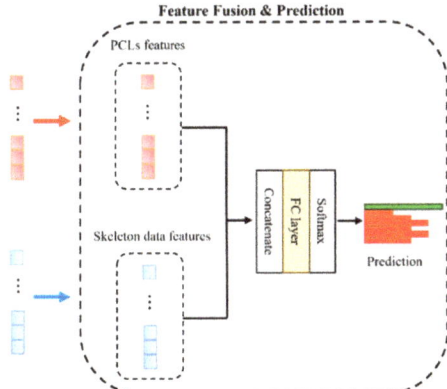

Figure 6. Multimodal feature fusion by concatenation. After features were extracted by three ST-GCN layers, the point cloud and skeleton data features were concatenated and fed into the fully connected layer. Then, a softmax classifier made a prediction.

Table 4 describes the results, which reveal the necessity of an attention mechanism. The best-performing MTGEA model achieved 98.14% accuracy, whereas the MTGEA model without attention that used the same multimodal two-stream framework achieved a lower accuracy of 96.27%. The weighted F1 score was also 1.9% lower than the MTGEA model with attention.

Table 4. Performance comparison of fusion models on the DGUHA dataset.

Model	Accuracy (%)	Weighted F_1 Score (%)
MTGEA (ZP + Skeleton) without attention	83.85	77.77
MTGEA (GN + Skeleton) without attention	94.41	94.40
MTGEA (AHC + Skeleton) without attention	96.27	96.24

In the case of the MTGEA model without attention that used the GN augmentation strategy, it had a 0.62% lower accuracy and a 0.73% lower weighted F1 score than the original MTGEA model with the same augmentation strategy. Similarly, the MTGEA model without attention that used the ZP augmentation strategy had a 1.24% lower accuracy and a 1.58% lower weighted F1 score than the original MTGEA model that used the ZP augmentation strategy.

One notable point is that the MTGEA model without an attention mechanism generally had higher score values than the single-modal models, except for one weighted F1 score, while displaying lower score values than the MTGEA model with an attention mechanism. This means that utilizing accurate skeletal features from the Kinect sensor was critical. Additionally, comparisons between models with the same multimodal two-stream framework but with and without an attention mechanism indicated the necessity of an attention mechanism.

6. Conclusions

This paper presented a radar-based human activity recognition system called MTGEA that does not cause an invasion of privacy or require strict lighting environments. The proposed MTGEA model can classify human activities in a 3D space. To improve the accuracy of human activity recognition using sparse point clouds only, MTGEA uses a multimodal two-stream framework with the help of accurate skeletal features obtained from Kinect models. We used an attention mechanism for efficient multimodal data alignment. Moreover, we provided a newly produced dataset, called the DGUHA, that contains human skeleton data from a Kinect V4 sensor and 3D coordinates from a mmWave radar sensor. MTGEA was evaluated extensively using the DGUHA dataset. The results obtained after training the MTGEA model show that the proposed MTGEA model successfully recognizes human activities using sparse point clouds alone. Training/test datasets, including the raw dataset of DGUHA, are provided on our GitHub page. An ablation study on the multimodal two-stream framework was conducted, and it showed that two-stream framework structures were better than single-modal framework structures for human activity recognition. A similar conclusion was drawn from the second ablation study. This is because even when comparing the results with the MTGEA model that did not consist of an attention mechanism, it showed better performance than the single-modal framework structure. The second ablation study shows the effectiveness of an attention mechanism, an alignment method we used to leverage accurate skeletal features. For the same augmented point clouds, the MTGEA model without an attention mechanism had lower score values than that with an attention mechanism. In this experiment, we chose concatenation as a feature fusion strategy. Our experimental evaluations show the efficiency and necessity of each component of our MTGEA model. The MTGEA uses a multimodal two-stream framework to address the sparse point clouds and an attention mechanism to consider efficient alignment for two multimodal datasets. The entire workflow diagram is shown in Figure 7. Although the model needs some improvement for distinguishing simple activities that do not have complex movements, it can be one of the first steps toward creating a smart home care system.

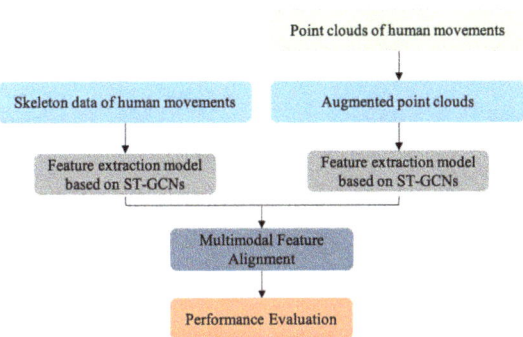

Figure 7. Proposed model diagram.

Author Contributions: Conceptualization, G.L. and J.K.; methodology, G.L. and J.K.; data collection, G.L.; experiment, G.L.; validation, G.L.; formal analysis, G.L.; Writing—original draft, G.L.; Writing—review & editing, G.L. and J.K.; visualization, G.L.; supervision, J.K.; project administration, J.K.; funding acquisition, J.K. All authors have read and agreed to the published version of the manuscript.

Funding: This research was supported by the MSIT (Ministry of Science, ICT), Korea, under the High-Potential Individuals Global Training Program (RS-2022-00155054) (50%) and under the ITRC (Information Technology Research Center) support program (IITP-2023-2020-0-01789) (50%) supervised by the IITP (Institute for Information & Communications Technology Planning & Evaluation).

Institutional Review Board Statement: Not applicable.

Informed Consent Statement: Not applicable.

Data Availability Statement: Publicly available datasets were used in this study. The datasets can be found here: (1) MMActivity (https://github.com/nesl/RadHAR accessed on 1 March 2023) and (2) ours: https://github.com/AIC-DGU/MTGEA (accessed on 1 March 2023).

Acknowledgments: Special thanks to Kyungeun Jung of the Department of Multimedia Engineering and Sejoon Park of the Department of Computer Engineering at the University of Dongguk for their help in data collection.

Conflicts of Interest: The authors declare no conflict of interest.

References

1. Vaiyapuri, T.; Lydia, E.L.; Sikkandar, M.Y.; Diaz, V.G.; Pustokhina, I.V.; Pustokhin, D.A. Internet of Things and Deep Learning Enabled Elderly Fall Detection Model for Smart Homecare. *IEEE Access* **2021**, *9*, 113879–113888. [CrossRef]
2. Ma, W.; Chen, J.; Du, Q.; Jia, W. PointDrop: Improving object detection from sparse point clouds via adversarial data augmentation. In Proceedings of the 2020 25th International Conference on Pattern Recognition (ICPR), Milan, Italy, 10–15 January 2021; pp. 10004–10009.
3. Xu, S.; Zhou, X.; Ye, W.; Ye, Q. Classification of 3D Point Clouds by a New Augmentation Convolutional Neural Network. *IEEE Geosci. Remote Sens. Lett.* **2022**, *19*, 7003405.
4. Kim, K.; Kim, C.; Jang, C.; Sunwoo, M.; Jo, K. Deep learning-based dynamic object classification using LiDAR point cloud augmented by layer-based accumulation for intelligent vehicles. *Expert Syst. Appl.* **2021**, *167*, 113861. [CrossRef]
5. Kulawiak, M. A Cost-Effective Method for Reconstructing City-Building 3D Models from Sparse Lidar Point Clouds. *Remote Sens.* **2022**, *14*, 1278. [CrossRef]
6. Singh, A.D.; Sandha, S.S.; Garcia, L.; Srivastava, M. Radhar: Human activity recognition from point clouds generated through a millimeter-wave radar. In Proceedings of the 3rd ACM Workshop on Millimeter-Wave Networks and Sensing Systems, Los Cabos, Mexico, 25 October 2019; pp. 51–56.
7. Palipana, S.; Salami, D.; Leiva, L.A.; Sigg, S. Pantomime: Mid-air gesture recognition with sparse millimeter-wave radar point clouds. *Proc. ACM Interact. Mob. Wearable Ubiquitous Technol.* **2021**, *5*, 1–27. [CrossRef]
8. Vonstad, E.K.; Su, X.; Vereijken, B.; Bach, K.; Nilsen, J.H. Comparison of a deep learning–based pose estimation system to marker–based and kinect systems in exergaming for balance training. *Sensors* **2020**, *20*, 6940. [CrossRef] [PubMed]

9. Radu, I.; Tu, E.; Schneider, B. Relationships between body postures and collaborative learning states in an Augmented Reality Study. In *International Conference on Artificial Intelligence in Education, Ifrane, Morocco, 6–10 July 2020*; Springer: Berlin/Heidelberg, Germany, 2020; pp. 257–262.
10. Shahroudy, A.; Liu, J.; Ng, T.-T.; Wang, G. NTU RGB+D: A Large Scale Dataset for 3D Human Activity Analysis. In Proceedings of the IEEE Conference on Computer Vision and Pattern Recognition 2016, Las Vegas Valley, NV, USA, 26 June–1 July 2016.
11. Liu, J.; Shahroudy, A.; Perez, M.; Wang, G.; Duan, L.Y.; Kot, A.C. NTU RGB+D 120: A large-scale benchmark for 3D human activity understanding. *IEEE Trans. Pattern Anal. Mach. Intell.* **2019**, *24*, 2684–2701. [CrossRef]
12. Haocong, R.; Shihao, X.; Xiping, H.; Jun, C.; Bin, H. Augmented skeleton based contrastive action learning with momentum LSTM for unsupervised action recognition. *Inf. Sci.* **2021**, *569*, 90–109.
13. Ryselis, K.; Blažauskas, T.; Damaševičius, R.; Maskeliūnas, R. Computer-aided depth video stream masking framework for human body segmentation in depth sensor images. *Sensors* **2022**, *22*, 3531. [CrossRef]
14. Wozniak, M.; Wieczorek, M.; Silka, J.; Polap, D. Body pose prediction based on motion sensor data and recurrent neural network. *IEEE Trans. Ind. Inform.* **2021**, *17*, 2101–2111. [CrossRef]
15. Weiyao, X.; Muqing, W.; Min, Z.; Ting, X. Fusion of skeleton and RGB features for RGB-D human action recognition. *IEEE Sens. J.* **2021**, *21*, 19157–19164. [CrossRef]
16. Zheng, C.; Feng, J.; Fu, Z.; Cai, Y.; Li, Q.; Wang, T. Multimodal relation extraction with efficient graph alignment. In Proceedings of the MM '21: ACM Multimedia Conference, Virtual Event, 20–24 October 2021; pp. 5298–5306.
17. Yang, W.; Zhang, J.; Cai, J.; Xu, Z. Shallow graph convolutional network for skeleton-based action recognition. *Sensors* **2021**, *21*, 452. [CrossRef] [PubMed]
18. Ogundokun, R.O.; Maskeliūnas, R.; Misra, S.; Damasevicius, R. Hybrid inceptionv3-svm-based approach for human posture detection in health monitoring systems. *Algorithms* **2022**, *15*, 410. [CrossRef]
19. Sengupta, A.; Cao, S. mmPose-NLP: A natural language processing approach to precise skeletal pose estimation using mmwave radars. *arXiv* **2021**, arXiv:2107.10327. [CrossRef]
20. Lee, G.; Kim, J. Improving human activity recognition for sparse radar point clouds: A graph neural network model with pre-trained 3D human-joint coordinates. *Appl. Sci.* **2022**, *12*, 2168. [CrossRef]
21. Pan, L.; Chen, X.; Cai, Z.; Zhang, J.; Liu, Z. Variational Relational Point Completion Network. In Proceedings of the 2021 IEEE/CVF Conference on Computer Vision and Pattern Recognition (CVPR), Nashville, TN, USA, 19–25 June 2021; pp. 8520–8529.
22. Zhang, R.; Cao, S. Real-time human motion behavior detection via CNN using mmWave radar. *IEEE Sens. Lett.* **2019**, *3*, 1–4. [CrossRef]
23. Yan, S.; Xiong, Y.; Lin, D. Spatial temporal graph convolutional networks for skeleton-based action recognition. *arXiv* **2018**, arXiv:1801.07455. [CrossRef]
24. Kipf, T.N.; Welling, M. Semi-supervised classification with graph convolutional networks. In Proceedings of the 5th International Conference on Learning Representations, Toulon, France, 24–26 April 2017.
25. Bahdanau, D.; Cho, K.H.; Bengio, Y. Neural machine translation by jointly learning to align and translate. In Proceedings of the 3rd International Conference on Learning Representations, San Diego, CA, USA, 7 May 2015.
26. Rashid, M.; Khan, M.A.; Alhaisoni, M.; Wang, S.-H.; Naqvi, S.R.; Rehman, A.; Saba, T. A Sustainable Deep Learning Framework for Object Recognition Using Multi-Layers Deep Features Fusion and Selection. *Sustainability* **2020**, *12*, 5037. [CrossRef]
27. Yen, C.-T.; Liao, J.-X.; Huang, Y.-K. Feature Fusion of a Deep-Learning Algorithm into Wearable Sensor Devices for Human Activity Recognition. *Sensors* **2021**, *21*, 8294. [CrossRef]
28. Wu, P.; Cui, Z.; Gan, Z.; Liu, F. Three-Dimensional ResNeXt Network Using Feature Fusion and Label Smoothing for Hyperspectral Image Classification. *Sensors* **2020**, *20*, 1652. [CrossRef]
29. Petrovska, B.; Zdravevski, E.; Lameski, P.; Corizzo, R.; Štajduhar, I.; Lerga, J. Deep learning for feature extraction in remote sensing: A case-study of aerial scene classification. *Sensors* **2020**, *20*, 3906. [CrossRef] [PubMed]
30. Vaswani, A.; Shazeer, N.; Parmar, N.; Uszkoreit, J.; Jones, L.; Gomez, A.N.; Kaiser, L.; Polosukhin, I. Attention is all you need. *arXiv* **2017**, arXiv:1706.03762.

Disclaimer/Publisher's Note: The statements, opinions and data contained in all publications are solely those of the individual author(s) and contributor(s) and not of MDPI and/or the editor(s). MDPI and/or the editor(s) disclaim responsibility for any injury to people or property resulting from any ideas, methods, instructions or products referred to in the content.

Article

Revisiting Consistency for Semi-Supervised Semantic Segmentation [†]

Ivan Grubišić [1], Marin Oršić [2] and Siniša Šegvić [1,*]

[1] Faculty of Electrical Engineering and Computing, University of Zagreb, Unska 3, 10000 Zagreb, Croatia
[2] Microblink Ltd., Strojarska Cesta 20, 10000 Zagreb, Croatia
* Correspondence: sinisa.segvic@fer.hr
[†] This paper is an extended version of our paper published in the proceedings of the 17th International Conference on Machine Vision and Applications, MVA 2021, Aichi, Japan, 25–27 July 2021.

Abstract: Semi-supervised learning is an attractive technique in practical deployments of deep models since it relaxes the dependence on labeled data. It is especially important in the scope of dense prediction because pixel-level annotation requires substantial effort. This paper considers semi-supervised algorithms that enforce consistent predictions over perturbed unlabeled inputs. We study the advantages of perturbing only one of the two model instances and preventing the backward pass through the unperturbed instance. We also propose a competitive perturbation model as a composition of geometric warp and photometric jittering. We experiment with efficient models due to their importance for real-time and low-power applications. Our experiments show clear advantages of (1) one-way consistency, (2) perturbing only the student branch, and (3) strong photometric and geometric perturbations. Our perturbation model outperforms recent work and most of the contribution comes from the photometric component. Experiments with additional data from the large coarsely annotated subset of Cityscapes suggest that semi-supervised training can outperform supervised training with coarse labels. Our source code is available at https://github.com/Ivan1248/semisup-seg-efficient.

Keywords: semi-supervised learning; semantic segmentation; dense prediction; one-way consistency; deep learning; scene understanding

1. Introduction

Most machine learning applications are hampered by the need to collect large annotated datasets. Learning with incomplete supervision [1,2] presents a great opportunity to speed up the development cycle and enable rapid adaptation to new environments. Semi-supervised learning [3–5] is especially relevant in the dense prediction context [6–8] since pixel-level labels are very expensive, whereas unlabeled images are easily obtained.

Dense prediction typically operates on high resolutions in order to be able to recognize small objects. Furthermore, competitive performance requires learning on large batches and large crops [9–11]. This typically entails a large memory footprint during training, which constrains model capacity [12]. Many semi-supervised algorithms introduce additional components to the training setup. For instance, training with surrogate classes [13] implies infeasible logit tensor size, while GAN-based approaches require an additional generator [6,14] or discriminator [7,15,16]. Some other approaches require multiple model instances [17–20] or accumulated predictions across the dataset [21]. Such designs are less appropriate for dense prediction since they constrain model capacity.

This paper studies semi-supervised approaches [3,5,18,21,22] that require consistent predictions over input perturbations. In the considered consistency objective, input perturbations affect only one of two model instances, while the gradient is not propagated towards the model instance which operates on the clean (weakly perturbed) input [4,5]. For brevity, we refer to the two model instances as the perturbed branch and the clean branch. If the gradient is not computed in a branch, we refer to it as the teacher, and otherwise as the student. Hence, we refer to the considered approach as a one-way consistency with the clean teacher.

Let x be the input, T a perturbation to which the ideal model should be invariant, h_θ the student, and $h_{\theta'}$ the teacher, where θ' denotes a frozen copy of the student parameters θ. Then, one-way consistency with clean teacher can be expressed as a divergence D between the two predictions:

$$L_\theta^{ct}(x, T) = D(h_{\theta'}(x), h_\theta(T(x))) . \tag{1}$$

We argue that the clean teacher approach is a method of choice in case of perturbations that are too strong for standard data augmentation. In this setting, perturbed inputs typically give rise to less reliable predictions than their clean counterparts. Figure 1 illustrates the advantage of the clean teacher approach in comparison with other kinds of consistency on the Two moons dataset. The clean student experiment (Figure 1b) shows that many blue data points get classified into the red class due to teacher inputs being pushed towards labeled examples of the opposite class. This aberration does not occur when the teacher inputs are clean (Figure 1c). Two-way consistency [21] (Figure 1d) can be viewed as a superposition of the two one-way approaches and works better than (Figure 1b), but worse than (Figure 1c). In our experiments, D corresponds to KL divergence.

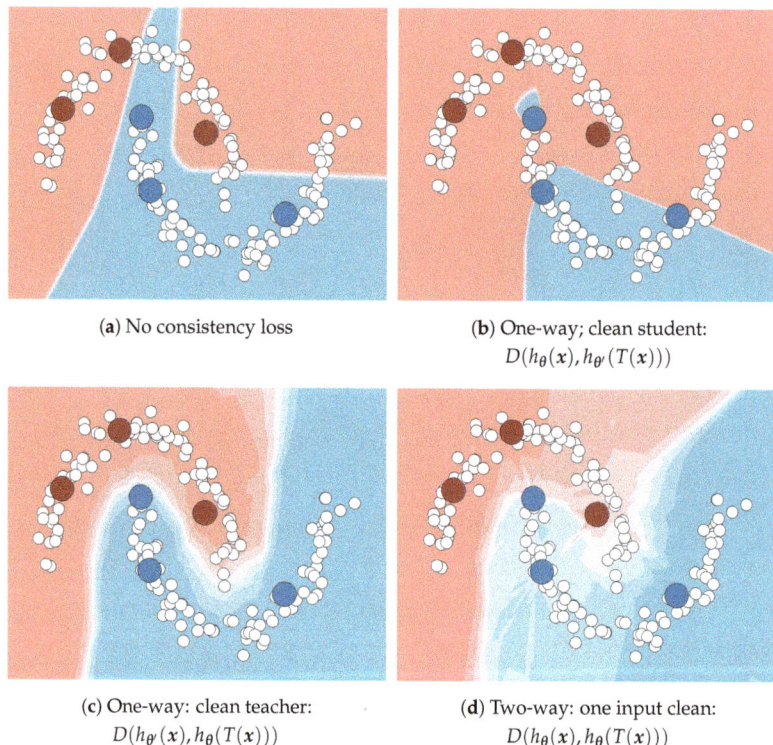

(a) No consistency loss

(b) One-way; clean student: $D(h_\theta(x), h_{\theta'}(T(x)))$

(c) One-way: clean teacher: $D(h_{\theta'}(x), h_\theta(T(x)))$

(d) Two-way: one input clean: $D(h_\theta(x), h_\theta(T(x)))$

Figure 1. A toy semi-supervised classification problem with six labeled (red, blue) and many unlabeled 2D datapoints (white). All setups involve 20,000 epochs of semi-supervised training with cross-entropy and default Adam optimization hyper-parameters. The consistency loss was set to none (**a**), one-way with clean student (**b**), one-way with clean teacher (**c**), and two-way with one input clean (**d**). One-way consistency with clean teacher outperforms all other formulations.

One-way consistency is especially advantageous in the dense prediction context since it does not require caching latent activations in the teacher. This allows for better training in many practical cases where model capacity is limited by GPU memory [12,23]. In comparison with two-way consistency [20,21], the proposed approach both improves generalization and approximately halves the training memory footprint.

This paper is an extended version of our preliminary conference report [24]. It exposes the elements of our method in much more detail and complements them with many new experiments. In particular, the most important additions are additional ablation and validation studies, full-resolution Cityscapes experiments, and a detailed analysis of a large-scale experiment that compares the contribution of coarse labels with semi-supervised learning on unlabeled images. The new experiments add more evidence in favor of one-way consistency with respect to other consistency variants, investigate the influence of particular components of our algorithm and various hyper-parameters, and investigate the behavior of the proposed algorithm in different data regimes (higher resolution; additional unlabeled images).

The consolidated paper proposes a simple and effective method for semi-supervised semantic segmentation. One-way consistency with clean teacher [4,5,25] outperforms the two-way formulation in our validation experiments. In addition, it retains the memory footprint of supervised training because the teacher activations depend on parameters that are treated as constants. Experiments with a standard convolutional architecture [26] reveal that our photometric and geometric perturbations lead to competitive generalization performance and outperform their counterpart from a recent related work [25]. A similar advantage can be observed in experiments with a recent efficient architecture [27], which offers a similar performance while requiring an order of magnitude of less computation. To our knowledge, this is the first account of the evaluation of semi-supervised algorithms for dense prediction with a model capable of real-time inference. This contributes to the goals of Green AI [28] by enabling competitive research with less environmental damage.

This paper proceeds as follows. Section 2 presents related work. Section 3 describes the one-way consistency objective adapted to dense prediction, our perturbation model, and a description of our memory-efficient consistency training procedure. Section 4 presents the experimental setup, which includes information about datasets and training details, as well as the performed experiments in semi-supervised semantic segmentation. Finally, Section 5 presents the conclusion.

2. Related Work

Our work spans the fields of dense prediction and semi-supervised learning. The proposed methodology is most related to previous work in semi-supervised semantic segmentation.

2.1. Dense Prediction

Image-wide classification models usually achieve efficiency, spatial invariance, and integration of contextual information by gradual downsampling of representations and use of global spatial pooling operations. However, dense prediction also requires location accuracy. This emphasizes the trade-off between efficiency and quality of high-resolution features in the model design. Some common designs use a classification backbone as a feature encoder and attach a decoder that restores the spatial resolution. Many approaches seek to enhance contextual information, starting with FCN-8s [29]. UNet [30] improves spatial details by directly using earlier representations of the encoder in a symmetric decoder. Further work improves the efficiency with lighter decoders [23,31]. Some models use context aggregation modules such as spatial pyramid pooling [32] and multi-scale inference [31,33]. DeepLab [26] increases the receptive field through dilated convolutions and improves spatial details through CRF post-processing. HRNet [34] maintains the full resolution throughout the whole model and incrementally introduces parallel lower-resolution branches that exchange information between stages. Semantic segmentation gains much from ImageNet pre-trained encoders [23,26].

2.2. Semi-Supervised Learning

Semi-supervised methods often rely on some of the following assumptions about the data distribution [35]: (1) similar inputs in high density regions correspond to similar outputs (smoothness assumption), (2) inputs form clusters separated by low-density regions and inputs within clusters are likely to correspond to similar outputs (cluster assumption), and (3) the data lies on a lower-dimensional manifold (manifold assumption). Semi-supervised methods devise various inductive biases that exploit such regularities for learning from unlabeled data.

Entropy minimization [36] encourages high confidence in unlabeled inputs. Such designs push decision boundaries towards low-density regions, under assumptions of clustered data and prediction smoothness. Pseudo-label training (or self-training) [37–39] also encourages high confidence (because of hard pseudo-labels) as well as consistency with a previously trained teacher. The basic forms of such algorithms do not achieve competitive performance on their own [40], but can be effective in conjunction with other approaches [4,41]. Pseudo-labels can be made very effective by confidence-based selection and other processing [37,38,42]. Note that some concurrent work [42] uses the term pseudo-label as a synonym for processed teacher prediction in one-way consistency, but we do not follow this practice.

Many approaches exploit the smoothness assumption by enforcing prediction consistency across different versions of the same input or different model instances. Introducing knowledge about equivariance has been studied for understanding and learning useful image representations [43,44] and improving dense prediction [45–47]. Exemplar training [13] associates patches with their original images (each image is a separate surrogate class). Temporal ensembling [21] enforces per-datapoint consistency between the current prediction and a moving average of past predictions. Mean Teacher [3] encourages consistency with a teacher whose parameters are an exponential moving average of the student's parameters.

Clusterization of latent representations can be promoted by penalizing walks which start in a labeled example, pass over an unlabeled example, and end in another example with a different label [48]. PiCIE [46] obtains semantically meaningful segmentation without labels by jointly learning clustering and representation consistency under photometric and geometric perturbations.

MixMatch [49] encourages consistency between predictions in different MixUp perturbations of the same input. The average prediction is used as a pseudo-label for all variants of the input. Deep co-training [19] produces complementary models by encouraging them to be consistent while each is trained on adversarial examples of the other one.

Consistency losses may encourage trivial solutions, where all inputs give rise to the same output. This is not much of a problem in semi-supervised learning since there the trivial solution is inhibited through the supervised objective. Interestingly, recent work shows that a variant of simple one-way consistency evades trivial solutions even in the context of self-supervised representation learning [50,51].

Virtual adversarial training (VAT) [4] encourages one-way consistency between predictions in original datapoints and their adversarial perturbations. These perturbations are recovered by maximizing a quadratic approximation of the prediction divergence in a small L^2 ball around the input. Better performance is often obtained by additionally encouraging low-entropy predictions [36]. Unsupervised data augmentation (UDA) [5] also uses a one-way consistency loss. FixMatch [52] shows that pseudo-label selection and processing can be useful in a one-way consistency. However, instead of adversarial additive perturbations, they use random augmentations generated by RandAugment. Different from all previous approaches, we explore an exhaustive set of consistency formulations.

2.3. Semi-Supervised Semantic Segmentation

In the classic semi-supervised GAN (SGAN) setup, the classifier also acts as a discriminator which distinguishes between real data (both labeled and unlabeled) and fake data produced by the generator [14]. This approach has been adapted for dense prediction by expressing the discriminator as a segmentation network that produces dense C+1-way logits [6]. KE-GAN [53] additionally enforces semantic consistency of neighbouring predictions by leveraging label-similarity recovered from a large text corpus (MIT ConceptNet). A semantic segmentation model can also be trained as a GAN generator (AdvSemSeg) [7]. In this setup, the discriminator guesses whether its input is ground truth or generated by the segmentation network. The discriminator is also used to choose better predictions for use as pseudo-labels for semi-supervised training. s4GAN + MLMT [8] additionally post-processes the recovered dense predictions by emphasizing classes identified by an image-wide classifier trained with Mean Teacher [3]. The authors note that the image-wide classification component is not appropriate for datasets such as Cityscapes, where almost all images contain a large number of classes.

A recent approach enforces consistency between outputs of redundant decoders with noisy intermediate representations [54]. Other recent work studies pseudo-labeling in the dense prediction context [55–57]. Zhu et al. [55] observe advantages in hard pseudo-labels. A recent approach [20] proposes a two-way consistency loss, which is related to the Π-model [21], and perturbs both inputs with geometric warps. However, we show that perturbing only the student branch generalizes better and has a smaller training footprint. A concurrent work [58] successfully applies a contrastive loss [59,60] between two branches which receive overlapping crops, and proposes a pixel-dependent consistency direction. Mean Teacher consistency with CutMix perturbations achieved state-of-the-art performance on half-resolution Cityscapes [25] prior to this work. Different than most presented approaches and similar to [25,55,56,61], our method does not increase the training footprint [12]. In comparison with [55,56,61], our teacher is updated in each training step, which eliminates the need for multiple training episodes. In comparison with [25], this work proposes a perturbation model which results in better generalization and shows that simple one-way consistency can be competitive with Mean Teacher. None of the previous approaches addresses semi-supervised training of efficient dense prediction models. We examine the simplest forms of consistency, explain advantages of perturbing only the student with respect to other forms of consistency, and propose a novel perturbation model. None of the previous approaches considered semi-supervised training of efficient dense-prediction models, nor studied composite perturbations of photometry and geometry.

3. Method

We formulate dense consistency as a mean pixel-wise divergence between corresponding predictions in the clean image and its perturbed version. We perturb images with a composition of photometric and geometric transformations. Photometric transformations do not disturb the spatial layout of the input image. Geometric transformations affect the spatial layout of the input image and the same kind of disturbance is expected at the model output. Ideally, our models should exhibit invariance to photometric transformations and equivariance [44] to the geometric ones.

3.1. Notation

We typeset vectors and arrays in bold, sets in blackboard bold, and we underline random variables. $P[\underline{y}|\underline{x} = x]$ denotes the distribution of a random variable $\underline{y}|x$, while $P(y|x)$ is a shorthand for the probability $P(\underline{y} = y|\underline{x} = x)$. We denote the expectation of a function of a random variable as e.g., $\mathbf{E}_{\underline{\tau}} f(\tau)$. We use similar notation to denote the average over a set: $\mathbf{E}_{x \in \mathbb{D}} f(x)$. We use the Iverson bracket notation: given a statement P, $[\![P]\!] = 1$ if P is true; 0 otherwise. We denote cross-entropy with $\mathrm{H}_y(y^*) := \mathbf{E}_{\underline{y} \sim y^*} \ln p(\underline{y} = y)$, and entropy with $\mathrm{H}(y)$ [62]. We use Python-like array indexing notation.

We denote the labeled dataset as $\mathbb{D}_l$, and the unlabeled dataset as $\mathbb{D}_u$. We consider input images $x \in [0,1]^{H \times W \times 3}$ and dense labels $y \in \{1 \ldots C\}^{H \times W}$. A model instance maps an image to per-pixel class probabilities: $h_\theta(x)_{[i,j,c]} = P(\underline{y}_{[i,j]} = c | x, \theta)$. For convenience, we identify output vectors of class probabilities with distributions: $h_\theta(x)_{[i,j]} \equiv P[\underline{y}_{[i,j]} | x, \theta]$.

3.2. Dense One-Way Consistency

We adapt one-way consistency [4,5] for dense prediction under our perturbation model $T_\tau = T_\gamma^G \circ T_\varphi^P$, where T_γ^G is a geometric warp, T_φ^P a per-pixel photometric perturbation, and $\tau = (\gamma, \varphi)$ perturbation parameters. T_γ^G displaces pixels with respect to a dense deformation field. The same geometric warp is applied to the student input and the teacher output. Figure 2 illustrates the computational graph of the resulting dense consistency loss. In simple one-way consistency, the teacher parameters θ' are a frozen copy of the student parameter θ. In Mean Teacher, θ' is a moving average of θ. In simple two-way consistency, both branches use the same θ and are subject to gradient propagation.

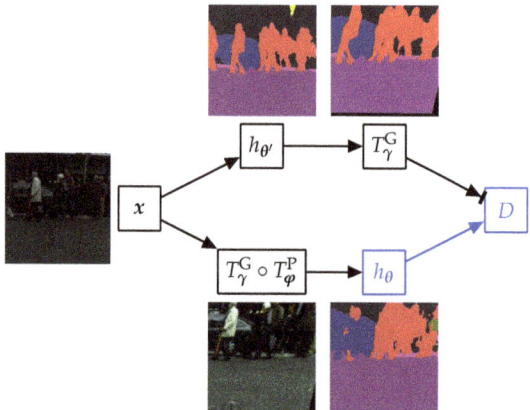

Figure 2. Dense one-way consistency with clean teacher. Top branch: the input is fed to the teacher $h_{\theta'}$. The resulting predictions are perturbed with geometric perturbations T_γ^G. Bottom branch: the input is perturbed with geometric and photometric perturbations and fed to the student h_θ. The loss D corresponds to the average pixel-wise KL divergence between the two branches. Gradients are computed only in the blue part of the graph.

A general semi-supervised training criterion $L(\theta; \mathbb{D}_l, \mathbb{D}_u)$ can be expressed as a weighted sum of a supervised term L_s over labeled data and an unsupervised consistency term L_c:

$$L(\theta; \mathbb{D}_l, \mathbb{D}_u) = \underset{(x,y) \in \mathbb{D}_l}{\mathbb{E}} L_s(\theta; x, y) + \alpha \underset{x \in \mathbb{D}_u}{\mathbb{E}} \underset{\tau}{\mathbb{E}} L_c(\theta; x, \tau). \qquad (2)$$

In our experiments, L_s is the usual mean per-pixel cross entropy with L^2 regularization. We stochastically estimate the expectation over perturbation parameters τ with one sample per training step.

We formulate the unsupervised term L_c at pixel (i,j) as a one-way divergence D between the prediction in the perturbed image and its interpolated correspondence in the clean image. The proposed loss encourages the trained model to be equivariant to T_γ^G and invariant to T_φ^P:

$$L_c^{i,j}(\theta; x, \tau) = D(T_\gamma^G(h_{\theta'}(x))_{[i,j]}, h_\theta((T_\gamma^G \circ T_\varphi^P)(x))_{[i,j]}). \qquad (3)$$

We use a validity mask $v^\gamma \in \{0,1\}^{H \times W}$, $v^\gamma_{[i,j]} = \left[T_\gamma^G(\mathbf{1}_{H \times W})_{[i,j]} = 1 \right]$ to ensure that the loss is unaffected by padding sampled from outside of $[1,H] \times [1,W]$. A vector produced by

$T_\gamma^G(h_\theta(x))_{[i,j]}$ represents a valid distribution wherever $v_{[i,j]}^\gamma = 1$. Finally, we express the consistency term L_c as a mean contribution over all pixels:

$$L_c(\theta; x, \tau) = \frac{1}{\Sigma(v^\gamma)} \sum_{i,j} v_{[i,j]}^\gamma L_c^{i,j}(\theta; x, \tau). \quad (4)$$

Recall that the gradient is not computed with respect to θ'. Consequently, L_c allows gradient propagation only towards the perturbed image. We refer to such training as one-way consistency with clean teacher (and perturbed student). Such formulation provides two distinct advantages over other kinds of consistency. First, predictions in perturbed images are pulled towards predictions in clean images. This improves generalization when the perturbations are stronger than data augmentations used in L_s (cf. Figures 1 and 3). Second, we do not have to cache teacher activations during training since the gradients propagate only towards the student branch. Hence, the proposed semi-supervised objective does not constrain model complexity with respect to the supervised baseline.

We use KL divergence as a principled choice for D:

$$D(y, \tilde{y}) := \mathbb{E}_y \ln \frac{P(y=y)}{P(\tilde{y}=y)} = H_{\tilde{y}}(y) - H(y). \quad (5)$$

Note that the entropy term $-H(y)$ does not affect parameter updates since the gradients are not propagated through θ'. Hence, one-way consistency does not encourage increasing entropy of model predictions in clean images. Several researchers have observed improvement after adding an entropy minimization term [36] to the consistency loss [4,5]. This practice did not prove beneficial in our initial experiments.

Note that two-way consistency [20,21] would be obtained by replacing θ' with θ. It would require caching latent activations for both model instances, which approximately doubles the training footprint with respect to the supervised baseline. This would be undesirable due to constraining the feasible capacity of the deployed models [12,63].

We argue that consistency with clean teacher generalizes better than consistency with clean student since strong perturbations may push inputs beyond the natural manifold and spoil predictions (cf. Figure 1). Moreover, perturbing both branches sometimes results in learning to map all perturbed pixels to similar arbitrary predictions (e.g., always the same class) [64]. Figure 3 illustrates that consistency training has the best chance to succeed if the teacher is applied to the clean image, and the student learns on the perturbed image.

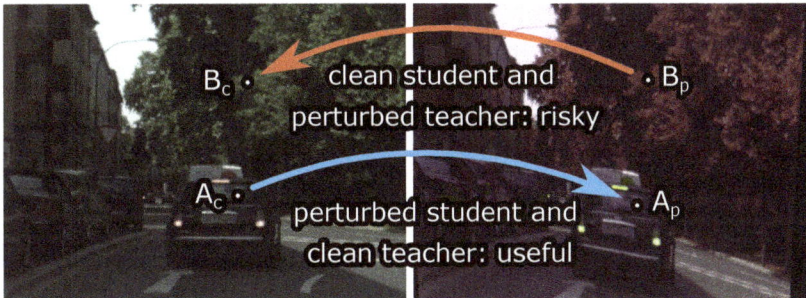

Figure 3. Two variants of one-way consistency training on a clean image (**left**) and its perturbed version (**right**). The arrows designate information flow from the teacher to the student. The proposed clean-teacher formulation trains in the perturbed pixels (A_p) according to the corresponding predictions in the clean image (A_c). The reverse formulation (training in B_c according to the prediction in B_p) worsens performance, since strongly perturbed images often give rise to less accurate predictions.

3.3. Photometric Component of the Proposed Perturbation Model

We express pixel-level photometric transformations as a composition of five simple perturbations with five image-wide parameters $\varphi = (b, s, h, c, \pi)$. These perturbations are applied in each pixel in the following order: (1) brightness is shifted by adding b to all channels, (2) saturation is multiplied with s, (3) hue is shifted by addition with h, (4) contrast is modulated by multiplying all channels with c, and (5) RGB channels are randomly permuted according to π. The resulting compound transformation T_φ^P is independently applied to all image pixels.

Our training procedure randomly picks image-wide parameters φ for each unlabeled image. The parameters are sampled as follows: $b \sim \mathcal{U}(-0.25, 0.25)$, $s \sim \mathcal{U}(0.25, 2)$, $h \sim \mathcal{U}(-36°, 36°)$, $c \sim \mathcal{U}(0.25, 2)$, and $\pi \sim \mathcal{U}(\mathbb{S}_3)$, where $\mathbb{S}_3$ represents the set of all 6 3-element permutations.

3.4. Geometric Component of the Proposed Perturbation Model

We formulate a fairly general class of parametric geometric transformations by leveraging thin plate splines (TPS) [65,66]. We consider the 2D TPS warp $f: \mathbb{R}^2 \to \mathbb{R}^2$, which maps each image coordinate pair q to the relative 2D displacement of its correspondence q':

$$f(q) = q' - q. \tag{6}$$

TPS warps minimize the bending energy (curvature) $\int_{\mathrm{dom}(f)} \left\| \frac{\partial^2 f(q)}{\partial q^2} \right\|_F^2 dq$ given a set of control points and their displacements $\{(c_i, d_i) : i = 1 \ldots n\} \subset \mathbb{R}^2 \times \mathbb{R}^2$. In simple words, a TPS warp produces a smooth deformation field which optimally satisfies all constraints $f(c_i) = d_i$. In the 2D case, the solution of the TPS problem takes the following form:

$$f(q) = A \begin{bmatrix} 1 \\ q \end{bmatrix} + W \left[\phi(\|q - c_i\|) \right]_{i=1\ldots n}^\top, \tag{7}$$

where q denotes a 2D coordinate vector to be transformed, A is a 2×3 affine transformation matrix, W is a $2 \times n$ control point coefficient matrix, and $\phi(r) = r^2 \ln(r)$. Such a 2D TPS warp is equivariant to rotation and translation [66]. That is, $f(T(q)) = T(f(q))$ for every composition of rotation and translation T.

TPS parameters A and W can be determined as a solution of a standard linear system which enforces deformation constraints (c_i, d_i), and square-integrability of second derivatives of f. When we determine A and W, we can easily transform entire images.

We first consider images as continuous domain functions and later return to images as arrays from $[0, 1]^{H \times W \times 3}$. Let $I : \mathrm{dom}(I) \to [0, 1]^3$ be the original image of size (W, H), where $\mathrm{dom}(I) = [0, W] \times [0, H]$. Then the transformed image I' can be expressed as

$$I'(q + f(q)) = \begin{cases} I(q), & q \in \mathrm{dom}(I), \\ 0, & \text{otherwise.} \end{cases} \tag{8}$$

The resulting formulation is known as forward warping [67] and is tricky to implement. We, therefore, prefer to recover the reverse transformation $\tilde{f}$, which can be conducted by replacing each control point c_i with $c'_i = c_i + d_i$. Then, the transformed image is:

$$\tilde{I}(q') = \begin{cases} I(q' - \tilde{f}(q')), & q' - \tilde{f}(q') \in \mathrm{dom}(I), \\ 0, & \text{otherwise.} \end{cases} \tag{9}$$

This formulation is known as backward warping [67]. It can be easily implemented for discrete images by leveraging bilinear interpolation. Contemporary frameworks already include the implementations for the GPU hardware. Hence, the main difficulty is to determine the TPS parameters by solving two linear systems with $(n + 3) \times (n + 3)$ variables [66].

In our experiments, we use $n = 4$ control points corresponding to the centers of the four image quadrants: $(c'_1, \ldots, c'_4) = \left(\left[\frac{1}{4}H, \frac{1}{4}W\right]^T, \ldots, \left[\frac{3}{4}H, \frac{3}{4}W\right]^T \right)$. The parameters of our geometric transformation are four 2D displacements $\gamma = (d_1, \ldots, d_4)$. Let f_γ denote the resulting TPS warp. Then, we can express our transformation as $T^G_\gamma(x) =$ backward_warp(x, f_γ).

Our training procedure picks a random γ for each unlabeled image. Each displacement is sampled from a bivariate normal distribution $\mathcal{N}(0_2, 0.05 \times H \times I_2)$, where H is the height of training crops.

3.5. Training Procedure

Algorithm 1 sketches a procedure for recovering gradients of the proposed semi-supervised loss (2) on a mixed batch of labeled and unlabeled examples. For simplicity, we make the following changes in notation here: x_l and y_l are batches of size B_l, x_u, γ and φ batches of size B_u, and all functions are applied to batches. The algorithm computes the gradient of the supervised loss, discards cached activations, computes the teacher predictions, applies the consistency loss (3), and finally accumulates the gradient contributions of the two losses. Backpropagation through one-way consistency with clean teacher requires roughly the same extent of caching as in the supervised baseline. Hence, our approach constrains the model complexity much less than the two-way consistency.

Algorithm 1. Evaluation of the gradient of the proposed semi-supervised loss given perturbation parameters (γ, φ) on a mixed batch of labeled (x_l, y_l) and unlabeled (x_u) examples. CE denotes mean cross entropy, while KL_masked denotes mean KL divergence over valid pixels.

```
1   # x_u, γ, φ, p_t, p_s, and v contain batches of B_u elements.
2   # x_l, y, and p_l contain batches of B_l elements.
3   procedure compute_loss_gradient(h, θ, x_l, y_l, x_u, γ, φ):
4       θ' ← frozen_copy(θ) # simple one-way
5
6       # supervised loss
7       p_l ← h_θ(x_l)
8       L_s ← CE(y_l, p_l)
9       g ← ∇_θ L_s # clears cached activations
10
11      # unsupervised loss
12      with no_grad(): # activations not cached here
13          p_t ← T^G_γ(h_θ'(x_u)) # clean teacher
14      p_s ← h_θ((T^G_γ ∘ T^P_φ)(x_u)) # perturbed student
15      v ← ⌊T^G_γ(1_{B_u × H × W})⌋ # validity mask
16      L_c ← α · KL_masked(p_t, p_s, v)
17      g ← g + ∇_θ L_c
18
19      return g
```

Figure 4 illustrates GPU memory allocation during a semi-supervised training iteration of a SwiftNet-RN34 model with one-way and two-way consistency. We recovered these measurements by leveraging the following functions of the torch.cuda package: max_memory_allocated, memory_allocated, reset_peak_memory_stats, and empty_cache. The training was carried out on a RTX A4500 GPU. Numbers on the x-axis correspond to lines of the pseudo-code in Algorithm 1. Line 9 backpropagates through the supervised loss and caches the gradients. The memory footprint briefly peaks due to temporary storage and immediately declines since PyTorch automatically releases all cached activations immediately after the backpropagation. Line 13 computes the teacher output. This step does not cache intermediate activations due to torch.no_grad. Line

16 computes the unsupervised loss, which requires the caching of activations on a large spatial resolution. The memory footprint briefly peaks since we delete perturbed inputs and teacher predictions immediately after line 16 (for simplicity, we omit opportunistic deletions from Algorithm 1). Line 17 triggers the backpropagation algorithm and accumulates the gradients of the consistency loss. The memory footprint briefly peaks due to temporary storage and immediately declines due to automatic deletion of the cached activations. At this point, the memory footprint is slightly greater than at line 4 since we still hold the supervised predictions in order to accumulate the recognition performance on the training dataset.

The ratio between memory allocations at lines 16 and 9 reveals the relative memory overhead of our semi-supervised approach. Note that the absolute overhead is model independent since it corresponds to the total size of perturbed inputs and predictions, and intermediate results of dense KL-divergence. On the other hand, the memory footprint of the supervised baseline is model dependent, since it reflects the computational complexity of the backbone. Consequently, the relative overhead approaches 1 as the model size increases, and is around 1.26 for SwiftNet-RN34.

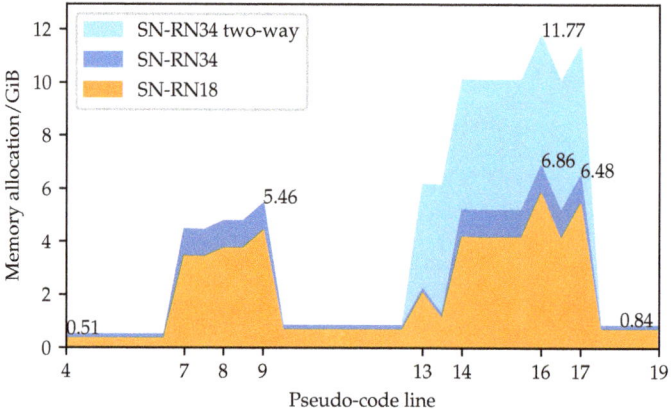

Figure 4. GPU memory allocation during and after execution of particular lines from Algorithm 1 during the 2nd iteration of training. Our PyTorch implementations involve SwiftNet-RN18 and SwiftNet-RN34 models with one-way and two-way consistency, 768 × 768 crops, and batch sizes $(B_l, B_u) = (8, 8)$. Line 9 computes the supervised gradient. Line 13 computes the teacher output (without caching interemediate activations). Lines 16 and 17 compute the consistency loss and its gradient.

4. Results

Our experiments evaluate one-way consistency with clean teacher and a composition of photometric and geometric perturbations ($T_\gamma^G \circ T_\phi^P$). We compare our approach with other kinds of consistency and the state of the art in semi-supervised semantic segmentation. We denote simple one-way consistency as "simple", Mean Teacher [3] as "MT", and our perturbations as "PhTPS". In experiments that compare consistency variants, "1w" denotes one-way, "2w" denotes two way, "ct" denotes clean teacher, "cs" denotes clean student, and "2p" denotes both inputs perturbed. We present semi-supervised experiments in several semantic segmentation setups as well as in image-classification setups on CIFAR-10. Our implementations are based on the PyTorch framework [68].

4.1. Experimental Setup

Datasets. We perform semantic segmentation on Cityscapes [9], and image classification on CIFAR-10. Cityscapes contains 2975 training, 500 validation and 1525 testing images with resolution 1024 × 2048. Images are acquired from a moving vehicle during daytime and fine weather conditions. We present half-resolution and full-resolution experiments. We use bilinear interpolation for images and nearest neighbour subsampling for labels. Some experiments on Cityscapes also use the coarsely labeled Cityscapes subset ("train-extra") that contains 19,998 images. CIFAR-10 consists of 50,000 training and 10,000 test images of resolution 32 × 32.

Common setup. We include both unlabeled and labeled images in $\mathbb{D}_u$, which we use for the consistency loss. We train on batches of B_l labeled and B_u unlabeled images. We perform $\lfloor |\mathbb{D}_l| / B_l \rfloor$ training steps per epoch. We use the same perturbation model across all datasets and tasks (TPS displacements are proportional to image size), which is likely suboptimal [69]. Batch normalization statistics are updated only in non-teacher model instances with clean inputs except for full-resolution Cityscapes, for which updating the statistics in the perturbed student performed better in our validation experiments (cf. Appendix B). The teacher always uses the estimated population statistics, and does not update them. In Mean Teacher, the teacher uses an exponential moving average of the student's estimated population statistics.

Semantic segmentation setup. Cityscapes experiments involve the following models: SwiftNet with ResNet-18 (SwiftNet-RN18) or ResNet-34 (SwiftNet-RN34), and DeepLab v2 with a ResNet-101 backbone. We initialize the backbones with ImageNet pre-trained parameters. We apply random scaling, cropping, and horizontal flipping to all inputs and segmentation labels. We refer to such examples as clean. We schedule the learning rate according to $e \mapsto \eta \cos(e\pi/2)$, where $e \in [0\ldots1]$ is the fraction of epochs completed. This alleviates the generalization drop at the end of training with standard cosine annealing [70]. We use learning rates $\eta = 4 \times 10^{-4}$ for randomly initialized parameters and $\eta = 10^{-4}$ for pre-trained parameters. We use Adam with $(\beta_1, \beta_2) = (0.9, 0.99)$. The L^2 regularization weight in supervised experiments is 10^{-4} for randomly initialized and 2.5×10^{-5} for pre-trained parameters [27]. We have found that such L^2 regularization is too strong for our full-resolution semi-supervised experiments. Thus, we use a 4× smaller weight there. Based on early validation experiments, we use $\alpha = 0.5$ unless stated otherwise. Batch sizes are $(B_l, B_u) = (8, 8)$ for SwiftNet-RN18 [27] and $(B_l, B_u) = (4, 4)$ for DeepLab v2 (ResNet-101 backbone) [26]. The batch size in corresponding supervised experiments is B_l.

In half-resolution Cityscapes experiments the size of crops is 448 × 448 and the logarithm of the scaling factor is sampled from $\mathcal{U}(\ln(1.5^{-1}), \ln(1.5))$. We train SwiftNet for $200 \times \frac{2975}{|\mathbb{D}_l|}$ epochs (200 epochs or 74,200 iterations when all labels are used), and DeepLab v2 for $100 \times \frac{2975}{|\mathbb{D}_l|}$ epochs (100 epochs or 74,300 iterations when all labels are used). In comparison with SwiftNet-RN18, DeepLab v2 incurs a 12-fold per-image slowdown during supervised training. However, it also requires less epochs since it has very few parameters with random initialization. Hence, semi-supervised DeepLab v2 trains more than 4× slower than SwiftNet-RN18 on RTX 2080Ti. Appendix A.2 presents more detailed comparisons of memory and time requirements of different semi-supervised algorithms.

Our full-resolution experiments only use SwiftNet models. The crop size is 768 × 768 and the spatial scaling is sampled from $\mathcal{U}(2^{-1}, 2)$. The number of epochs is 250 when all labels are used. The batch size is 8 in supervised experiments, and $(B_l, B_u) = (8, 8)$ in semi-supervised experiments.

Appendix A.1 presents an overview and comparison of hyper-parameters with other consistency-based methods that are compared in the experiments.

Classification setup. Classification experiments target CIFAR-10 and involve the Wide ResNet model WRN-28-2 with standard hyper-parameters [71]. We augment all training images with random flips, padding and random cropping. We use all training images (including labeled images) in $\mathbb{D}_u$ for the consistency loss. Batch sizes are $(B_l, B_u) = (128, 640)$. Thus, the number of iterations per epoch is $\lfloor \frac{|\mathbb{D}_l|}{128} \rfloor$. For example, only one iteration is performed if $|\mathbb{D}_l| = 250$. We run $1000 \times \frac{4000}{|\mathbb{D}_l|}$ epochs in semi-supervised, and 100 epochs in supervised training. We use default VAT hyper-parameters $\xi = 10^{-6}$,

$\epsilon = 10$, $\alpha = 1$ [4]. We perform photometric perturbations as described, and sample TPS displacements from $\mathcal{N}(\mathbf{0}, 3.2 \times \mathbf{I}_2)$.

Evaluation. We report generalization performance at the end of training. We report sample means and sample standard deviations (with Bessel's correction) of the corresponding evaluation metric (mIoU or classification accuracy) of 5 training runs, evaluated on the corresponding validation dataset.

4.2. Semantic Segmentation on Half-Resolution Cityscapes

Table 1 compares our approach with the previous state of the art. We train using different proportions of training labels and evaluate mIoU on half-resolution Cityscapes val. The top section presents the previous work [7,8,25,57]. The middle section presents our experiments based on DeepLab v2 [26]. Note that here we outperform some previous work due to more involved training (as described in Section 4.1). since that would be a method of choice in all practical applications. Hence, we get consistently greater performance. We perform a proper comparison with [25] by using our training setup in combination with their method. Our MT-PhTPS outperforms MT-CutMix with L2 loss and confidence thresholding when 1/4 or more labels are available, while underperforming with 1/8 labels.

The bottom section involves the efficient model SwiftNet-RN18. Our perturbation model outperforms CutMix both with simple consistency, as well as with Mean Teacher. Overall, Mean Teacher outperforms simple consistency. We observe that DeepLab v2 and SwiftNet-RN18 get very similar benefits from the consistency loss. SwiftNet-RN18 comes out as a method of choice due to about 12× faster inference than DeepLab v2 with ResNet-101 on RTX 2080Ti (see Appendix A.2 for more details). Experiments from the middle and the bottom section use the same splits to ensure a fair comparison.

Table 1. Semantic segmentation performance (mIoU/%) on half-resolution Cityscapes val after training with different proportions of labeled data. The top section reviews experiments from previous work. The middle section presents our experiments with DeepLab v2. The bottom section presents our experiments with SwiftNet-RN18. We run experiments across 5 different dataset splits and report mean mIoUs with standard deviations. The subscript "~[25]" denotes training with L^2 loss, confidence thresholding, and $\alpha = 1$, as proposed in [25]. The best results overall are bold, and best results within sections are underlined.

Method	Label Proportion			
	1/8	1/4	1/2	1/1
DLv2-RN101 supervised [8,25]	56.2	60.2	64.6^1	66.0
DLv2-RN101 s4GAN+MLMT [8]	59.3	61.9	–	65.8
DLv2-RN101 supervised [7]	55.5	59.9	64.1	66.4
DLv2-RN101 AdvSemSeg [7]	58.8	62.3	65.7	67.7
DLv2-RN101 supervised [57]	56.0	60.5	–	66.0
DLv2-RN101 ECS [57]	<u>60.3</u>	<u>63.8</u>	–	<u>67.7</u>
DLv2-RN101 MT-CutMix [25]	$\underline{60.3}_{1.2}$	$\underline{63.9}_{0.7}$	–	$\underline{67.7}_{0.4}$
DLv2-RN101 supervised	$56.4_{0.4}$	$61.9_{1.1}$	$66.6_{0.6}$	$69.8_{0.4}$
DLv2-RN101 MT-CutMix~[25]	$\mathbf{63.2}_{1.4}$	$65.6_{0.8}$	$67.6_{0.4}$	$70.0_{0.3}$
DLv2-RN101 MT-PhTPS	$61.5_{1.0}$	$\mathbf{66.4}_{1.1}$	$\mathbf{69.0}_{0.6}$	$\mathbf{71.0}_{0.7}$
SN-RN18 supervised	$55.5_{0.9}$	$61.5_{0.5}$	$66.9_{0.7}$	$70.5_{0.6}$
SN-RN18 simple-CutMix	$59.8_{0.5}$	$63.8_{1.2}$	$67.0_{1.4}$	$69.3_{1.1}$
SN-RN18 simple-PhTPS	$60.8_{1.6}$	$64.8_{1.5}$	$68.8_{0.7}$	$71.1_{0.9}$
SN-RN18 MT-CutMix~[25]	$61.6_{0.9}$	$64.6_{0.5}$	$67.6_{0.7}$	$69.9_{0.6}$
SN-RN18 MT-CutMix	$59.3_{1.3}$	$63.3_{1.0}$	$66.8_{0.6}$	$69.7_{0.5}$
SN-RN18 MT-PhTPS	$\underline{62.0}_{1.3}$	$\mathbf{66.0}_{1.0}$	$\mathbf{69.1}_{0.5}$	$\mathbf{71.2}_{0.7}$

Now, we present ablation and hyper-parameter validation studies for simple-PhTPS consistency with SwiftNet-RN18. Table 2 presents ablations of the perturbation model, and also includes supervised training with PhTPS augmentations in one half of each mini-batch in addition to standard jittering. Perturbing the whole mini-batch with PhTPS in supervised training did not improve upon the baseline. We observe that perturbing half of each mini-batch with PhTPS in addition to standard jittering improves the supervised performance, but quite less than semi-supervised training. Semi-supervised experiments suggest that photometric perturbations (Ph) contribute most, and that geometric perturbations (TPS) are not useful when there is 1/2 or more of the labels.

Table 2. Ablation experiments on half-resolution Cityscapes val (mIoU/%) with SwiftNet-RN18. Subscripts denote the difference from the supervised baseline. The label "supervised PhTPS-aug" denotes supervised training where half of each mini-batch is perturbed with PhTPS. The bottom three rows compare PhTPS with Ph (only photometric) and TPS (only geometric) under simple one-way consistency. We present means of experiments on 5 different dataset splits. Numerical subscripts are differences with respect to the supervised baseline.

Method	Label Proportion			
	1/8	1/4	1/2	1/1
SN-RN18 supervised	55.5	61.5	66.9	70.5
SN-RN18 supervised PhTPS-aug	$56.2_{+1.5}$	$62.2_{+0.7}$	$67.4_{+0.5}$	$70.4_{-0.1}$
SN-RN18 simple-Ph	$59.2_{+3.7}$	$64.9_{+3.4}$	$68.3_{+1.4}$	$71.8_{+1.3}$
SN-RN18 simple-TPS	$58.2_{+3.1}$	$63.4_{+1.9}$	$66.7_{-0.2}$	$70.1_{-0.4}$
SN-RN18 simple-PhTPS	$60.8_{+5.3}$	$64.8_{+3.3}$	$68.8_{+1.9}$	$71.1_{+0.6}$

Figure 5 shows perturbation strength validation using 1/4 of the labels. Rows correspond to the factor that multiplies the standard deviation of control point displacements s_G defined at the end of Section 3.4. Columns correspond to the strength of the photometric perturbation s_P. The photometric strength s_P modulates the random photometric parameters according to the following expression:

$$(\underline{b}, \underline{s}, \underline{h}, \underline{c}) \mapsto (s_P \cdot \underline{b}, \exp(s_P \cdot \ln(\underline{s})), s_P \cdot \underline{h}, \exp(s_P \cdot \ln(\underline{c})) \, . \quad (10)$$

We set the probability of choosing a random channel permutation as $\min\{s_P, 1\}$. Hence, $s_P = 0$ corresponds to the identity function. Note that the "1/4" column in Table 2 uses the same semi-supervised configurations with strengths $s_G, s_P \in \{0, 1\}$. Moreover, note that the case $(s_G, s_P) = (0, 0)$ is slightly different from supervised training in that batch normalization statistics are still updated in the student. The differences in results are due to variance—the estimated standard error of the mean of 5 runs is between 0.35 and 0.5. We can observe that the photometric component is more important, and that a stronger photometric component can compensate for a weaker geometric component. Our perturbation strength choice $(s_G, s_P) = (1, 1)$ is close to the optimum, which the experiments suggests to be at $(1, 0.5)$.

Figure 6 shows our validation of the consistency loss weight α with SN-RN18 simple-PhTPS. We observe the best generalization performance for $\alpha \in [0.25 \ldots 0.75]$. We do not scale the learning rate with $(1 + \alpha)^{-1}$ because we use a scale-invariant optimization algorithm.

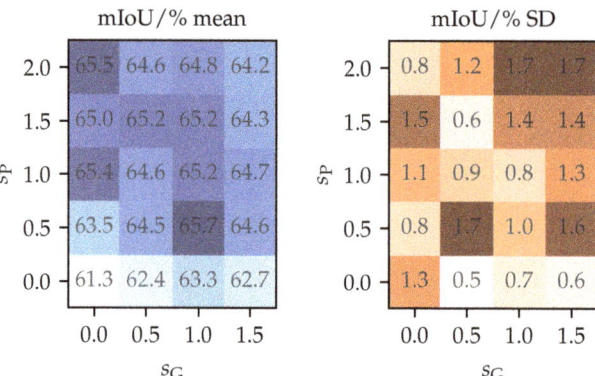

Figure 5. Validation of perturbation strength hyper-parameters on Cityscapes val (mIoU/%). We use 5 different subsets with 1/4 of the total number of training labels. The hyper-parameters s_P (photometric) and s_G (geometric) are defined in the main text. SD denotes the sample standard deviation.

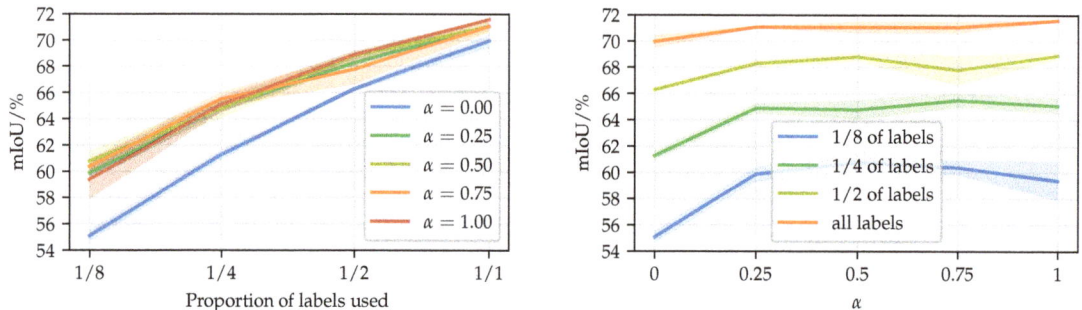

Figure 6. Validation of the consistency loss weight α on Cityscapes val (mIoU/%). We present the same results in two plots with different x-axes: the proportion of labels (**left**), and the consistency loss weight α (**right**).

Appendix B presents experiments that quantify the effect of updating batch normalization statistics when the inputs are perturbed.

Figure 7 shows qualitative results on the first few validation images with SwiftNet-RN18 trained with 1/4 of labels. We observe that our method displays a substantial resilience to heavy perturbations, such as those used during training.

Figure 7. Qualitative results on the first few validation images with SwiftNet-RN18 trained with 1/4 of half-resolution Cityscapes labels. Odd rows contain unperturbed inputs, and even rows contain PhTPS perturbed inputs. The columns are (left to right): ground truth segmentations, predictions of simple-PhTPS consistency training, and predictions of supervised training.

4.3. Semantic Segmentation on Full-Resolution Cityscapes

Table 3 presents our full resolution experiments in setups such as in Table 1, and comparison with previous work, but with full-resolution images and labels. In comparison with KE-GAN [53] and ECS [57], we underperform with 1/8 labeled images, but outperform with 1/2 labeled images. Note that KE-GAN [53] also trains on a large text corpus (MIT ConceptNet) as well as that ECS DLv3$^+$-RN50 requires 22 GiB of GPU memory with batch size 6 [57], while our SN-RN18 simple-PhTPS requires less than 8 GiB of GPU memory with batch size 8 and can be trained on affordable GPU hardware. Appendix A.2 presents more detailed memory and execution time comparisons with other algorithms.

We note that the concurrent approach DLv3$^+$-RN50 CAC [58] outperforms our method with 1/8 and 1/1 labels. However, ResNet-18 has significantly less capacity than ResNet-50. Therefore, the bottom section applies our method to the SwiftNet model with a ResNet-34 backbone, which still has less capacity than ResNet-50. The resulting model outperforms DLv3$^+$-RN50 CAC across most configurations. This shows that our method consistently improves when more capacity is available.

We note that training DLv3$^+$-RN50 CAC requires three RTX 2080Ti GPUs [58], while our SN-RN34 simple-PhTPS setup requires less than 9 GiB of GPU memory and fits on a single such GPU. Moreover, SN-RN34 has about 4× faster inference than DLv3$^+$-RN50 on RTX 2080Ti.

Table 3. Semi-supervised semantic segmentation performance (mIoU/%) on full-resolution Cityscapes val with different proportions of labeled data. We compare simple-PhTPS and MT-PhTPS (ours) with supervised training and previous work. DLv3$^+$-RN50 stands for DeepLab v3$^+$ with ResNet-50, and SN for SwiftNet. We run experiments across 5 different dataset splits and report mean mIoUs with standard deviations. Best results overall are bold, and best results within sections are underlined.

Method	Label Proportion			
	1/8	1/4	1/2	1/1
KE-GAN [53]	66.9	70.6	72.2	75.3
DLv3$^+$-RN50 supervised [57]	63.2	68.4	72.9	74.8
DLv3$^+$-RN50 ECS [57]	67.4	70.7	72.9	74.8
DLv3$^+$-RN50 supervised [58]	63.9	68.3	71.2	76.3
DLv3$^+$-RN50 CAC [58]	**69.7**	72.7	–	77.5
SN-RN18 supervised	61.1$_{0.4}$	67.3$_{1.1}$	71.9$_{0.1}$	75.4$_{0.4}$
SN-RN18 simple-PhTPS	66.3$_{1.0}$	71.0$_{0.5}$	74.3$_{0.7}$	75.8$_{0.4}$
SN-RN18 MT-PhTPS	68.6$_{0.6}$	72.0$_{0.3}$	73.8$_{0.4}$	75.0$_{0.4}$
SN-RN34 supervised	64.9$_{0.8}$	69.8$_{1.0}$	73.8$_{1.4}$	76.1$_{0.8}$
SN-RN34 simple-PhTPS	69.2$_{0.8}$	73.1$_{0.7}$	**76.3$_{0.7}$**	**77.9$_{0.2}$**
SN-RN34 MT-PhTPS	**70.8$_{1.5}$**	**74.3$_{0.5}$**	76.0$_{0.5}$	77.2$_{0.4}$

Finally, we present experiments in the large-data regime, where we place the whole fine subset into $\mathbb{D}_l$. In some of these experiments, we also train on the large coarsely labeled subset. We denote the extent of supervision with subscripts "l" (labeled) and "u" (unlabeled). Hence, C_u in the table denotes the coarse subset without labels. Table 4 investigates the impact of the coarse subset on the SwiftNet performance on the full-resolution Cityscapes val. We observe that semi-supervised learning brings considerable improvement with respect to fully supervised learning on fine labels only (columns F_l vs. $F_l \cup C_u$). It is also interesting to compare the proposed semi-supervised setup ($F_l \cup C_u$) with classic fully supervised learning on both subsets (F_l, C_l). We observe that semi-supervised learning with SwiftNet-RN18 comes close to supervised learning with coarse labels. Moreover, semi-supervised learning prevails when we plug in the SwiftNet-RN34. These experiments suggest that semi-supervised training represents an attractive alternative to coarse labels and large annotation efforts.

Table 4. Effects of an additional large dataset on supervised and semi-supervised learning on full-resolution Cityscapes val (mIoU/%). Tags F and C denote fine and coarse subsets, respectively. Subset indices denote whether we train with labels (l) or one-way consistency (u).

Method	F_l	(F_l, F_u)	$(F_l, F_u \cup C_u)$	(F_l, C_l)
SN-RN18 simple-PhTPS	$75.4_{0.4}$	$75.8_{0.4}$	$76.5_{0.3}$	$76.9_{0.3}$
SN-RN18 MT-PhTPS		$75.0_{0.4}$	$75.5_{0.3}$	
SN-RN34 simple-PhTPS	$76.1_{0.8}$	$77.9_{0.2}$	$\mathbf{78.5}_{0.4}$	$77.7_{0.4}$

4.4. Validation of Consistency Variants

Table 5 presents experiments with supervised baselines and four variants of semi-supervised consistency training. All semi-supervised experiments use the same PhTPS perturbations on CIFAR-10 (4000 labels and 50,000 images) and half-resolution Cityscapes (the SwiftNet-RN18 setups with 1/4 labels from Table 1). We investigate the following kinds of consistency: one-way with clean teacher (1w-ct, cf. Figure 1c), one-way with clean student (1w-cs, cf. Figure 1b), two-way with one clean input (2w-c1, cf. Figure 1d), and one-way with both inputs perturbed (1w-p2). Note that two-way consistency is not possible with Mean Teacher. Moreover, when both inputs are perturbed (1w-p2), we have to use the inverse geometric transformation on dense predictions [20]. We achieve that by forward warping [72] with the same displacement field. Two-way consistency with both inputs perturbed (2w-p2) is possible as well. We expect it to behave similarly to 1w-2p because it could be observed as a superposition of two opposite one-way consistencies, and our preliminary experiments suggest as much.

We observe that 1w-ct outperforms all other variants, while 2w-c1 performs in-between 1w-ct and 1w-cs. This confirms our hypothesis that predictions in clean inputs make better consistency targets. We note that 1w-p2 often outperforms 1w-cs, while always underperforming with respect to 1w-ct. A closer inspection suggests that 1w-p2 sometimes learns to cheat the consistency loss by outputting similar predictions for all perturbed images. This occurs more often when batch normalization uses the batch statistics estimated during training. A closer inspection of 1w-cs experiments on Cityscapes indicates the consistency cheating combined with severe overfitting to the training dataset.

Table 5. Comparison of 4 consistency variants under PhTPS perturbations: one-way with clean teacher (1w-ct), one-way with clean student (1w-cs), two-way with one input clean (2w-c1), and one-way with both inputs perturbed (1w-p2). Algorithms are evaluated on CIFAR-10 test (accuracy/%) while training on 4000 out of 50,000 labels (CIFAR-10, 2/25) and half-resolution Cityscapes val (mIoU/%) while training on 1/4 of labels from Cityscapes train with SwiftNet-RN18 (CS-half, 1/4).

Dataset	Method	sup.	1w-ct	1w-cs	2w-c1	1w-p2
CIFAR-10, 4k	WRN-28-2 simple-PhTPS	$80.8_{0.4}$	$\mathbf{90.8}_{0.3}$	$69.3_{4.2}$	$72.9_{2.6}$	$73.3_{7.0}$
CIFAR-10, 4k	WRN-28-2 MT-PhTPS	$80.8_{0.4}$	$\mathbf{90.8}_{0.4}$	$80.5_{0.5}$	-	$73.4_{1.4}$
CS-half, 1/4	SN-RN18 simple-PhTPS	$61.5_{0.5}$	$65.3_{1.9}$	$1.6_{1.0}$	$16.7_{3.0}$	$61.6_{0.5}$
CS-half, 1/4	SN-RN18 MT-PhTPS	$61.5_{0.5}$	$\mathbf{66.0}_{1.0}$	$61.5_{1.4}$	-	$62.0_{1.1}$

4.5. Image Classification on CIFAR-10

Table 6 evaluates the image classification performance of two supervised baselines and 4 semi-supervised algorithms on CIFAR-10. The first supervised baseline uses only labeled data with standard data augmentation. The second baseline additionally uses our perturbations for data augmentation. The third algorithm is VAT with entropy minimization [4]. The simple-PhTPS approach outperforms supervised approaches and VAT. Again, two-way consistency results in the worst generalization performance. Perturbing the teacher input results in accuracy below 17% for 4000 or less labeled examples, and is not displayed. Note that somewhat better performance can be achieved by complementing consistency with other techniques that are either suitable for dense prediction or out of the scope of this paper [5,49,69].

Table 6. Classification accuracy [%] on CIFAR-10 test with WRN-28-2. We compare two supervised approaches (top), VAT with entropy minimization [4] (middle), and two-way and one-way consistency with our perturbations (bottom three rows). We report means and standard deviations of 5 runs. The label "supervised PhTPS-aug" denotes the supervised training, where half of each mini-batch is perturbed with PhTPS.

Method	Number of Labeled Examples			
	250	1000	4000	50,000
supervised	$31.8_{0.6}$	$59.3_{1.4}$	$81.0_{0.2}$	94.7
supervised PhTPS-aug	$48.7_{0.9}$	$67.2_{0.5}$	$81.7_{0.2}$	95.0
VAT + entropy minimization	$41.0_{2.5}$	$73.2_{1.5}$	$84.2_{0.4}$	$90.5_{0.2}$
1w-cs simple-PhTPS	$27.7_{5.9}$	$51.7_{3.5}$	$69.3_{4.2}$	$91.6_{1.5}$
2w-c1 simple-PhTPS	$30.3_{1.8}$	$54.8_{1.5}$	$72.9_{2.6}$	$95.9_{0.2}$
1w-ct simple-PhTPS	**$68.8_{5.4}$**	**$84.2_{0.4}$**	**$90.6_{0.4}$**	**$96.2_{0.2}$**

5. Discussion

We have presented the first comprehensive study of one-way consistency for semi-supervised dense prediction, and proposed a novel perturbation model, which leads to the competitive generalization performance on Cityscapes. Our study clearly shows that one-way consistency with clean teacher outperforms other forms of consistency (e.g., clean student or two-way) both in terms of generalization performance and training footprint. We explain this by observing that predictions in perturbed images tend to be less reliable targets.

The proposed perturbation model is a composition of a photometric transformation and a geometric warp. These two kinds of perturbations have to be treated differently, since we desire invariance to the former and equivariance to the latter. Our perturbation model outperforms CutMix both in standard experiments with DeepLabv2-RN101 and in combination with recent efficient models (SwiftNet-RN18 and SwiftNet-RN34).

We consider two teacher formulations. In the simple formulation, the teacher is a frozen copy of the student. In the Mean Teacher formulation, the teacher is a moving average of student parameters. Mean Teacher outperforms simple consistency in low data regimes (half resolution; few labels). However, experiments with more data suggest that the simple one-way formulation scales significantly better.

To the best of our knowledge, this is the first account of semi-supervised semantic segmentation with efficient models. This combination is essential for many practical real-time applications where there is a lack of large datasets with suitable pixel-level groundtruth. Many of our experiments are based on SwiftNet-RN18, which behaves similarly to DeepLabv2-RN101, while offering about $9\times$ faster inference on half-resolution images, and about $15\times$ faster inference on full-resolution images on RTX 2080Ti. Experiments on Cityscapes coarse reveal that semi-supervised learning with one-way consistency can come close and exceed full supervision with coarse annotations. Simplicity, competitive performance and speed of training make this approach a very attractive baseline for evaluating future semi-supervised approaches in the dense prediction context.

Author Contributions: Conceptualization, M.O., I.G. and S.Š.; methodology, I.G., M.O. and S.Š.; software, I.G. and M.O.; validation, I.G. and M.O.; formal analysis, I.G.; investigation, I.G., M.O. and S.Š.; resources, S.Š.; data curation, I.G. and M.O.; writing—original draft preparation, I.G., M.O. and S.Š.; writing—review and editing, I.G. and S.Š.; visualization, I.G., M.O. and S.Š.; supervision, S.Š.; project administration, S.Š.; funding acquisition, S.Š. All authors have read and agreed to the published version of the manuscript.

Funding: This research was funded by the Croatian Science Foundation, grant IP-2020-02-5851 ADEPT. This research was also funded by European Regional Development Fund, grant KK.01.1.1.01.0009 DATACROSS.

Data Availability Statement: Data are available in a publicly accessible repository that does not issue DOIs and are provided by third parties. The datasets that we use are available at https://www.cityscapes-dataset.com/ and https://www.cs.toronto.edu/~kriz/cifar.html (accessed on 8 Jan 2023). Source code that enables reproducibility of our experiments is available at https://github.com/Ivan1248/semisup-seg-efficient.

Acknowledgments: This research was supported by VSITE College for Information Technologies, and NVIDIA Academic Hardware Grant Program.

Conflicts of Interest: The authors declare no conflict of interest.

Abbreviations

The following abbreviations are used in this manuscript:

DLv2	DeepLab v2
DLv3$^+$	DeepLab v3$^+$
CS	Cityscapes
MT	Mean Teacher
PhTPS	Our composition of photometric and geometric perturbations
pp	Percentage point
RN	ResNet
simple-X	Simple one-way consistency with clean teacher, with perturbation model X
1w-ct	One-way consistency with clean teacher
1w-cs	One-way consistency with clean student
2w-c1	Two-way consistency with one input perturbed
1w-p2	One-way consistency with both inputs perturbed
SD	Sample standard deviation with Bessel's correction
SN	SwiftNet
TPS	Thin plate spline
WRN	Wide ResNet

Appendix A. Additional Algorithm Comparisons

Appendix A.1. Hyper-Parameters

Tables A1 and A2 review hyper-parameters of consistency-based semi-supervised algorithms from Tables 1 and 3.

Table A1. Overview of hyper-parameters of consistency-based algorithms for semi-supervised semantic segmentation. We denote the setup that we use in our experiments including supervised training with "our". We use the following symbols: y is the teacher prediction, $\hat{y}$ is the student prediction, hard is a function that maps a vector representing a distribution to the closest one-hot vector ($\text{hard}_{[c]} = [\![c = \arg\max_k y_{[k]}]\!]$), e is the proportion of completed epochs, α is the consistency loss weight, and η is the base learning rate. The if-clause in the "Consistency loss" column represents confidence thresholding, i.e., determines whether the pixel is included in loss computation.

Model	Method	Crop Size	Jitter. Scale	Iterations	Epochs	B_l	B_u	Consistency Loss	α	Learning Rate Schedule
DLv2	MT-CutMix [25]	321	[0.5...1.5]	40,000	135*	10	10	$\|\text{hard}(y) - \tilde{y}\|_2^2$ if $\max_c y_{[c]} > 0.97$	0.5	$\eta(1-e)^{0.9}$
DLv2	MT-CutMix~[25]	321	[0.5...1.5]	37,100	100	4	4	$\|\text{hard}(y) - \tilde{y}\|_2^2$ if $\max_c y_{[c]} > 0.97$	1	$\eta(1-e)^{0.9}$
DLv2	ours	448	[0.5...2]	74,200	200	4	4	$D(y,\tilde{y})$	0.5	$\eta(1-e)^{0.9}$
SN	ours	448	[0.5...2]	74,200	200	8	8	$D(y,\tilde{y})$	0.5	$\eta\cos(e\pi/2)$
SN	ours	768	[0.5...2]	92,750	250	8	8	$D(y,\tilde{y})$	0.5	$\eta\cos(e\pi/2)$
DLv3+	CAC [58]	720	[0.5...1.5]	92,560	249*	8	8	see [58]	$[\![e > 5/80]\!] \times 0.1$	$\eta(1-e)^{0.9}$
DLv3+	ours	720	[0.5...1.5]	92,560	249*	8	8	$D(y,\tilde{y})$	0.5	$\eta(1-e)^{0.9}$

* Some authors [25,58] use the word "epoch" to refer to a fixed number of iterations — 1000 and 1157 iterations, respectively.

Table A2. Optimizer hyper-parameter configurations for Cityscapes semantic segmentation experiments, represented in a PyTorch-like style.

Model	Method	Main	Backbone (Difference)
DLv2	MT-CutMix [25]	SGD, lr = 3×10^{-5}, momentum = 0.9, weight_decay = 5×10^{-4}	
DLv2	MT-CutMix~[25]	SGD, lr = 3×10^{-5}, momentum = 0.9, weight_decay = 5×10^{-4}	
DLv2	ours	Adam, betas = (0.9, 0.99), lr = 4×10^{-4}, weight_decay = 1×10^{-4}	lr = 1×10^{-4}, weight_decay = 2.5×10^{-5}
SN	ours	Adam, betas = (0.9, 0.99), lr = 4×10^{-4}, weight_decay = 2.5×10^{-5}	lr = 1×10^{-4}, weight_decay = 6.25×10^{-6}
DLv3+	CAC [58]	SGD, lr = 1×10^{-1}, momentum = 0.9	lr = 1×10^{-2}, weight_decay = 1×10^{-4}
DLv3+	ours	SGD, lr = 1×10^{-1}, momentum = 0.9	lr = 1×10^{-2}, weight_decay = 1×10^{-4}

Appendix A.2. Time and Memory Performance Characteristics

Table A3 shows memory requirements and training times of methods from Tables 1 and 3. The times include data loading and processing, and do not include the evaluation on the validation set. The memory measurements are based on the `max_memory_allocated` and `reset_peak_memory_stats` procedures from the `torch.cuda` package. Note that some overhead that is required by PyTorch is not included in this measurement (see PyTorch memory management documentation for more information: https://pytorch.org/docs/master/notes/cuda.html (accessed on 8 Jan 2023). Some algorithms did not fit into the memory of GTX 2080Ti. The memory allocations are higher than in Figure 4 because the supervised prediction, perturbed inputs, and perturbed outputs are unnecessarily kept in memory.

For DeepLabv3$^+$-RN50, we use the number of iterations, batch size, and crop size from [58]. Note, however, that the method from [58] has the memory requirements of two-way consistency because of per-pixel directionality.

Table A3. Half resolution Cityscapes (top section) and Cityscapes (bottom section) maximum memory allocation and training time on two GPUs.

Model	Method	Crop Size	Iterations	B_l	B_u	Memory /MiB	Duration /min A4500	2080Ti
DLv2-RN101	MT-CutMix [25]	321	40,000	10	10	16,289	1067	–
	MT-CutMix~[25]	321	37,100	4	4	7037	794	1314
DLv2-RN101	supervised	448	74,300	4	–	6611	338	602
	MT-PhTPS	448	74,300	4	4	7021	816	1397
SN-RN18	supervised	448	74,200	4	–	1646	119	161
	simple-PhTPS	448	74,200	8	8	2398	228	279
	MT-PhTPS	448	74,200	8	8	2456	234	297
SN-RN18	supervised	768	92,750	8	–	4444	321	432
	simple-PhTPS	768	92,750	8	8	6683	732	963
	MT-PhTPS	768	92,750	8	8	6727	768	965
SN-RN34	supervised	768	92,750	8	–	5500	422	570
	simple-PhTPS	768	92,750	8	8	7737	994	1268
	MT-PhTPS	768	92,750	8	8	7818	1013	1276
DLv3+-RN50	supervised	720	92,560	8	–	11,645	1229	–
	simple-PhTPS	720	92,560	8	8	13,384	1884	–
	CAC [58]	720	92,560	8	8	25,165 †	>3000 *	–

† Estimated by running on NVidia A100. * The original implementation requires 36,005 MiB. Approximately 10.6 GiB can be saved by accumulating gradients as in Algorithm 1.

Table A4 shows the numbers of model parameters, and Table A5 shows inference speeds of models from Tables 1 and 3.

Table A4. Number of model parameters.

Model	Number of Parameters
DeepLabv2-RN101	43.80×10^6
DeepLabv3$^+$-RN50	40.35×10^6
SwiftNet-RN34	21.91×10^6
SwiftNet-RN18	11.80×10^6

Table A5. Model inference speed (number of iterations per second) on three different GPUs and two input resolutions. Inputs are processed one by one, without overlap in computation. The measurements include the computation of the cross-entropy loss, but do not include data loading and preparation.

	1024 × 2048			512 × 1024		
	A4500	2080Ti	1080Ti	A4500	2080Ti	1080Ti
DeepLabv2-RN101	5.1	3.0	1.5	19.6	12.2	6.3
DeepLabv3+-RN50	16.1	9.6	5.2	54.2	30.7	23.5
SwiftNet-RN34	39.2	30.5	23.6	93.4	86.1	73.5
SwiftNet-RN18	56.5	45.3	34.6	139.5	115.8	98.4

Appendix B. Effect of Updating Batch Normalization Statistics in the Perturbed Student

Batch normalization layers estimate population statistics during training as exponential moving averages. Training on perturbed images can adversely affect the suitability of these estimates. Consequently, we explore the design choice to update the statistics only on clean inputs (when the supervised loss is computed), instead of both on clean and on perturbed inputs. Note that this configuration difference does not affect parameter optimization because the training always relies on mini-batch statistics.. Figure A1 shows the effect of disabling the updates of batch normalization statistics when the model instance (student) receives perturbed inputs in our semi-supervised training (one way consistency with clean teacher). The experiments are conducted according to the corresponding descriptions in Section 4. In case of half-resolution Cityscapes, disabling updates in the perturbed student (blue) increased the validation mIoU by between 0.3 and 1.4 pp, depending on the proportion of labels. However, in case of full-resolution Cityscapes, an opposite effect occured—mIoU decreased by between 0.1 and 1.1 pp. In CIFAR-10 experiments, the effect is mostly neutral.

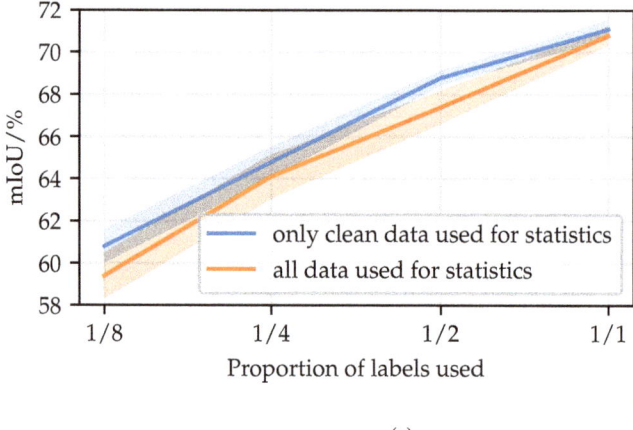

(a)

Figure A1. *Cont.*

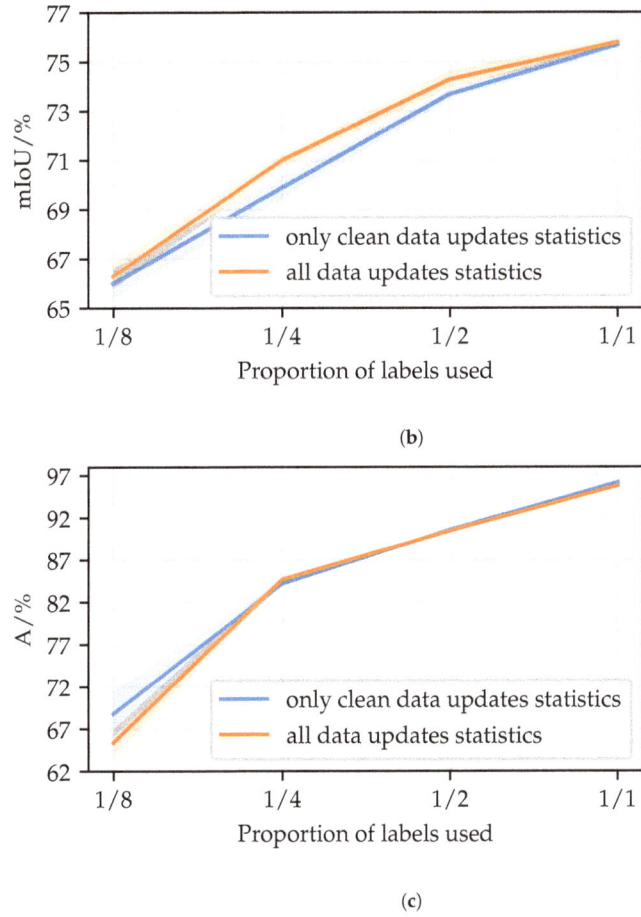

Figure A1. Effect of updating batch normalization statistics in the perturbed student. (**a**) Half-resolution Cityscapes val. (**b**) Cityscapes val. (**c**) CIFAR-10 validation set.

References

1. Kolesnikov, A.; Zhai, X.; Beyer, L. Revisiting Self-Supervised Visual Representation Learning. In Proceedings of the IEEE/CVF Conference on Computer Vision and Pattern Recognition, CVPR 2019, Long Beach, CA, USA, 16–20 June 2019; pp. 1920–1929.
2. Lee, J.; Kim, E.; Lee, S.; Lee, J.; Yoon, S. FickleNet: Weakly and Semi-Supervised Semantic Image Segmentation Using Stochastic Inference. In Proceedings of the IEEE/CVF Conference on Computer Vision and Pattern Recognition, CVPR 2019, Long Beach, CA, USA, 16–20 June 2019; pp. 5267–5276.
3. Tarvainen, A.; Valpola, H. Mean teachers are better role models: Weight-averaged consistency targets improve semi-supervised deep learning results. In Proceedings of the Advances in Neural Information Processing Systems, Long Beach, CA, USA, 4–9 December 2017; pp. 1195–1204.
4. Miyato, T.; Maeda, S.; Koyama, M.; Ishii, S. Virtual Adversarial Training: A Regularization Method for Supervised and Semi-Supervised Learning. *IEEE Trans. Pattern Anal. Mach. Intell.* **2019**, *41*, 1979–1993. [CrossRef] [PubMed]
5. Xie, Q.; Dai, Z.; Hovy, E.H.; Luong, T.; Le, Q. Unsupervised Data Augmentation for Consistency Training. In Proceedings of the Advances in Neural Information Processing Systems 33: Annual Conference on Neural Information Processing Systems 2020, NeurIPS 2020, Virtual, 6–12 December 2020; Larochelle, H., Ranzato, M., Hadsell, R., Balcan, M., Lin, H., Eds.; Curran Associates Inc.: Red Hook, NY, USA, 2020.
6. Souly, N.; Spampinato, C.; Shah, M. Semi Supervised Semantic Segmentation Using Generative Adversarial Network. In Proceedings of the IEEE International Conference on Computer Vision, ICCV 2017, Venice, Italy, 22–29 October 2017; pp. 5689–5697.

7. Hung, W.; Tsai, Y.; Liou, Y.; Lin, Y.; Yang, M. Adversarial Learning for Semi-supervised Semantic Segmentation. In Proceedings of the BMVC, Newcastle, UK, 3–6 September 2018; p. 65.
8. Mittal, S.; Tatarchenko, M.; Brox, T. Semi-Supervised Semantic Segmentation with High- and Low-level Consistency. *IEEE Trans. Pattern Anal. Mach. Intell.* **2019**, *43*, 1369–1379. [CrossRef] [PubMed]
9. Cordts, M.; Omran, M.; Ramos, S.; Scharwächter, T.; Enzweiler, M.; Benenson, R.; Franke, U.; Roth, S.; Schiele, B. The Cityscapes Dataset. In Proceedings of the CVPRW, Boston, MA, USA, 7–12 June 2015.
10. Neuhold, G.; Ollmann, T.; Rota Bulò, S.; Kontschieder, P. Mapillary Vistas Dataset for Semantic Understanding of Street Scenes. In Proceedings of the ICCV, Venice, Italy, 22–29 October 2017; pp. 5000–5009.
11. Maggiori, E.; Tarabalka, Y.; Charpiat, G.; Alliez, P. Can semantic labeling methods generalize to any city? The inria aerial image labeling benchmark. In Proceedings of the 2017 IEEE International Geoscience and Remote Sensing Symposium, IGARSS 2017, Fort Worth, TX, USA, 23–28 July 2017; pp. 3226–3229.
12. Rota Bulò, S.; Porzi, L.; Kontschieder, P. In-Place Activated BatchNorm for Memory-Optimized Training of DNNs. In Proceedings of the 2018 IEEE/CVF Conference on Computer Vision and Pattern Recognition, CVPR 2018, Salt Lake City, UT, USA, 18–22 June 2018.
13. Dosovitskiy, A.; Springenberg, J.T.; Riedmiller, M.; Brox, T. Discriminative unsupervised feature learning with convolutional neural networks. In Proceedings of the Advances in Neural Information Processing Systems, hlMontreal, QC, Canada, 8–13 December 2014; pp. 766–774.
14. Salimans, T.; Goodfellow, I.J.; Zaremba, W.; Cheung, V.; Radford, A.; Chen, X. Improved Techniques for Training GANs. In Proceedings of the Advances in Neural Information Processing Systems 29: Annual Conference on Neural Information Processing Systems 2016, Barcelona, Spain, 5–10 December 2016; pp. 2226–2234.
15. Tsai, Y.; Hung, W.; Schulter, S.; Sohn, K.; Yang, M.; Chandraker, M. Learning to Adapt Structured Output Space for Semantic Segmentation. In Proceedings of the 2018 IEEE/CVF Conference on Computer Vision and Pattern Recognition, CVPR 2018, Salt Lake City, UT, USA, 18–22 June 2018; pp. 7472–7481.
16. Gao, H.; Yao, D.; Wang, M.; Li, C.; Liu, H.; Hua, Z.; Wang, J. A Hyperspectral Image Classification Method Based on Multi-Discriminator Generative Adversarial Networks. *Sensors* **2019**, *19*, 3269. [CrossRef]
17. Rasmus, A.; Berglund, M.; Honkala, M.; Valpola, H.; Raiko, T. Semi-supervised Learning with Ladder Networks. In Proceedings of the Advances in Neural Information Processing Systems 28: Annual Conference on Neural Information Processing Systems 2015, Montreal, QC, Canada, 7–12 December 2015; pp. 3546–3554.
18. Sajjadi, M.; Javanmardi, M.; Tasdizen, T. Mutual exclusivity loss for semi-supervised deep learning. In Proceedings of the Advances in Neural Information Processing Systems 29: Annual Conference on Neural Information Processing Systems 2016, Barcelona, Spain, 5–10 December 2016; pp. 1163–1171.
19. Qiao, S.; Shen, W.; Zhang, Z.; Wang, B.; Yuille, A. Deep co-training for semi-supervised image recognition. In Proceedings of the European Conference on Computer Vision (ECCV), Munich, Germany, 8–14 September 2018; pp. 135–152.
20. Bortsova, G.; Dubost, F.; Hogeweg, L.; Katramados, I.; de Bruijne, M. Semi-supervised Medical Image Segmentation via Learning Consistency Under Transformations. In Proceedings of the MICCAI, Shenzhen, China, 13–17 October 2019.
21. Laine, S.; Aila, T. Temporal Ensembling for Semi-Supervised Learning. In Proceedings of the 5th International Conference on Learning Representations, ICLR 2017, Toulon, France, 24–26 April 2017.
22. Zheng, S.; Song, Y.; Leung, T.; Goodfellow, I.J. Improving the Robustness of Deep Neural Networks via Stability Training. In Proceedings of the 2016 IEEE Conference on Computer Vision and Pattern Recognition, CVPR 2016, Las Vegas, NV, USA, 27–30 June 2016; pp. 4480–4488.
23. Krešo, I.; Krapac, J.; Šegvić, S. Efficient ladder-style densenets for semantic segmentation of large images. *IEEE Trans. Intell. Transp. Syst.* **2020**, *40*, 1369–1379.
24. Grubišić, I.; Oršić, M.; Šegvić, S. A baseline for semi-supervised learning of efficient semantic segmentation models. In Proceedings of the 17th International Conference on Machine Vision and Applications, MVA 2021, Aichi, Japan, 25–27 July 2021; pp. 1–5. [CrossRef]
25. French, G.; Laine, S.; Aila, T.; Mackiewicz, M.; Finlayson, G. Semi-supervised semantic segmentation needs strong, varied perturbations. In Proceedings of the BMVC, Virtual, 7–10 September 2020.
26. Chen, L.C.; Papandreou, G.; Kokkinos, I.; Murphy, K.; Yuille, A.L. DeepLab: Semantic Image Segmentation with Deep Convolutional Nets, Atrous Convolution, and Fully Connected CRFs. *IEEE Trans. Pattern Anal. Mach. Intell.* **2018**, *40*, 834–848. [CrossRef]
27. Oršić, M.; Krešo, I.; Bevandić, P.; Šegvić, S. In defense of pre-trained imagenet architectures for real-time semantic segmentation of road-driving images. In Proceedings of the IEEE Conference on Computer Vision and Pattern Recognition, Long Beach, CA, USA, 15–20 June 2019; pp. 12607–12616.
28. Schwartz, R.; Dodge, J.; Smith, N.A.; Etzioni, O. Green AI. *Commun. ACM* **2020**, *63*, 54–63. [CrossRef]
29. Long, J.; Shelhamer, E.; Darrell, T. Fully convolutional networks for semantic segmentation. In Proceedings of the 2015 IEEE Conference on Computer Vision and Pattern Recognition, CVPR 2015, Boston, MA, USA, 7–12 June 2015; pp. 3431–3440. [CrossRef]

30. Ronneberger, O.; Fischer, P.; Brox, T. U-Net: Convolutional Networks for Biomedical Image Segmentation. In *Medical Image Computing and Computer-Assisted Intervention—MICCAI 2015, Proceedings of the 18th International Conference, Munich, Germany, 5–9 October 2015*; Navab, N., Hornegger, J., III, Wells, M.W., Frangi, A.F., Eds.; Lecture Notes in Computer Science; Proceedings, Part III; Springer: Cham, Switzerland, 2015; Volume 9351, pp. 234–241.
31. Zhao, H.; Qi, X.; Shen, X.; Shi, J.; Jia, J. ICNet for Real-Time Semantic Segmentation on High-Resolution Images. In Proceedings of the ECCV, Munich, Germany, 8–14 September 2018; Volume 11207, pp. 418–434.
32. Zhao, H.; Shi, J.; Qi, X.; Wang, X.; Jia, J. Pyramid Scene Parsing Network. In Proceedings of the 2017 IEEE Conference on Computer Vision and Pattern Recognition, CVPR 2017, Honolulu, HI, USA, 22–25 July 2017.
33. Oršić, M.; Šegvić, S. Efficient semantic segmentation with pyramidal fusion. *Pattern Recognit.* **2021**, *110*, 107611. [CrossRef]
34. Sun, K.; Xiao, B.; Liu, D.; Wang, J. Deep High-Resolution Representation Learning for Human Pose Estimation. In Proceedings of the IEEE/CVF Conference on Computer Vision and Pattern Recognition, CVPR 2019, Long Beach, CA, USA, 18–20 June 2019.
35. Chapelle, O.; Schlkopf, B.; Zien, A. *Semi-Supervised Learning*, 1st ed.; The MIT Press: Cambridge, MA, USA, 2010.
36. Grandvalet, Y.; Bengio, Y. Semi-supervised Learning by Entropy Minimization. In *Proceedings of the Advances in Neural Information Processing Systems*; Saul, L.K., Weiss, Y., Bottou, L., Eds.; MIT Press: Cambridge, MA, USA, 2005; pp. 529–536.
37. Yarowsky, D. Unsupervised Word Sense Disambiguation Rivaling Supervised Methods. In Proceedings of the 33rd Annual Meeting of the Association for Computational Linguistics, Cambridge, MA, USA, 26–30 June 1995; Association for Computational Linguistics: Stroudsburg, PA, USA, 1995; pp. 189–196. [CrossRef]
38. McClosky, D.; Charniak, E.; Johnson, M. Effective Self-Training for Parsing. In Proceedings of the Human Language Technology Conference of the NAACL, Main Conference; New York, NY, USA, 4–9 June 2006; Association for Computational Linguistics: Stroudsburg, PA, USA, 2006; pp. 152–159.
39. hyun Lee, D. Pseudo-Label: The Simple and Efficient Semi-Supervised Learning Method for Deep Neural Networks. In Proceedings of the ICML 2013 Workshop: Challenges in Representation Learning (WREPL), Atlanta, GA, USA, 16–21 June 2013.
40. Oliver, A.; Odena, A.; Raffel, C.A.; Cubuk, E.D.; Goodfellow, I. Realistic Evaluation of Deep Semi-Supervised Learning Algorithms. In Proceedings of the Advances in Neural Information Processing Systems 32, NeurIPS 2018, Montréal, QC, Canada, 2–8 December 2018; Bengio, S., Wallach, H., Larochelle, H., Grauman, K., Cesa-Bianchi, N., Garnett, R., Eds.; Curran Associates, Inc.: Red Hook, NY, USA, 2018; Volume 31.
41. Xie, Q.; Luong, M.; Hovy, E.H.; Le, Q.V. Self-Training with Noisy Student Improves ImageNet Classification. In Proceedings of the IEEE/CVF Conference on Computer Vision and Pattern Recognition, CVPR 2020, Seattle, WA, USA, 13–19 June 2020; pp. 10684–10695.
42. Wang, Y.; Wang, H.; Shen, Y.; Fei, J.; Li, W.; Jin, G.; Wu, L.; Zhao, R.; Le, X. Semi-Supervised Semantic Segmentation Using Unreliable Pseudo-Labels. In Proceedings of the IEEE/CVF Conference on Computer Vision and Pattern Recognition, CVPR 2022, New Orleans, LA, USA, 18–24 June 2022; pp. 4238–4247.
43. Gerken, J.E.; Aronsson, J.; Carlsson, O.; Linander, H.; Ohlsson, F.; Petersson, C.; Persson, D. Geometric Deep Learning and Equivariant Neural Networks. *arXiv* **2021**, arXiv:2105.13926.
44. Lenc, K.; Vedaldi, A. Understanding Image Representations by Measuring Their Equivariance and Equivalence. *Int. J. Comput. Vis.* **2019**, *127*, 456–476. [CrossRef] [PubMed]
45. Wang, Y.; Zhang, J.; Kan, M.; Shan, S.; Chen, X. Self-supervised Equivariant Attention Mechanism for Weakly Supervised Semantic Segmentation. In Proceedings of the 2020 IEEE/CVF Conference on Computer Vision and Pattern Recognition, CVPR 2020, Seattle, WA, USA, 13–19 June 2020.
46. Cho, J.H.; Mall, U.; Bala, K.; Hariharan, B. PiCIE: Unsupervised Semantic Segmentation using Invariance and Equivariance in Clustering. In Proceedings of the 2021 IEEE/CVF Conference on Computer Vision and Pattern Recognition, CVPR 2021, Virtual, 19–25 June 2021.
47. Patel, G.; Dolz, J. Weakly supervised segmentation with cross-modality equivariant constraints. *Med. Image Anal.* **2022**, *77*, 102374. [CrossRef]
48. Häusser, P.; Mordvintsev, A.; Cremers, D. Learning by Association—A Versatile Semi-Supervised Training Method for Neural Networks. In Proceedings of the 2017 IEEE Conference on Computer Vision and Pattern Recognition, CVPR 2017, Honolulu, HI, USA, 21–26 July 2017; pp. 626–635.
49. Berthelot, D.; Carlini, N.; Goodfellow, I.; Papernot, N.; Oliver, A.; Raffel, C.A. MixMatch: A Holistic Approach to Semi-Supervised Learning. In Proceedings of the Advances in Neural Information Processing Systems 33, NeurIPS 2019, Vancouver, BC, Canada, 8–14 December 2019; pp. 5049–5059.
50. Chen, X.; He, K. Exploring Simple Siamese Representation Learning. In Proceedings of the IEEE/CVF Conference on Computer Vision and Pattern Recognition, CVPR 2021, Nashville, TN, USA, 20–25 June 2021; Computer Vision Foundation/IEEE: New York, NY, USA, 2021; pp. 15750–15758.
51. Tian, Y.; Chen, X.; Ganguli, S. Understanding self-supervised learning dynamics without contrastive pairs. In Proceedings of the 38th International Conference on Machine Learning, ICML 2021, Virtual, 18–24 July 2021; Meila, M., Zhang, T., Eds.; Proceedings of Machine Learning Research; PMLR: Brookline, MA, USA, 2021; Volume 139, pp. 10268–10278.

52. Kurakin, A.; Li, C.L.; Raffel, C.; Berthelot, D.; Cubuk, E.D.; Zhang, H.; Sohn, K.; Carlini, N.; Zhang, Z. FixMatch: Simplifying Semi-Supervised Learning with Consistency and Confidence. In Proceedings of the Advances in Neural Information Processing Systems 33: Annual Conference on Neural Information Processing Systems 2020, NeurIPS 2020, Virtual, 6–12 December 2020; Larochelle, H., Ranzato, M., Hadsell, R., Balcan, M., Lin, H., Eds.; Curran Associates Inc.: Red Hook, NY, USA, 2020.
53. Qi, M.; Wang, Y.; Qin, J.; Li, A. KE-GAN: Knowledge Embedded Generative Adversarial Networks for Semi-Supervised Scene Parsing. In Proceedings of the 2019 IEEE/CVF Conference on Computer Vision and Pattern Recognition, CVPR 2019, Long Beach, CA, USA, 18–20 June 2019; pp. 5237–5246.
54. Ouali, Y.; Hudelot, C.; Tami, M. Semi-Supervised Semantic Segmentation With Cross-Consistency Training. In Proceedings of the 2020 IEEE/CVF Conference on Computer Vision and Pattern Recognition, CVPR 2020, Seattle, WA, USA, 13–19 June 2020.
55. Zhu, Y.; Zhang, Z.; Wu, C.; Zhang, Z.; He, T.; Zhang, H.; Manmatha, R.; Li, M.; Smola, A.J. Improving Semantic Segmentation via Self-Training. *arXiv* **2020**, arXiv:2004.14960.
56. Chen, L.; Lopes, R.G.; Cheng, B.; Collins, M.D.; Cubuk, E.D.; Zoph, B.; Adam, H.; Shlens, J. Naive-Student: Leveraging Semi-Supervised Learning in Video Sequences for Urban Scene Segmentation. In Proceedings of the Computer Vision—ECCV 2020—16th European Conference, Glasgow, UK, 23–28 August 2020; Vedaldi, A., Bischof, H., Brox, T., Frahm, J., Eds.; Lecture Notes in Computer Science; Proceedings, Part IX; Springer: London, UK, 2020; Volume 12354, pp. 695–714.
57. Mendel, R.; Souza, L., Jr.; Rauber, D.; Papa, J.; Palm, C. Semi-Supervised Segmentation based on Error-Correcting Supervision. In Proceedings of the Computer Vision—ECCV 2020—16th European Conference, Glasgow, UK, 23–28 August 2020.
58. Lai, X.; Tian, Z.; Jiang, L.; Liu, S.; Zhao, H.; Wang, L.; Jia, J. Semi-supervised Semantic Segmentation with Directional Context-aware Consistency. In Proceedings of the IEEE Conference on Computer Vision and Pattern Recognition, CVPR 2021, Virtual, 19–25 June 2021; Computer Vision Foundation/IEEE: New York, NY, USA, 2021.
59. van den Oord, A.; Li, Y.; Vinyals, O. Representation Learning with Contrastive Predictive Coding. *arXiv* **2018**, arXiv:1807.03748.
60. He, K.; Fan, H.; Wu, Y.; Xie, S.; Girshick, R. Momentum Contrast for Unsupervised Visual Representation Learning. In Proceedings of the 2020 IEEE/CVF Conference on Computer Vision and Pattern Recognition, CVPR 2020, Seattle, WA, USA, 13–19 June 2020; pp. 9726–9735. [CrossRef]
61. Yang, L.; Zhuo, W.; Qi, L.; Shi, Y.; Gao, Y. ST++: Make Self-training Work Better for Semi-supervised Semantic Segmentation. In Proceedings of the 2020 IEEE/CVF Conference on Computer Vision and Pattern Recognition, CVPR 2020, Seattle, WA, USA, 13–19 June 2022.
62. Olah, C. Visual Information Theory, 2015. Available online: https://colah.github.io/posts/2015-09-Visual-Information/ (accessed on 8 January 2023).
63. Huang, G.; Liu, Z.; Pleiss, G.; van der Maaten, L.; Weinberger, K.Q. Convolutional Networks with Dense Connectivity. *IEEE Trans. Pattern Anal. Mach. Intell.* **2022**, *44*, 8704–8716. [CrossRef] [PubMed]
64. Grill, J.B.; Strub, F.; Altché, F.; Tallec, C.; Richemond, P.; Buchatskaya, E.; Doersch, C.; Avila Pires, B.; Guo, Z.; Gheshlaghi Azar, M.; et al. Bootstrap Your Own Latent—A New Approach to Self-Supervised Learning. In Proceedings of the Advances in Neural Information Processing Systems 33: Annual Conference on Neural Information Processing Systems 2020, NeurIPS 2020, Virtual, 6–12 December 2020; Larochelle, H., Ranzato, M., Hadsell, R., Balcan, M., Lin, H., Eds.; Curran Associates Inc.: Red Hook, NY, USA, 2020.
65. Duchon, J. Splines minimizing rotation-invariant semi-norms in Sobolev spaces. In *Constructive Theory of Functions of Several Variables*; Springer: Berlin/Heidelberg, Germany, 1977; pp. 85–100.
66. Bookstein, F.L. Principal Warps: Thin-Plate Splines and the Decomposition of Deformations. *IEEE Trans. Pattern Anal. Mach. Intell.* **1989**, *11*, 567–585. [CrossRef]
67. Szeliski, R. *Computer Vision: Algorithms and Applications*; Springer Science & Business Media: Berlin, Germany, 2010.
68. Paszke, A.; Gross, S.; Massa, F.; Lerer, A.; Bradbury, J.; Chanan, G.; Killeen, T.; Lin, Z.; Gimelshein, N.; Antiga, L.; et al. PyTorch: An Imperative Style, High-Performance Deep Learning Library. In Proceedings of the Advances in Neural Information Processing Systems 33, NeurIPS 2019, Vancouver, BC, Canada, 8–14 December 2019; pp. 8024–8035.
69. Cubuk, E.D.; Zoph, B.; Shlens, J.; Le, Q. RandAugment: Practical Automated Data Augmentation with a Reduced Search Space. In Proceedings of the Advances in Neural Information Processing Systems 33: Annual Conference on Neural Information Processing Systems 2020, NeurIPS 2020, Virtual, 6–12 December 2020; Larochelle, H., Ranzato, M., Hadsell, R., Balcan, M., Lin, H., Eds.; Curran Associates Inc.: Red Hook, NY, USA, 2020.
70. Loshchilov, I.; Hutter, F. SGDR: Stochastic Gradient Descent with Warm Restarts. In Proceedings of the 5th International Conference on Learning Representations, ICLR 2017, Toulon, France, 24–26 April 2017.
71. Zagoruyko, S.; Komodakis, N. Wide Residual Networks. In Proceedings of the British Machine Vision Conference (BMVC) 2016, York, UK, 19–22 September 2016. [CrossRef]
72. Niklaus, S.; Liu, F. Softmax Splatting for Video Frame Interpolation. In Proceedings of the 2020 IEEE/CVF Conference on Computer Vision and Pattern Recognition, CVPR 2020, Seattle, WA, USA, 13–19 June 2020.

Disclaimer/Publisher's Note: The statements, opinions and data contained in all publications are solely those of the individual author(s) and contributor(s) and not of MDPI and/or the editor(s). MDPI and/or the editor(s) disclaim responsibility for any injury to people or property resulting from any ideas, methods, instructions or products referred to in the content.

Article

Adversarial Patch Attacks on Deep-Learning-Based Face Recognition Systems Using Generative Adversarial Networks

Ren-Hung Hwang [1,*], Jia-You Lin [2], Sun-Ying Hsieh [2], Hsuan-Yu Lin [3] and Chia-Liang Lin [3]

1. College of Artificial Intelligence, National Yang Ming Chiao Tung University, Tainan 71150, Taiwan
2. Computer Science and Information Engineering Department, National Chung Cheng University, Chiayi 62102, Taiwan
3. Telecom Technology Center, Kao-Hsiung 82151, Taiwan
* Correspondence: rhhwang@nycu.edu.tw

Abstract: Deep learning technology has developed rapidly in recent years and has been successfully applied in many fields, including face recognition. Face recognition is used in many scenarios nowadays, including security control systems, access control management, health and safety management, employee attendance monitoring, automatic border control, and face scan payment. However, deep learning models are vulnerable to adversarial attacks conducted by perturbing probe images to generate adversarial examples, or using adversarial patches to generate well-designed perturbations in specific regions of the image. Most previous studies on adversarial attacks assume that the attacker hacks into the system and knows the architecture and parameters behind the deep learning model. In other words, the attacked model is a white box. However, this scenario is unrepresentative of most real-world adversarial attacks. Consequently, the present study assumes the face recognition system to be a black box, over which the attacker has no control. A Generative Adversarial Network method is proposed for generating adversarial patches to carry out dodging and impersonation attacks on the targeted face recognition system. The experimental results show that the proposed method yields a higher attack success rate than previous works.

Keywords: deep learning; face recognition; adversarial attack; perturbation; adversarial examples; adversarial patches; Generative Adversarial Network

1. Introduction

Face recognition technology has undergone significant advances in recent years through the application of deep learning models. Meanwhile, the COVID-19 pandemic has brought about many lifestyle changes, including a desire for non-contact business opportunities wherever possible [1]. As a result, face recognition now plays a significant role in improving security and convenience in all manner of fields and applications. For example, face recognition is widely used throughout manufacturing and warehousing, banking and financial insurance, smart offices, smart homes, retail, public transportation and airports, medical scenes, schools and education institutions, hotels, and many other service industries. MarketsandMarkets [2] estimated that the global output value of face recognition would grow at an annual rate of 17.2% from 2020 onwards and would reach a global market value of USD 13.87 billion in 2028. Thus, face recognition offers significant business opportunities in the coming years and decades.

However, despite the many benefits of deep learning technology, it is not without risk. For example, the authors in [3] showed that image classification systems built on deep learning models can be easily attacked by adding a small perturbation to the original image to form an adversarial example, which is subsequently misclassified. Similar misclassification errors can be induced by applying adversarial patches [4] to the original image to produce local perturbations. In such cases, the reliability of the image classification system is significantly impaired. Consequently, the problem of improving the robustness of deep learning models, and the applications which rely on these models (face recognition systems, for example), is a crucial concern in real-world environments.

Based on the above issue, we propose a method that can make the face recognition model misclassified, and the method can achieve attack effectiveness in the physical world as well. Moreover, we also explore adding adversarial images to the face recognition model as a training dataset to improve the model's robustness.

Adversarial attacks against deep learning models can be divided into many types. For example, depending on the adversarial capacity, they can be classified as either white-box or black-box attacks, where in the former case, the attacker knows the parameters and architecture of the deep learning model, whereas in the latter case, they do not. Black-box attacks are thus generally more challenging than white box attacks. A second class of attack is that of poisoning attacks, in which adversarial images are injected into the model during the training stage in order to affect the learning performance; or input or evasion attacks, in which the input images are deliberately perturbed in order to produce misclassification errors. Depending on the space in which they are launched, adversarial attacks can also be classified as either physical world attacks [5,6] or digital world attacks. Finally, depending on whether or not the attack has a specific target, adversarial attacks can be categorized as either targeted attacks or non-targeted attacks, where such attacks are generally referred to as dodging attacks or impersonation attacks, respectively, in the face recognition field. Dodging attacks aim to cause the input face image to be identified as any other individual in the face database. By contrast, impersonation attacks aim to cause the individual to be identified as a specific person (i.e., the attack target) in the database.

However, for the attack method, Bhambri et al. [7] surveyed the relevant literature, in which Deb et al. [8] proposed a GAN-based [9] adversarial attack method that generates perturbations for the human face to achieve an attack in a digital environment. That perturbation cannot be examined with the human eye, nor can physical cameras. Therefore, based on this method, we can conduct an attack on a face recognition system in the physical world by generating perturbation for the glasses of a specific person. When an attacker wears attack glasses to attack a face recognition system, it can cause misidentification.

As aforementioned, the majority of face recognition systems are built in the real world, the present study focuses on the challenging problem of black-box input attacks using GAN-based adversarial patches in the physical world. For the sake of robustness, the study considers both dodging attacks and impersonation attacks. In short, our contributions are summarized as follows.

- We propose the adversarial patches method for face recognition attacks applicable to the physical world. It does not require knowledge of the parameters of the deep learning model (black box) to achieve attack effectiveness.
- For the reliability of our approach, we performed a comprehensive attack test for all one-to-one combinations. Based on testing quantities, the number of subjects and the number of testing by each subject is higher than the previous literature, which results show that the success rate of dodging attacks is 57.99%, and the impersonation attack success rate is 46.78% in the digital world. The success rate of dodging attacks is 81.77%, impersonation attack success rate reached 63.85% in the physical world.
- The proposed attack method utilizes the adversarial patch, which occupies only a small area of the face, instead of the adversarial example, which occupies the whole face. Therefore, the attacker can adjust the noise region according to the requirements. In our case, we hide the adversarial perturbation in the glasses to achieve the effectiveness of being judged as someone else. As a result, it is difficult for the layperson to know our attack intent and, therefore, poses a significant threat to the face recognition system.
- Based on the method proposed in this study, we found that the number of people with two face databases of different numbers of people, the number of people will further affect the attack's success rate. The attack success rate increases when the number of people in the database increases. In short, the chances of being hacked increase.

- We propose a novel defense mechanism to counter the GAN-based adversarial patch method. The results show that the proposed mechanism can detect almost all dodging attacks and more than half of the impersonation attacks with high defense effectiveness.
- We explored the relationship between thresholds and attack success and proved that both are relative. In addition, we attack different models by the no-box attack, showing that our attack method is transferable.

2. Background

2.1. Face Recognition

Various face recognition methods have been proposed in the past, such as SVM-based, subspace learning-based, and deep-learning-based methods. We summarize and compare the previous works in Table 1.

Table 1. Previous works of face recognition.

Previous Works	Database	Method	Accuracy	Year
[10]	Tufts face	MvRDTSVM	91.55%	2022
		MvFRDTSVM	88.82 %	2022
[11]	AT&T face	DWT + PCA + SVM	96%	2018
		DWT + LDA + SVM	96%	2018
		DWT + ICA + SVM	94.5%	2018
[12]	PubFig83	CSV-DML	84.6%	2022
[13]	LFW	DeepFace-ensemble	97.35%	2014
[14]	LFW	Siamese Network (ZFNet + Inception-v1)	99.63%	2015
[15]	VGGFace2	ResNet-50	99.6%	2018
[16]	LFW	LightCNN-v29	98.98%	2020
[17]	VGGFace2-FP	PDA	95.32%	2020
[18]	VGGFace2-FP	HOG + Autoencoders	99.60%	2017
[19]	CASIA NIR-VIS2.0	CpGAN	96.63%	2020
[20]	LFW	FI-GAN	99.6%	2020
[21]	IJB-A	DAC	0.976 ± 0.01%	2020
[22]	YTF	ADRL	96.52 ± 0.54%	2017

In many classification issues, the samples of one class are usually surrounded by the samples of the other classes. To address this issue, Ye et al. [10] proposed multiview robust double-sided twin SVM(MvRDTSVM) and a fast version of MvRDTSVM (named MvFRDTSVM). In the Tufts face database, MvRDTSVM and MvFRDTSVM achieved an accuracy of 91.55% and 88.82%, respectively. The previous method used in the two classification problems is not suited for face recognition that has many people. In general, face recognition algorithms consist of three steps: pre-processing images, feature extraction, and face classification. Lahaw et al. [11] proposed combining Two Dimensional Discrete Wavelet Transform (2D-DWT), which can capture localized information of images, with Principal Component Analysis (PCA), Linear Discriminant Analysis (LDA), or Independent Component Analysis (ICA) to extract face feature. Finally, the SVM algorithm combined with the 2D-DWT method has led to the increase of the performance of PCA + SVM, LDA + SVM, and ICA + SVM from 90.24%, 93.9%, and 91% to 96%, 96%, and 94.5%, respectively. Most of the existing distance metric-based(DML) methods are kNN DML methods. The disadvantage of kNN DML method is that the classification result is affected by the setting of the nearest neighbor number k. Ruan et al. [12] proposed a convex model for support vector DML (CSV-DML), which increased the accuracy of the CSV-DML to 84.6%, better than the existing kNN DML and support vector DML methods.

Furthermore, many well-known face recognition studies are based on deep learning approaches, including Deepface [13], FaceNet [14], VGG-Face [15], and ArcFace [23]. Deepface [13], proposed by Facebook in 2014, uses a nine-layer neural network with Softmax in the loss function, and achieved a recognition accuracy of 97.35% when applied to the LFW (Labeled Faces in the Wild) dataset. FaceNet [14] was proposed by Google in 2015 and uses ZFNet and Inception-v1 as the Siamese network [24] architecture and a triplet loss in the loss function. The model achieved an accuracy of 99.63% on the LFW dataset in the validation stage. The Visual Geometry Group (VGG) proposed VGG-Face [15] in 2017, which is a neural network for large-scale image recognition based on a small number of VGG [25] training samples and the Softmax loss function. It was shown that the accuracy of the proposed network reached 99.6% when using the triplet loss proposed in FaceNet [14] for inference purposes.

ArcFace [23], proposed in 2018, is based on the ResNet deep neural network architecture [26], but employs a novel loss referred to as Additive Angular Margin Loss. The model achieved an accuracy of 99.83% when applied to the LFW dataset.

In recent studies, Fuad et al. [27] surveyed many deep learning (DL) methods for face recognition (FR). The authors explored them in several parts. For the CNN-based method, Chen et al. [16] considered the angle discrepancy and magnitude gap between high-resolution and corresponding low-resolution faces. It successfully identified faces with fewer than 32×32 pixels, resulting in LightCNN-v29 achieving a 98.98% success rate. Wang et al. [17] proposed a pyramid-diverse attention framework to avoid the model focusing on fixed blocks by extracting features in multiple layers so that the model can extract facial features more comprehensively. For the Autoencoder-based method, Autoencoder combines generated and learned properties, but it still learns irrelevant features. Therefore, Pidhorskyi et al. [28] proposed an Adversarial Latent Autoencoder (ALAE) to solve this issue and improve the training procedure of GAN. Additionally, Usman et al. [18] used multiple levels of hidden layers for feature extraction and dimension reduction for expression recognition. For GAN-based methods, Iranmanesh et al. [19] proposed the CpGAN method, which processes visible and non-visible spectra separately through two sub-networks of independent GANs. The CpGAN was then used for heterogeneous face recognition. In addition, Rong et al. [20] used GAN to solve the issue of failing recognition when the identified person has a large pose change. For the Reinforcement Learning-based method, Liu et al. [21] and Rao et al. [22] applied reinforcement learning to find the attention of videos in a heterogeneous collection of unordered images and videos, and both achieve rich results.

To sum up the above methods, despite the many novel architectures proposed in the recent literature, FaceNet has a unique architecture and employs a triplet loss to process the data. FaceNet continues to be one of the most commonly used and accurate models for face recognition purposes. Although the accuracy of ArcFace is slightly higher than that of FaceNet, its performance advantage is obtained at the expense of a higher computational cost. Consequently, the present study deliberately adopts FaceNet to build the face recognition systems used for evaluation purposes.

2.2. Adversarial Attack

2.2.1. Adversarial Example

Adversarial examples are produced by adding small perturbations to the original input sample. Many methods are available for generating adversarial examples, including the Fast Gradient Sign Method (FGSM) [3], the Basic Iterative Method (BIM) [5], the Projected Gradient Descent (PGD) [29], and the Carlini & Wagner attack [30]. All of these methods have a high attack success rate and are widely used in digital attack scenarios. However, FGSM is not suitable for black-box attacks since they require knowledge of the model parameters when training. BIM and PGD are based on FGSM, albeit with a smaller step size, and is thus equally inapplicable to black-box attacks.

Consequently, among these methods, only the Carlini and Wagner attack model can be applied to black-box attacks. Most attack methods are based on loss functions and add a gradient value to the image pixels as noise. The target model is then queried repeatedly until model convergence. However, face recognition systems are generally implemented in the real world and, provided that the face recognition system is not hacked, the likelihood of an attack succeeding simply by directly modifying the image pixels is rather low. Furthermore, if the attack queries the model many times in an attempt to deceive the system, it is likely to trigger a security mechanism and will thus similarly fail. In other words, adversarial attack methods, which add noise to the entire image, have only a limited effectiveness against face recognition applications.

2.2.2. Adversarial Patch

Adversarial patches [4] differ from adversarial examples in that they add noise only to certain regions of the image, rather than the entire image. Adversarial patch attacks can be easily applied in the real world and require no knowledge of the parameters or architecture of the model. The authors in [31] demonstrated the feasibility for fooling automated surveillance cameras by applying adversarial T-shirts to the subject. Similarly, the authors in [32] generated adversarial patches for road signs using a GAN-based [9] method and showed that the patches prevented the classifier from identifying the road signs correctly.

2.3. Attention Area of Face Recognition

The accuracy of face recognition systems based on deep learning methods is significantly higher than that of earlier image-processing-based or statistical methods. However, besides the prediction accuracy of such methods, there is growing interest in the interpretability of the prediction results. Castanon and Bryne [33] used a heat map to quantify the relative importance of each feature in the classification model. The results indicated that the prediction outcome was determined mainly by the features extracted from the eyes, nose and mouth regions of the image. Deb et al. [8] also showed that the success rate of adversarial patch attacks against face images was enhanced when applying noise mainly to the eyes and nose regions of the face.

3. Related Works

The literature contains many studies on attack methods against deep-learning-based face recognition systems. However, many of these studies assume that the attacker somehow gains access to the face recognition system, or consider only attacks in the digital world. By contrast, the present study aims to explore a more realistic attack scenario, in which the attack occurs in the physical world and the attacker has no information of the model parameters and architectures. Thus, in considering the previous work in the field, the present study focuses mainly on the works shown in Table 2 where we distinguish related works by attack method, attack situation, generate object and attack capacity.

Table 2. Related Works.

Related Works	Attack Method	Situation	Generate Object	Adversarial Capacity
[34]	L-BFGS	Physical	Patches	White-Box
[35]	LED	Physical	LED Infrared	White-Box
[36]	FGSM	Physical	Patches	White-Box
[37]	MI-FGSM	Physical	Patches	White-Box
[38]	VLA	Physical	Visible Light	Black-Box
[39]	Transformation-Invariant Adversarial Pattern	Physical	Visible Light	Black-Box

Table 2. *Cont.*

Related Works	Attack Method	Situation	Generate Object	Adversarial Capacity
[8]	GAN-based	Digital	Digital Face Image	Black-Box
[40]	GAN-based	Physical	Adv-Makeup	Black-Box
[41]	GAN-based	Digital	Digital Image	Black-Box
[42]	FACESEC	Physical	Eyeglass	Black-box
[43]	GAN-based	Physical	Sticker	Black-box

As shown in Table 2, many of the adversarial attack methods reported in the literature are white-box attacks and use adversarial patches. For example, the method in [34] uses the L-BFGS method to print 2D or 3D images with adversarial glasses. The attack success rate was found to be as much as 97.22% in dodging attacks and 75% in impersonation attacks. However, the test dataset was limited to just three individuals. In [35], an LED is added to the hat, an infrared light is projected onto the face which is adjusted according to the attack target. It is noticed that the attack is unstable which indicates some limitation on using infrared light attack. Moreover, the attack method is white-box attack. The studies in [36,37] conduct white-box attacks using adversarial stickers attached to hats and the nose, forehead and eyes regions of the face image, respectively. It was shown in [36], that the adversarial patches effectively reduce the similarity between the input image and the target image and therefore is able to attack the face recognition system. In [37], although the similarity of the patch face image with the ground truth class was only just slightly lower that that with the targeted class, the attack is successful as the neural network classifies the patch face image as the targeted class.

Although the methods in [34–37] are capable of deceiving the face recognition model, their success rate is relatively low. Moreover, the adversarial attacks are launched using white-box attack methods. As described above, white-box attacks require a knowledge of the parameters and architecture of the face recognition model. Thus, white-box attacks are generally ineffective in real-world scenarios, where such information is carefully guarded. Accordingly, the authors in [38,39] proposed black-box attacks, in which light produced by a projector was used to generate attack noise. The study in [38] attacked the FaceNet face recognition system and achieved an average dodging success rate of 85.7% for nine test subjects and an average impersonation success rate of 32.4%. In [39], projected light was generated as noise using an update gradient method and was used to conduct attacks against a commercial face recognition model. The dodging attack and impersonation attack success rates were shown to be 70% and 60%, respectively. However, the impersonation attack considered only one test subject. The attack success rates of the methods in [38,39] are generally higher than those of previous methods. However, in both cases, it is necessary to query the system multiple times during training. Moreover, it is impractical to carry and use the light projection equipment in real-world situations, and the projection angle and light intensity must be carefully considered and managed.

Deb et al. [8] used a GAN to generate adversarial noise, which was added to the original face to form an adversarial face in the digital world. In the original GAN architecture, the aim is to generate an image which is as similar to the original image as possible. In [8], however, the performance of the GAN in generating adversarial noise was improved by extending the loss function to include not only the original loss L_{GAN}, but also two new losses, namely $L_{perturbation}$ and $L_{identity}$, respectively. The aim of the $L_{perturbation}$ loss was to control the amount of noise generated, while that of the $L_{identity}$ loss was to control the generated noise such that the image was classified into a specified class. The GAN model was used against the FaceNet face recognition model and achieved a success rate of 97.22% for obfuscation attacks and 24.30% for impersonation attacks. Thus, even though the attack model considered the more realistic scenario of a black-box attack, the impersonation success rate was still rather low. Bin et al. [40] used a GAN-based method to add makeup around the eyes as adversarial noise. The experimental results showed that the proposed method achieved an average success rate of 33.17% for impersonation attacks against FaceNet and a maximum success rate of 52.92% for impersonation attacks against a commercial face recognition model. Xiao et al. [41] proposed another GAN-based method (advGAN) for generating adversarial examples. However, it was limited only to attacks in the digital world and was aimed at image classification systems rather than face recognition systems. Tong et al. [42] proposed the FACESEC method based on gradient l_0-norm to generate stickers, eyeglass frames, and face masks, which in turn attack FaceNet and VGGFACE. The attack success rate of the eyeglass frame on the FaceNet model is 54%. In addition, [42] explored the effect of knowing the parameter and architecture of the attack model on the attack success rate. Shen et al. [43] proposed a GAN-based adversarial attack to generate stickers that can be adhered to the ciliary arches, nasal bones, and two nasolabial folds on both sides. This study attacked Arcface, CosFace, FaceNet, and VGGFace models. For the dodging attack on FaceNet in the physical world, the work could achieve an attack success rate of 100.0%. For the impersonation attack, the attack success rate is 55.32%. Finally, [43] also explored the effects of camera distance, sticker size, and head pose.

In summary, other papers use Gradient-based (e.g., L-BFGS [34], FGSM [36,37], FACESEC [42]), Visible Light-based [35,38,39], and GAN-based [8,40,41,43] methods. In the physical world, the model architecture and parameters are mandatory knowledge for the traditional Gradient-based method, which means it is only applicable to white-box attacks. This approach is not realistic for practical applications. The Visible Light-based method uses visible light projection to change the face feature pixels. Besides requiring many resources and projectors, this method is easily vulnerable to external environmental factors, such as an infrared cut-off lens leading to the attack's failure. In contrast, we use a GAN-based approach to generate attack glasses or face patches, which is convenient. We also restrict the noise to the frame of the glasses (small area, not modified face features), which achieves a high success rate of attacks in both the physical and digital worlds. In addition, our method does not result in a "this person does not exist" warning in real-world face recognition systems. It is difficult for a layman such as a security guard to know the intent of our attack.

4. Proposed Method

The present study proposes a GAN-based attack method based on the generation of adversarial patches. The attack method assumes the use of a black-box model and is applicable to both the digital world and the physical world. It is shown that the proposed method achieves a success rate of 57.99% in digital dodging attacks and 48.78% in digital impersonation attacks. Moreover, the success rates for physical dodging and impersonation attacks are 81.77% and 63.85%, respectively. In other words, the attack performance of the proposed method is significantly better than that of previous methods reported in Section 3.

The present study refers to the model architecture used in [8,44], in which the training images comprise a face data set, and adversarial noise is added to each face using the conventional GAN method. In contrast to the method in [8], however, the generator in the present study generates adversarial noise only on glasses rather than on the entire face, and then adds these adversarial glasses to the face prior to judgement by the discriminator. The proposed method is thus more easily applied for attacks in the physical world.

On the whole, our proposed GAN architecture is illustrated in Figure 1, which comprises three main components: the generator G, the discriminator D, and the face matcher F. For the execution process of the proposed architecture, first, the glasses are the generator's input for generating the perturbation. Second, the perturbation will be combined with the glasses and the person to form the merged image, which will be used as the input to the discriminator (D) and the face matcher (F). Third, the generator, discriminator, and face matcher will calculate the losses, $L_{perturbation}$, L_{adv}, and $L_{identity}$, respectively. The pseudo codes for generating adversarial glasses and patches are shown in Algorithms 1 and 2. The details of these components will be presented below.

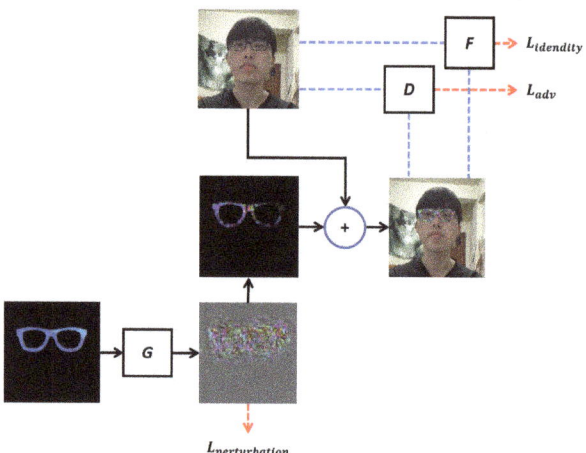

Figure 1. Proposed architecture.

Algorithm 1 Training AdvFace in dodging attack

Input:
  $\mathcal{X}$ Training Glasses Dataset
  f Training Face Dataset
  $\mathcal{F}$ Cosine similarity between an image pair obtained by face matcher
  $\mathcal{G}$ Generator with weight $\mathcal{G}_\theta$
  $\mathcal{D}$ Discriminator with $\mathcal{D}_\theta$
  m Batch size
  α Learning size
1: **for** number of training iterations **do**
2:   Sample a batch of probes $\{x^{(i)}\}_{i=1}^m \sim \mathcal{X}$
3:   Sample a batch of origin face images $\{y^{(i)}\}_{i=1}^m \sim f$
4:   $\delta^{(i)} = \mathcal{G}(x^{(i)})$
5:   $x_{adv}^{(i)} = x^{(i)} + \delta^{(i)}$
6:   $\mathcal{L}_{perturbation} = \frac{1}{m}[\sum_{i=1}^m max(P, ||\delta^{(i)}||_2)]$
7:   $\mathcal{L}_{identity} = \frac{1}{m}[\sum_{i=1}^m E[(\mathcal{F}(y^{(i)}, x_{adv}^{(i)}))]]$
8:   $\mathcal{L}^{\mathcal{D}} = \frac{1}{m}[\sum_{i=1}^m \log(1 - \mathcal{D}(x_{adv}^{(i)}))]$
9:   $\mathcal{L}_{adv} = \frac{1}{m}[\sum_{i=1}^m \log(\mathcal{D}(x^{(i)})) + \log(1 - \mathcal{D}(x_{adv}^{(i)}))]$
10: $\mathcal{L}^{\mathcal{G}} = \mathcal{L}_{adv} + \lambda_i \mathcal{L}_{identity} + \lambda_p \mathcal{L}_{perturbation}$
11: $\mathcal{G}_\theta = \text{Adam}(\bigtriangledown_\mathcal{G} \mathcal{L}^{\mathcal{G}}, \mathcal{G}_\theta, \beta_1, \beta_2)$
12: $\mathcal{D}_\theta = \text{Adam}(\bigtriangledown_\mathcal{D} \mathcal{L}^{\mathcal{D}}, \mathcal{G}_\theta, \beta_1, \beta_2)$
13: **end for**

Algorithm 2 Training AdvFace in impersonation attack

Input:
- $\mathcal{X}$ Training Face Dataset
- f Target Face Dataset
- $\mathcal{F}$ Cosine similarity between an image pair obtained by face matcher
- $\mathcal{G}$ Generator with weight $\mathcal{G}_\theta$
- $\mathcal{D}$ Discriminator with $\mathcal{D}_\theta$
- m Batch size
- α Learning size

1: **for** number of training iterations **do**
2:     Sample a batch of probes $\{x^{(i)}\}_{i=1}^{m} \sim \mathcal{X}$
3:     Sample a batch of target images $\{y^{(i)}\}_{i=1}^{m} \sim f$
4:     $\delta^{(i)} = \mathcal{G}(x^{(i)}, y^{(i)})$
5:     $x_{adv}^{(i)} = x^{(i)} + \delta^{(i)}$
6:     $\mathcal{L}_{perturbation} = \frac{1}{m}[\sum_{i=1}^{m} max(P, ||\delta^{(i)}||_2)]$
7:     $\mathcal{L}_{identity} = \frac{1}{m}[\sum_{i=1}^{m} E[1 - \mathcal{F}(y^{(i)}, x_{adv}^{(i)})]]$
8:     $\mathcal{L}^D = \frac{1}{m}[\sum_{i=1}^{m} \log(1 - \mathcal{D}(x_{adv}^{(i)}))]$
9:     $\mathcal{L}_{adv} = \frac{1}{m}[\sum_{i=1}^{m} \log(\mathcal{D}(y^{(i)})) + \log(1 - \mathcal{D}(x_{adv}^{(i)}))]$
10:    $\mathcal{L}^\mathcal{G} = \mathcal{L}_{adv}^\mathcal{G} + \lambda_i \mathcal{L}_{identity} + \lambda_p \mathcal{L}_{perturbation}$
11:    $\mathcal{G}_\theta = \text{Adam}(\nabla_\mathcal{G} \mathcal{L}^\mathcal{G}, \mathcal{G}_\theta, \beta_1, \beta_2)$
12:    $\mathcal{D}_\theta = \text{Adam}(\nabla_\mathcal{D} \mathcal{L}^D, \mathcal{G}_\theta, \beta_1, \beta_2)$
13: **end for**

4.1. Generator

The aim of the generator is to generate an adversarial image that causes the face recognition system to misclassify the input. A glasses image x is input to the generator, and the generator randomly generates $G(x)$ from the multi-dimensional space. The noises generated by the generator are then added to the original glasses image to produce $x + G(x)$. (That is, $x + G(x)$ denotes that only noises within the glasses are remained on the glasses.) As shown in Figure 1, the generated noise, $G(x)$, can be controlled by the $L_{perturbation}$ loss. In particular, the L_2 norm of $G(x)$, designated as $||G(x)||_2$, is taken and compared with a predefined noise threshold, P. The loss function, $L_{perturbation}$, is then assigned based on the outcome of this comparison, as shown in the following:

$$L_{perturbation} = E[max(P, ||G(x)||_2)] \tag{1}$$

With the $L_{perturbation}$, the generator will be trained to generate as much noise as possible with the constraint that the L_2 norm of $G(x)$ is close to but not larger than the predefined noise threshold, P.

The proposed architecture incorporates an additional loss, designated as $L_{identity}$, which aims to encourage the generator (G) to generate noise specifically intended to cause the face recognition system to misjudge the input face as that of another individual. The adversarial image $x + G(x)$ is first attached to a face image f through image processing to produce $f + x + G(x)$. The matcher (F) then compares $f + x + G(x)$ with the original face image f and calculates the difference between them as $L_{identity}$. The aim of the generator is to minimize the output value of the face matcher, F. The $L_{identity}$ here has different definitions for non-targeted attack and targeted attack. When generating adversarial glasses for a non-targeted attack, $L_{identity}$ is calculated as follows:

$$L_{identity} = E[F(f, (f + x + G(x))], \tag{2}$$

where f is the face image of the person performing adversarial attack.

However, when generating adversarial faces ($(x + G(x)$, where $x = f_t$) to carry out targeted attacks, the aim is to attack a specific target, and hence the input image patched with the adversarial face image, $f + x + G(x)$, is compared to the face image of the attack target, f_t. Thus, in this case, $L_{identity}$ is computed as

$$L_{identity} = E[1 - F(f_t, f + x + G(x))]. \tag{3}$$

For the layer structure of the generator, Table 3 shows the structural parameters of the design in detail.

Table 3. Structural parameters of the generator.

Layer	Type	Filters/Neurons	Stride	Padding
1	Conv	64 (kernel size = 7)	1	3
2	Conv	128 (kernel size = 4)	2	-
3	Conv	256 (kernel size = 4)	3	-
-	residual block	kernel size = 3	-	-
-	residual block	kernel size = 3	-	-
-	residual block	kernel size = 3	-	-
4	Unsampling	128 (kernel size = 5)	1	2
5	Unsampling	64 (kernel size = 5)	1	2
6	Conv	3 (kernel size = 7)	1	3

4.2. Discriminator

The purpose of the discriminator is to compare the original face f and the face image with glasses $f + x + G$ generated by the generator. The original face f is input to the discriminator, and the generator then attempts to minimize the difference between f and $f + x + G$ such that the discriminator is unable to distinguish between them. In the present architecture, the discriminator is implemented using the loss function described in [9]. That is,

$$L_{adv} = E[\log D(f)] + [\log(1 - D(f + x + G(x)))] \tag{4}$$

For the layer structure of the discriminator, Table 4 shows the structural parameters of the design in detail.

Table 4. Structural parameters of the discriminator.

Layer	Type	Filters/Neurons	Stride	Padding
1	Conv	32 (kernel size = 4)	2	-
2	Conv	64 (kernel size = 4)	2	-
3	Conv	128 (kernel size = 4)	2	-
4	Conv	256 (kernel size = 4)	2	-
5	Conv	512 (kernel size = 4)	2	-
6	Conv	1 (kernel size = 1)	1	-

4.3. Face Matcher

The propose of the face matcher is to quantify the similarity between two faces. In other words, F can be regarded as a form of face recognition system. To compare the similarity between two faces, the face matcher receives the two face images and outputs two corresponding feature vectors. The distance between the two feature vectors is then taken as a measure of the similarity between them, where a smaller distance indicates a greater similarity, and vice versa.

The total loss function of the proposed GAN architecture thus has the form

$$L_{total} = L_{adv} + x_1 L_{perturbation} + x_2 L_{identity}, \tag{5}$$

where x_1 and x_2 are weighting values used to control the relative contributions of the Lperturbation and Lidentity losses, respectively. The total loss, L_{total}, is then fed back to the generator G for further training. In particular, the generator produces a new adversarial image, which is reevaluated by the discriminator D and face matcher F. The resulting loss, L_{total}, is then returned to the generator as feedback once again. The training process continues iteratively in this way until the generator produces a set of high-quality adversarial images which are virtually indistinguishable from the original face images by both the discriminator and the face matcher.

5. Experimental Results

The present study constructed two system environments: one in the digital world and another in the physical world. The former system implemented a face recognition system for the digital world and a generator for producing adversarial patches. The system was implemented on the Ubuntu 18.04 operating system with 256 GB of memory space, a Tesla V100 GPU, 32 GB PCIe (NVIDIA Corp., San Jose, CA, USA), and an Dell PowerEdge R740 with Intel® Xeon® Silver 4116 CPU @ 2.10 GHz. The system was programmed in Python 3.6 using a variety of deep learning tools, including TensorFlow 1.14.0, Keras 2.3.1, Pytorch 1.9, and CUDA version 11.0.

The second system implemented a face recognition system in the physical world. For testing convenience, and to reproduce a realistic face recognition system environment, the system was implemented on an ASUS X556UR laptop (ASUSTek Computer Inc., Taipei, Taiwan) under an Anaconda virtual environment. The face recognition system was run on a NVIDIA GeForce 940MX (NVIDIA Corp., San Jose, CA, USA) 2 GB graphics card and was programmed in Python 3.6 with TensorFlow 1.14.0, Keras 2.3.1 and Pytorch 1.9. The laptop camera had a poor resolution of just 480 p. Thus, to enhance the face recognition process, the laptop was interfaced with a Logitech C925e (Logitech International S.A., Lausanne, Switzerland, and Newark, CA, USA) webcam with an improved resolution of 1080 p.

5.1. Evaluation Metric

The performance of the proposed GAN-based attack method was quantified by evaluating the attack success rate in both the digital world and the physical world. For both worlds, the attack success rate was investigated for both dodging attacks and impersonation attacks. In the case of dodging attacks in the digital world, the success rate was computed as follows:

$$\frac{\sum_{i \in N}(\hat{y}_i \neq y_i) \text{ and } (d(\hat{y}_i) < threshold)}{|N|}, \qquad (6)$$

where y_i is the original class of the input image i; $\hat{y}_i$ is the image class (of the image i) predicted by the face recognition model; $d(\hat{y}_i)$ is the similarity (e.g., cosine distance) of the input image i and an image in class $\hat{y}_i$; $|N|$ is the total number of input images; and threshold is a threshold parameter used for classification judgement purposes. Note that the threshold parameter was assigned a value of 0.4, where a similarity less than this threshold was taken to indicate a valid classification result.

The success rate of impersonation attacks in the digital world was evaluated using Equation (7), in which $\tilde{y}$ is the class of the target, $\hat{y}_i$ is the class predicted by the face recognition model.

$$\frac{\sum_{i \in N}(\hat{y}_i = \tilde{y}) \text{ and } (d(\hat{y}_i) < threshold)}{|N|} \qquad (7)$$

In the physical world, the face images were read directly through the webcam. Thus, the adversarial patches produced by the generator were printed and worn by the subjects. For each subject, images were collected over a 10 s period with head motion allowed. For the dodging attacks, the attack was considered to be successful when the system identified three consecutive face images as belonging to an individual other than the subject. Similarly, for the impersonation attacks, the attack was considered to be a success when the system identified three consecutive face images as belonging to the target individual.

5.2. Datasets

Three datasets were used to construct the face recognition systems, namely one dataset consisting of 3000 face images chosen from the LFW open-source face dataset, and two self-collected small face datasets. The first self-collected database contained 10 subjects (6 male and 4 female) between the age of 22 and 27 years old, with five face images for each subject. The second self-collected database added an additional 12 individuals to the first dataset, giving a total of 22 individuals (16 male and 6 female) between the ages of 20 and 34 years old. Again, the dataset contained five face images for each individual. For both self-collected databases, the face images were captured in a well-lit indoor environment using a mobile phone camera.

When performing dodging attacks against the face recognition systems, adversarial patches were produced by adding noise to the glasses dataset in [34], which contains various styles of glasses, each with multiple colors, giving a total of 16,833 images. Meanwhile, the impersonation attacks were conducted using the 10 individuals in the first self-collected dataset as training subjects, where each individual wore printed adversarial glasses and face patches.

5.3. Face Recognition Systems in Digital World

The experiments commenced by evaluating the attack performance of the proposed GAN-based adversarial patch method against the face recognition systems constructed in the digital world. The attack performance was evaluated for both the face recognition system built using the LFW dataset and the systems built using the two self-collected databases, respectively. In all three cases, face recognition was performed using the Deepface [45] open-source model, with a similarity (cosine distance) threshold set as 0.4 in order to achieve a False Accept Rate (FAR) of 0.1%. Two different adversarial patches were generated to carry out dodging attacks and impersonation attacks, respectively. The dodging attacks were conducted using the adversarial glasses patches shown in Figure 2.

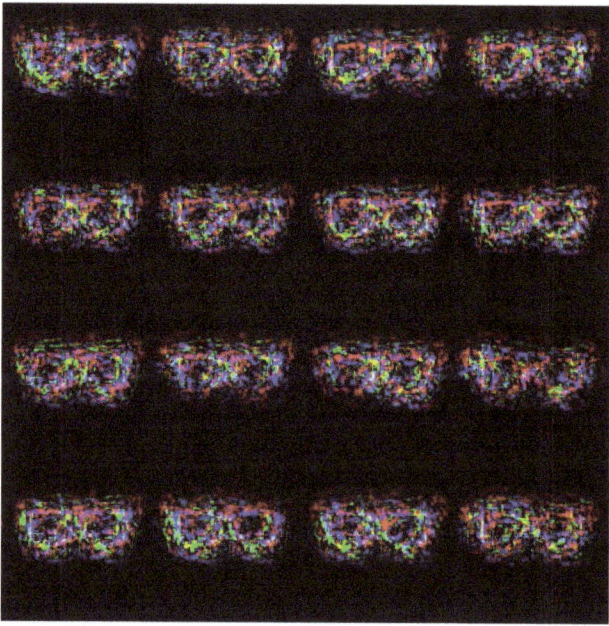

Figure 2. Generated adversarial glasses.

The adversarial glasses were passed through the mask shown in Figure 3, and the resulting frames were applied to the face images one-by-one to perform adversarial attacks, as shown in Figure 4.

Figure 3. Glasses mask.

Figure 4. Addition of adversarial glasses in the digital world.

To carry out the impersonation attacks, adversarial faces were generated by the proposed method, as shown in Figure 5. The adversarial faces were extracted as circles (see Figure 6) and were then attached to the forehead regions of the face images with adversarial glasses, as shown in Figure 7.

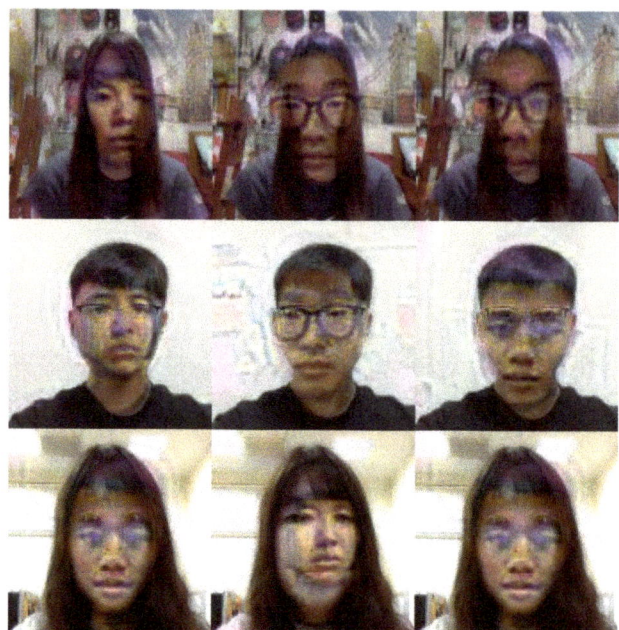

Figure 5. Generated adversarial faces.

Figure 6. Adversarial faces cut into circles.

Figure 7. Addition of adversarial faces to face images with adversarial glasses in the digital world.

5.4. Face Recognition Systems in Physical World

To evaluate the performance of the proposed model in the physical world, the face recognition system was constructed using the self-collected dataset containing 22 individuals. In implementing the face recognition system, the similarity threshold was set as 0.1 to achieve a FAR of 0.1%. For the dodging attacks, the adversarial glasses were printed on cardboard and worn around the eyes, as shown in Figure 8. For the impersonation attacks, the adversarial faces were additionally printed and attached to the forehead region, as shown in Figure 9.

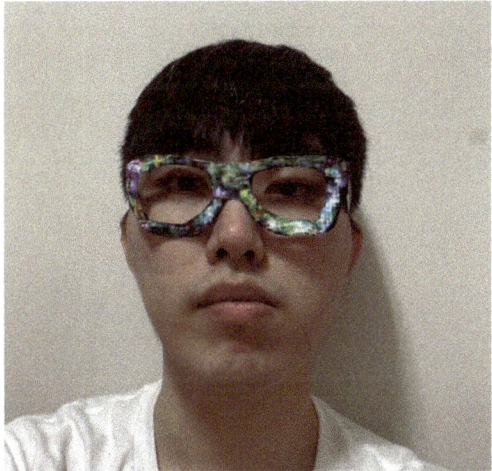

Figure 8. Wearing of adversarial glasses in the physical world.

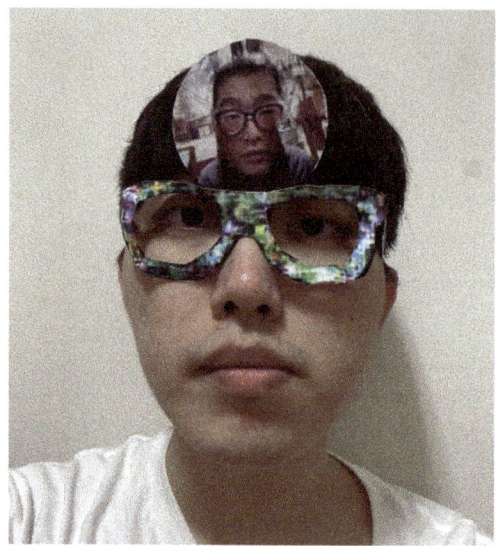

Figure 9. Wearing of adversarial glasses and an adversarial face in the physical world.

5.5. Results

5.5.1. Dodging Attacks in Digital World

The dodging attacks in the digital world were conducted against all three face recognition systems based on the LFW dataset (3000 images), Self-Collected Dataset 1 (10 subjects), and Self-Collected Dataset 2 (22 subjects), respectively. The intention of using the three different face recognition systems was to evaluate the relationship between the number of faces in the face recognition database and the attack success rate. The experiments commenced by evaluating the attack performance against the LBW face recognition system. The attack was conducted using 10 face images chosen at random from the LBW dataset, where each image wore 64 different adversarial glasses in turn. The corresponding attack results are shown in Table 5.

Table 5. Digital dodging attacks against LFW database.

Number	Attack Success Rate
No.1	92.18%
No.2	1.56%
No.3	10.93%
No.4	4.68%
No.5	35.93%
No.6	32.81%
No.7	62.5%
No.8	85.93%
No.9	100%
No.10	93.75%
Average	52.02%

Overall, the results presented in Table 5 show that the adversarial glasses result in a high attack success rate for some individuals (e.g., #1, #9 and #10), but a low attack success rate for others (e.g., #2 and #4). A detailed analysis revealed two main reasons for the low success rate in these cases: (1) even after wearing adversarial glasses, the adversarial face was still very similar to the original; and (2) the LFW dataset contained no faces similar to the adversarial face with glasses, i.e., the dataset contained no target to attack. Accordingly, the dodging attacks were repeated against the face recognition systems built using the two self-collected datasets, respectively. The corresponding results are presented in Tables 6 and 7.

Table 6. Digital Dodging Attack in 10 people's database.

Number	Attack Success Rate
No.1	0%
No.2	31.25%
No.3	78.13%
No.4	0%
No.5	37.5%
No.6	0%
No.7	0%
No.8	28.13%
No.9	34%
No.10	32.81%
Average	24.18%

Table 7. Digital Dodging Attack in 22 people's database.

Number	Attack Success Rate	Number	Attack Success Rate
No.1	10.9%	No.12	0%
No.2	100%	No.13	98.4%
No.3	100%	No.14	1.5%
No.4	43.75%	No.15	100%
No.5	57.8%	No.16	100%
No.6	0%	No.17	100%
No.7	1.5%	No.18	96.8%
No.8	100%	No.19	39%
No.9	56.2%	No.20	95.3%
No.10	54.6%	No.21	34.3%
No.11	35.9%	No.22	50%
Average			57.99%

As shown, the average attack success rates against the 10-person and 22-person databases are 24.18% and 57.99%, respectively. In general, in conducting successful dodging attacks, the aim is for the face recognition system not only to identify the original face image after the addition of adversarial glasses, but also to match the face with another face. For the face recognition system constructed with a larger number of faces, there exist more targets which can be matched by the adversarial face. Consequently, as the size of the database used by the face recognition system increases, the vulnerability of the system to adversarial glasses attacks also increases.

5.5.2. Impersonation Attacks in Digital World

The impersonation attacks were conducted against the face recognition system built using the self-collected database of 22 individuals. Ten individuals were chosen randomly from the database for testing purposes. In addition to wearing adversarial glasses, each face also wore the adversarial faces of the other nine individuals in turn. That is, each individual attacked nine targets. The corresponding attack results are shown in Table 8.

Table 8. Digital impersonation attacks against 22-person face recognition system.

		Attack Number										
		No.1	No.2	No.3	No.4	No.5	No.6	No.7	No.8	No.9	No.10	Average
Origin Number	No.1		90%	10%	30%	70%	90%	60%	0%	30%	70%	50%
	No.2	50%		10%	30%	60%	90%	60%	30%	50%	60%	50%
	No.3	30%	80%		10%	60%	80%	60%	10%	60%	50%	48.9%
	No.4	70%	90%	30%		80%	90%	60%	10%	30%	70%	58.9%
	No.5	40%	70%	10%	30%		80%	60%	20%	40%	60%	45.6%
	No.6	10%	50%	0%	20%	40%		40%	30%	10%	10%	23.3%
	No.7	50%	80%	20%	30%	50%	70%		0%	10%	50%	40%
	No.8	10%	90%	10%	20%	60%	90%	50%		30%	70%	47.8%
	No.9	70%	90%	10%	40%	40%	80%	40%	10%		80%	51.1%
	No.10	30%	100%	20%	30%	70%	90%	60%	20%	50%		52.2%
											Total	48.78%

It is seen in Table 8 that the overall average success rate is 48.78%. However, it is also noted that some of the targets (e.g., #3 and #8) are less easily impersonated than others. It is speculated that when adversarial face stickers are generated by the method of this study, some faces are more difficult to attack. Furthermore, some of the adversarial images also have a lower attack success rate than others. For example, the successful attack rate of adversarial image #6 is just 23.3%. In other words, the face recognition system matches the adversarial face image with the original image rather than the target individual. To investigate this phenomenon further, the method proposed in [8] was used to generate adversarial noise for all of the faces in the self-collected database, as shown in Figure 10.

It can be seen that some of the faces are attacked by generating noise in the eye region of the image, while in other cases, the noise is distributed over the entire face. However, the method proposed in the present study generates adversarial glasses and stickers which are applied only to certain regions of the face, and it cannot attack those faces with the noise is distributed over the entire face.

5.5.3. Dodging Attacks in Physical World

The dodging attack success rate in the physical world was evaluated for both self-collected databases. For each database, 10 subjects were selected for testing purposes, where each subject wore 11 adversarial glasses in turn. The corresponding attack success rates are shown in Tables 9 and 10, respectively.

In the attacks performed in the physical world, the subjects were allowed to turn their face during the detection process. Thus, compared to the digital case, in which the detection process was limited to only a single face input image, the detection process in the physical world was less constrained. As shown in Tables 9 and 10, the average dodging attack success rates against the 10-person database and 22-person database were 39.94% and 81.77%, respectively. In other words, as for the attacks performed in the digital world, the attack success rate in the physical world also increases as the size of the database increases. The results also imply that when the subject turns the face to a non-frontal angle, the likelihood of the face recognition system misclassifying the input face image also increases.

Figure 10. Generated adversarial masks.

Table 9. Physical dodging attacks against 10-person face recognition system.

Number	Attack Success Rate
No.1	45.4%
No.2	54.5%
No.3	36.3%
No.4	27.2%
No.5	36.3%
No.6	54.5%
No.7	18.1%
No.8	45.4%
No.9	27.2%
No.10	54.5%
Average	39.94%

Table 10. Physical dodging attacks against 22-person face recognition system.

Number	Attack Success Rate
No.1	81.8%
No.2	72.7%
No.3	63.6%
No.4	100%
No.5	72.7%
No.6	81.8%
No.7	81.8%
No.8	90.9%
No.9	81.8%
No.10	90.9%
Average	81.77%

5.5.4. Impersonation Attacks in Physical World

The performance of the impersonation attacks in the physical world was evaluated using the 10 subjects in the first self-collected database as test subjects and the 22 individuals in the second self-collected database as targets. Each of the test subjects wore adversarial glasses and the adversarial faces of all the other subjects in the first self-collected database in turn. The aim of the attack was to cause the face recognition system to recognize the target on the adversarial face sticker rather than the original subject. The corresponding attack results are presented in Table 11.

Table 11. The attack success rate of a single target for impersonation attacks in the physical world on a face recognition system with 22 persons in the database.

		Attack Number										
		No.1	No.2	No.3	No.4	No.5	No.6	No.7	No.8	No.9	No.10	Average
Origin Number	No.1		90%	0%	50%	90%	90%	80%	40%	90%	30%	62.2%
	No.2	50%		0%	40%	80%	90%	60%	50%	90%	90%	63.3%
	No.3	70%	80%		50%	90%	90%	90%	40%	90%	30%	70%
	No.4	60%	100%	20%		100%	90%	70%	40%	100%	80%	73.3%
	No.5	40%	80%	20%	40%		80%	90%	40%	90%	20%	55.5%
	No.6	40%	80%	10%	50%	80%		80%	40%	90%	20%	54.4%
	No.7	50%	90%	30%	50%	90%	90%		30%	90%	50%	63.3%
	No.8	40%	70%	10%	50%	90%	90%	90%		80%	60%	64.4%
	No.9	20%	100%	40%	50%	90%	90%	80%	40%		70%	64.4%
	No.10	40%	100%	30%	50%	80%	90%	90%	40%	90%		67.7%
											Total	63.85%

As shown, the average success rate of the impersonation attacks is 63.85%. Interestingly, the results show that even though the same adversarial face patch of a given target is added to all of the test subjects, the attack success rates are different for different test subjects. On the other hand, it is also evident that some subjects (e.g., #5, #6, #7) are easier to attack than others (e.g., #3). Notably, we achieve more than 70% attack success rate in the physical world on most of the targets, even if we set a threshold of 0.4, corresponding to a False Accept Rate (FAR) of 0.01%. Moreover, since the above table is a comprehensive attack success rate test for each person against others, the existing literature mainly discusses the attack success rate of a single target. Therefore, for a more intuitive comparison with the existing literature, we show in Table 12 that the average attack success result for a single target is 78%.

Table 12. In 22-person face recognition system, the attack success rate of a single target for impersonation attacks in the physical world.

Original No.	Attack Target	Attack Success Rate
No.1	No.3	70%
No.2	No.10	100%
No.3	No.9	40%
No.4	No.6	50%
No.5	No.4	100%
No.6	No.7	90%
No.7	No.5	90%
No.8	No.2	50%
No.9	No.4	100%
No.10	No.2	90%
Average		78%

5.6. Comparison of Dodging Attack Success Rates of Different Methods

Table 13 compares the dodging attack success rate of the proposed method in the physical world with that of several other attack methods proposed in the literature.

Table 13. Comparison of physical dodging attack success rates of different methods.

Literature	Generate Object	Face Recognition Model	Adversarial Capacity	Number of Subjects	Number of People in Database	Dodging Attack's Success Rate
[34]	Patches	VGG-Face	White-Box	3	10	97.22%
[38]	Visible Light	FaceNet	Black-Box	9	5749 (LFW)	85.7%
[39]	Visible Light	Commercial	Black-Box	10	50	70%
[42]	Eyeglass	FaceNet	Black-Box	10	5749 (LFW)	54%
Ours	Patches	FaceNet	Black-Box	10	22	81.77%

The method proposed in [34] applied adversarial glasses to the test images and achieved an average attack success rate of 97.22%. However, the adversary attack was a white-box attack, which is unrealistic in practical attack environments. The premise of the white-box attack is to know the model architecture and parameters, which is worlds apart from our black-box attack. At the same time, through the experiments of cosine similarity in the [36], we can observe some interesting variations of cosine similarity. In [36], cosine similarity varies between white box and black box attacks (note that when the cosine similarity is larger, the more similar it is). First, experiments in [36] are based on semi-white-box attacks, and we can see that the cosine similarity ranges from 0.15 to 0.2 in the Final similarity. Second, when the authors transfer the adversarial attack noise generated on the white-box attack to other unknown models, the architecture and parameters of the unknown model are unknown for the adversarial attack noise. In this case, it can be considered a no-box attack (more rigorous black box attack), which yields a cosine similarity of about 0.4. However, the no-box's cosine similarity increases from 0.2 to 0.4, which means that the similarity between the original image and the adversarial image is affected. That is, the attack success rate may be slightly decreased. In other words, the reason why our proposed attack success rate is lower than [34] is due to the difference between white-box and black-box. Moreover, the performance evaluation was conducted using just three test subjects in [34], which had slightly fewer test subjects. The methods in [38,39] adopted a black-box attack model and considered a greater number of test subjects (9 and 10 subjects, respectively). They were thus more representative of real-world attack scenarios than the method in [34]. Moreover, they achieved reasonable attack success rates of 85.7% and 70%, respectively. However, both methods require the use of visible light projection systems when conducting the attack and need careful consideration of the face angle and mask conditions. Thus, neither method is practical in real-world physical attack situations. In addition, we further compared our work with [38]. The work [38] did not set a similarity threshold. As aforementioned, the attack success rate will be higher without a similarity threshold. However, it is not practical as real-world face recognition systems will set the threshold properly. Therefore, we conducted an experiment by varying the threshold from 0 to 1 with an interval of 0.05. The purpose of this experiment is to illustrate the relationship between attack success rate and threshold. We discuss the results in Section 5.7. Tong et al. [42] proposed the FACESEC method based on gradient l_0-norm for adversarial patch attacks. However, this type of attack is significantly affected by the adversarial patch's wearing position, shape, and scale. Although in [42], whose attack method (patch) and position (eyeglass frame) are the same as ours, the attack success rate in [42] is lower than ours, at only 54%. The reason is the difference between Gradient-based and GAN-based. More interestingly, for the advantages and disadvantages of Gradient-based and GAN-based approaches, it has been shown in the experimental results of [40] that the performance of gradient-based attacks will be slightly lower than that of GAN-based approaches. Clearly, our results are consistent with the results of [42]. By contrast, the method proposed in this study not only considers a black-box attack model and achieves a relatively high success rate of 81.77% over 10 test subjects, but also requires only the use of simple temporary adversarial glasses stickers to deceive the face recognition system. It is thus more convenient and practical than the other methods presented in Table 13, while retaining a similar (if not better) attack performance.

Table 14 compares the impersonation attack success rate of the proposed method in the physical world with that of four other attack methods reported in the literature.

Table 14. Comparison of physical impersonation attack success rates of different methods.

Literature	Generate Object	Face Recognition Model	Attack Subjects	Number of Subjects	Number of Attack Target	Number of People in Database	Impersonation Attack's Success Rate
[34]	Patches	VGG-Face	White-Box	3	1	10	75%
[38]	Visible Light	FaceNet	Black-Box	9	60	5749 (LFW)	32.4%
[39]	Visible Light	Commercial	Black-Box	25	1	50	60%
[40]	Adv-Makeup	Commercial	Black-Box	1	1	20	52.92%
[43]	Sticker	FaceNet	Black-Box	20	3	20 (VolFace)	55.32%
Ours	Patches	FaceNet	Black-Box	10	10	22	63.85%
				10	1	22	78%

The method in [34] used a white-box attack model to generate adversarial glasses and achieved an average attack success rate of 75%. It is noted that the attack rate is slightly higher than that of the present study (63.85%). However, the present study is based on the more realistic assumption of a black-box model and, moreover, considers a greater number of test subjects (10 vs. 3 subjects), with the consequence that the results are expected to be more reliable. Although the average attack success rate of the combined experimental results is slightly lower than that of [34], the reason lay in the fact that we performed a comprehensive attack test on several targets and averaged the resulting attack success rates. For the above reasons, we compared the attack success rate of the single target with [34], as shown in Table 14, which resulted in 78% and had better results than that of Sharif et al. [34].

The method in [38] also considered a black-box attack model and used visible light to produce noise. Moreover, the attack evaluation considered a relatively large number of test subjects (9 subjects). However, the attack success rate against the FaceNet recognition system was just 32.4%, i.e., around half that of the present study. The authors in [39] also used projected light to produce noise in order to deceive the face recognition system. The attack success rate was 60%, and is thus close to that obtained in the present study. However, the evaluation process in [39] considered only 1 subject tested 10 times, and hence the evaluation results may not be reliable. Furthermore, as for the method in [39], the attack requires the use of visible light projection equipment, which is impractical for most real-world situations. By contrast, the present method requires only the use of adversarial stickers (glasses and face), which can be easily removed once the face recognition system has been fooled. The method in [40] requires only the application of makeup to the face, and is thus also convenient in real-world attack scenarios. However, the attacks in [40] were performed only once for seven different face angles, and hence the reliability of the evaluation results cannot be guaranteed. Moreover, the attack success rate was just 52.82%, and is hence lower than that of the present study (63.85%). Finally, in [43], the GAN-based adversarial stickers were crafted and put on five regions near to the facial organs (i.e., two superciliary arches, two nasolabial sulcus, and the nasal bone). Notably, these regions are critical regions for face recognition [8,33]. As a result, [43] was able to achieve 100% attack success rate of the dodging attack in the physical world which are higher than our results. However, the success rate for the impersonation attack is only 55.32%, which is slightly lower than ours.

Defense Mechanism

The attack method proposed in the present study exploits adversarial patches, which occupy only small regions of the face, rather than adversarial examples, which occupy the entire face. Many defense mechanisms based on face recognition rely on the detection of live subjects through temperature measurements [46], or the detection of adversarial samples [47]. These methods thus have only a limited ability to counter the adversarial patch-based method proposed in the present study. Accordingly, this study also proposes a new defense method, in which it is assumed that the defender already knows the attack method employed by the adversary. A new class, referred to as "Defense", is added to the output label. In particular, photos of each subject wearing adversarial glasses are added to the face database to counter dodging attacks, while photos of each subject wearing adversarial glasses and adversarial faces are also added to the face database to thwart impersonation attacks. All these inputs are labeled as "Defense" during training of the face recognition system. Tables 15 and 16 show the dodging attack success rate and dodging attack defense rate, respectively, following the implementation of the proposed defense mechanism.

Table 15. Dodging attack success rate after defense.

Number	Attack Success Rate
No.1	0%
No.2	0%
No.3	6.25%
No.4	0%
No.5	0%
No.6	0%
No.7	0%
No.8	0%
No.9	0%
No.10	0%
Average	0.06%

Table 16. Dodging defense rate after defense.

Number	Defense Rate
No.1	100%
No.2	81.25%
No.3	85.93%
No.4	78.12%
No.5	100%
No.6	56.25%
No.7	100%
No.8	100%
No.9	100%
No.10	95.3%
Average	89.69%

The results presented in Table 15 show that the dodging attack success rate reduces significantly from 57.99% to 0.06% following the implementation of the proposed defense mechanism. Moreover, the average defense rate is 89.69%. In other words, the proposed defense mechanism significantly improves the robustness of the face recognition system against dodging attacks.

Tables 17 and 18 show the equivalent results for impersonation attacks.

As shown, the implementation of the defense mechanism reduces the average impersonation attack success rate from 48.33% to 28.33% and achieves an average defense rate of 41.1%. It is noted that the defense rate is lower than that for dodging attacks. Nonetheless, the defense mechanism still reduces the original attack success rate by around 60%.

It can be shown that, the attack success rate attack success rate dropped from 48.33% to 28.33%, and the defense rate is 41.1%, the defense is not as effective as dodging attack, but it still can reduce the original attack by about 60%, and defense about 40% attack.

Table 17. Impersonation attack success rate after defense.

		\multicolumn{10}{c}{Attack Number}	
		No.1	No.2	No.3	No.4	No.5	No.6	No.7	No.8	No.9	No.10	Average
Origin Number	No.1		50%	10%	30%	0%	80%	30%	0%	0%	70%	30%
	No.2	50%		0%	30%	10%	90%	10%	0%	0%	60%	27.7%
	No.3	20%	80%		10%	0%	80%	20%	0%	0%	80%	32.2%
	No.4	50%	70%	20%		10%	90%	20%	0%	0%	70%	37.8%
	No.5	30%	50%	10%	30%		80%	10%	0%	0%	60%	30%
	No.6	10%	30%	0%	20%	0%		10%	0%	0%	10%	8.9%
	No.7	10%	50%	20%	30%	0%	70%		0%	0%	50%	25.6%
	No.8	10%	70%	10%	10%	0%	90%	0%		0%	70%	28.9%
	No.9	10%	70%	0%	40%	0%	80%	0%	0%		80%	31.1%
	No.10	30%	70%	20%	30%	0%	90%	30%	10%	0%		31.1%
											Total	28.33%

Table 18. Impersonation defense rate after defense.

		\multicolumn{10}{c}{Attack Number}	
		No.1	No.2	No.3	No.4	No.5	No.6	No.7	No.8	No.9	No.10	Average
Origin Number	No.1		40%	0%	0%	70%	10%	30%	40%	80%	0%	34.4%
	No.2	30%		10%	20%	80%	0%	90%	90%	10%	10%	37.8%
	No.3	10%	10%		0%	60%	0%	50%	90%	90%	0%	33.3%
	No.4	20%	20%	10%		80%	10%	70%	90%	10%	0%	34.4%
	No.5	60%	50%	60%	20%		20%	80%	80%	10%	20%	44.4%
	No.6	90%	70%	10%	70%	90%		90%	10%	10%	90%	58.9%
	No.7	90%	50%	70%	30%	90%	30%		90%	10%	40%	55.6%
	No.8	0%	20%	10%	0%	60%	0%	50%		80%	0%	24.4%
	No.9	80%	30%	70%	20%	90%	20%	10%	80%		20%	46.7%
	No.10	40%	30%	40%	20%	80%	0%	60%	90%	10%		41.1%
											Total	41.1%

5.7. Threshold

In our study, we additionally discuss two issues, which are (1) the relationship between threshold and attack success rate and (2) the portability of attacking other models by no-box.

First, about the relationship between threshold and attack success rate, we can observe from Figure 11 that the threshold directly affects the attack success rate. When the threshold is set smaller, the attack success rate is lower. Based on the observation, setting a proper threshold is necessary for the face recognition system. This also shows that our attack method is more realistic than [38] by properly setting the threshold.

Second, according to the definition of [48], a no-box attack does not query the face recognition system. That is, when generating adversarial samples, it does not refer to the confidence scores of the target face recognition system. Therefore, when our adversarial glasses attack other face recognition systems, that forms a no-box attack. Based on the experimental results shown in Figure 11, we can observe that our attack method exhibits transferability.

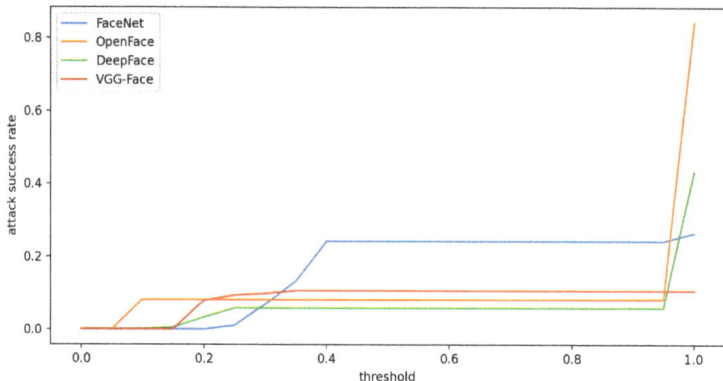

Figure 11. Compare the relationship between threshold and attack success rate.

5.8. Time Efficiency

For the analysis of techniques, other papers adopted Gradient-based (e.g., L-BFGS [34,42]), Visible-light-based [38,39], and GAN-based [40,43] methods. For Gradient-based, [34] took 4.39 h to output 35 attack images. The paper [42] did not mention its time efficiency. For the Visible Light-based approach, the paper [38] showed that VLA took less than 3 s on average to generate a frame pair containing a perturbation frame and a concealing frame. However, the paper [39] did not describe its time efficiency. For GAN-based, papers [40] neither discussed their time efficiency. [43] took 26 min to generate 8100 stickers. For our approach, first, one iteration could generate 64 pairs of glasses, which took 75 s. In all experiments, we ran 50 iterations. The total running time is 62.5 min. Second, for adversarial patches, each iteration could generate 32 adversarial patches in 20 s. It took 66.67 min to run 200 iterations. For the GAN model (including the Facenet) we used, the number of parameters is 31529204, and the FLOPs reach 58332210.

6. Conclusions

This work proposes an attack method based on GAN that generates noise and restricts the adversarial glasses and patches generated in the face region. This method can achieve black-box attacks in both digital and physical worlds. Among the attacks in the physical world, our method is more representative of the real-world attacks in the physical world by wearing adversarial glasses or patches on the face during face recognition. Furthermore, due to the assumption of a black-box model, the adversary requires no knowledge of the parameters and structure of the deep learning model employed by the face recognition system to conduct an attack.

On the other hand, the reason why the traditional method of dodging attacks fails is that "this person does not exist" attracts the attention of the guards, which is a severe problem. However, in dodging attacks, our method can make the face recognition system mistakenly identify the attacker as someone else, thus posing a significant threat to the face recognition system. Moreover, we take advantage of the fact that people often use glasses as a fashion accessory to hide the adversarial perturbation in the glasses. As such, it is difficult for laypersons to know our attack intentions. In impersonation attacks, we can wear glasses and patches that disguise a person as other people without tampering with focused features on the face. Namely, we can be who we want to be in front of the face recognition system. Furthermore, face recognition is used in various fields nowadays. Through our experiments, we have verified that when the database of the face recognition system is larger, the chances of being hacked will increase. In this case, serious security concerns still need to be considered and improvements need to be made.

In another exploration, we introduce a defense mechanism to counter the GAN-based adversarial patch method. The results show that the proposed mechanism detects almost all dodging attacks and more than half of the impersonation attacks. In impersonation attacks, although the adversarial patches applied in this study occupy only a small region of the face, which is still easily recognized by the supervisor monitoring the face recognition system. In this case, the attack is easy to fail. Therefore, future works will generate less obvious adversarial patches to improve the attack's success rate in the supervisor's presence.

Author Contributions: Conceptualization, R.-H.H., J.-Y.L. and S.-Y.H.; methodology, R.-H.H., J.-Y.L. and S.-Y.H.; software, S.-Y.H.; validation, J.-Y.L., R.-H.H., H.-Y.L. and C.-L.L.; formal analysis, R.-H.H. and S.-Y.H.; investigation, R.-H.H. and S.-Y.H.; resources, R.-H.H., H.-Y.L. and C.-L.L.; data curation, J.-Y.L. and S.-Y.H.; writing—original draft preparation, S.-Y.H. and J.-Y.L.; writing—review and editing, R.-H.H., H.-Y.L. and C.-L.L.; visualization, S.-Y.H.; supervision, R.-H.H.; project administration, R.-H.H., H.-Y.L. and C.-L.L.; funding acquisition, R.-H.H., H.-Y.L. and C.-L.L.; All authors have read and agreed to the published version of the manuscript.

Funding: This research was funded by Telecom Technology Center (TTC), Taiwan: TTC CCU 202205.

Institutional Review Board Statement: The study was conducted in accordance with the Declaration of Helsinki, and approved by the Human Research Ethics Committee of National Chung Cheng University (protocol code CCUREC111120101 Version 1 and date of approval 13 December 2022) for studies involving humans.

Informed Consent Statement: Informed consent was obtained from all subjects involved in the study.

Data Availability Statement: Due to the involvement of personal privacy (faces), these data are not publicly available.

Conflicts of Interest: The authors declare no conflict of interest.

Abbreviations

The following abbreviations are used in this manuscript:

DL	Deep Learning
FR	Face Recognition
ABC	Automatic Border Control
GAN	Generative Adversarial Network
VGG	Visual Geometry Group
LFW	Labeled Faces in the Wild
FGSM	Fast Gradient Sign Method
BIM	Basic Iterative Method
PGD	Projected Gradient Descent
FAR	False Accept Rate

References

1. Hariri, W. Efficient masked face recognition method during the COVID-19 pandemic. *Signal Image Video Process. (SIViP)* **2022**, *16*, 605–612. [CrossRef] [PubMed]
2. Facial Recognition Market Worth $8.5 Billion by 2025. Available online: https://www.marketsandmarkets.com/PressReleases/facial-recognition.asp (accessed on 19 September 2022).
3. Goodfellow, I.J.; Shlens, J.; Szegedy, C. Explaining and Harnessing Adversarial Examples. In Proceedings of the 3rd International Conference on Learning Representations (ICLR), San Diego, CA, USA, 7–9 May 2015.
4. Brown, T.B.; Mané, D.; Roy, A.; Abadi, M.; Gilmer, J. Adversarial Patch. Available online: https://arxiv.org/abs/1712.09665 (accessed on 19 September 2022).
5. Kurakin, A.; Goodfellow, I.J.; Bengio, S. Adversarial examples in the physical world. In *Artificial Intelligence Safety and Security*; Yampolskiy, R.V., Ed.; Chapman & Hall/CRC: New York, NY, USA, 2018; pp. 99–112.
6. Li, J.; Schmidt, F.; Kolter, Z. Adversarial camera stickers: A physical camera-based attack on deep learning systems. In Proceedings of the 36th International Conference on Machine Learning, Long Beach, CA, USA, 9–15 June 2019.
7. Bhambri, S.; Muku, S.; Tulasi, A.; Buduru, A.B. A Survey of Black-Box Adversarial Attacks on Computer Vision Models. Available online: https://arxiv.org/abs/1912.01667 (accessed on 9 November 2022).
8. Deb, D.; Zhang, J.; Jain, A.K. AdvFaces: Adversarial Face Synthesis. In Proceedings of the 2020 IEEE International Joint Conference on Biometrics (IJCB), Houston, TX, USA, 28 June–1 October 2020.

9. Goodfellow, I.; Pouget-Abadie, J.; Mirza, M.; Xu, B.; Warde-Farley, D.; Ozair, S.; Courville, A.; Bengio, Y. Generative Adversarial Nets. In Proceedings of the Neural Information Processing Systems (NIPS), Montreal, QC, Canada, 8–13 December 2014.
10. Ye, Q.; Huang, P.; Zhang, Z.; Zheng, Y.; Fu, L.; Yang, W. Multiview Learning With Robust Double-Sided Twin SVM. *IEEE Trans. Cybern.* **2022**, *52*, 12745–12758. [CrossRef] [PubMed]
11. Lahaw, Z.B.; Essaidani, D.; Seddik, H. Robust Face Recognition Approaches Using PCA, ICA, LDA Based on DWT, and SVM Algorithms. In Proceedings of the 2018 41st International Conference on Telecommunications and Signal Processing (TSP), Athens, Greece, 4–6 July 2018.
12. Ruan, Y.; Xiao, Y.; Hao, Z.; Liu, B. A Convex Model for Support Vector Distance Metric Learning. *IEEE Trans. Neural Netw. Learn. Syst.* **2022** *33*, 3533–3546. [CrossRef]
13. Taigman, Y.; Yang, M.;Ranzato, M.; Wolf, L. DeepFace: Closing the Gap to Human-Level Performance in Face Verification. In Proceedings of the 2014 IEEE Conference on Computer Vision and Pattern Recognition, Columbus, OH, USA, 23–28 June 2014.
14. Schroff, F.; Kalenichenko, D.; Philbin, J. FaceNet: A Unified Embedding for Face Recognition and Clustering. In Proceedings of the 2015 IEEE Conference on Computer Vision and Pattern Recognition (CVPR), Boston, MA, USA, 7–12 June 2015.
15. Cao, Q.; Shen, L.; Xie, W.; Parkhi, O.M.; Zisserman, A. VGGFace2: A Dataset for Recognising Faces across Pose and Age. In Proceedings of the 2018 13th IEEE International Conference on Automatic Face & Gesture Recognition (FG 2018), Los Alamitos, CA, USA, 15–19 May 2018.
16. Chen, J.; Chen, J.; Wang, Z.; Liang, C.; Lin, C.-W. Identity-Aware Face Super-Resolution for Low-Resolution Face Recognition. *IEEE Signal Process. Lett.* **2020**, *27*, 645–649. [CrossRef]
17. Wang, Q.; Wu, T.; Zheng, H.; Guo, G. Hierarchical Pyramid Diverse Attention Networks for Face Recognition. In Proceedings of the 2020 IEEE/CVF Conference on Computer Vision and Pattern Recognition (CVPR), Seattle, WA, USA, 13–19 June 2020.
18. Usman, M.; Latif, S.; Qadir, J. Using deep autoencoders for facial expression recognition. In Proceedings of the 2017 13th International Conference on Emerging Technologies (ICET), Islamabad, Pakistan, 27–28 December 2017.
19. Iranmanesh, S.M.; Riggan, B.; Hu, S.; Nasrabadi, N.M. Coupled generative adversarial network for heterogeneous face recognition *Image Vis. Comput.* **2020**, *94*, 1–10.
20. Rong, C.; Zhang, X.; Lin, Y. Feature-Improving Generative Adversarial Network for Face Frontalization. *IEEE Access* **2020**, *8*, 68842–68851. [CrossRef]
21. Liu, X.; Guo, Z.; You, J.; Vijaya Kumar, B.V. K. Dependency-Aware Attention Control for Image Set-Based Face Recognition. *IEEE Trans. Inf. Forensics Secur.* **2020**, *15*, 1501–1512. [CrossRef]
22. Rao, Y.; Lu, J.; Zhou, J. Attention-Aware Deep Reinforcement Learning for Video Face Recognition. In Proceedings of the 2017 IEEE International Conference on Computer Vision (ICCV), Venice, Italy, 22–29 October 2017.
23. Deng, J.; Guo, J.; Xue, N.; Zafeiriou, S. ArcFace: Additive Angular Margin Loss for Deep Face Recognition. In Proceedings of the 2019 IEEE/CVF Conference on Computer Vision and Pattern Recognition (CVPR), Long Beach, CA, USA, 15–20 June 2019.
24. Bromley, J.; Guyon, I.; LeCun, Y.; Säckinger, E.; Shah, R. Signature Verification using a "Siamese" Time Delay Neural Network. *Int. J. Pattern Recognit. Artif. Intell.* **1993**, *7*, 669–688. [CrossRef]
25. Simonyan, K.; Zisserman, A. Very Deep Convolutional Networks for Large-Scale Image Recognition. Available online: https://arxiv.org/abs/1409.1556 (accessed on 21 September 2022).
26. He, K.; Zhang, X.; Ren, S.; Sun, J. Deep Residual Learning for Image Recognition. In Proceedings of the 2016 IEEE Conference on Computer Vision and Pattern Recognition (CVPR), Los Alamitos, CA, USA, 27–30 June 2016.
27. Fuad, M.T.H.; Fime, A.A.; Sikder, D.; Iftee, M.A.R.; Rabbi, J.; Al-Rakhami, M.S.; Gumaei, A.; Sen, O; Fuad, M.; Islam, M.N. Recent Advances in Deep Learning Techniques for Face Recognition. *IEEE Access* **2021**, *9*, 99112–99142. [CrossRef]
28. Pidhorskyi, S.; Adjeroh, D.; Doretto, G. Adversarial Latent Autoencoders. Available online: https://arxiv.org/abs/2004.04467 (accessed on 19 November 2022).
29. Madry, A.; Makelov, A.; Schmidt, L.; Tsipras, D.; Vladu, A. Towards Deep Learning Models Resistant to Adversarial Attacks. Available online: https://arxiv.org/abs/1706.06083 (accessed on 22 September 2022).
30. Carlini, N.; Wagner, D. Towards Evaluating the Robustness of Neural Networks. In Proceedings of the 2017 IEEE Symposium on Security and Privacy (SP), San Jose, CA, USA, 22–24 May 2017.
31. Thys, S.; Ranst, W.V.; Goedeme, T. Fooling Automated Surveillance Cameras: Adversarial Patches to Attack Person Detection. In Proceedings of the IEEE/CVF Conference on Computer Vision and Pattern Recognition (CVPR) Workshops, Long Beach, CA, USA, 16–20 June 2019.
32. Liu, A.; Liu, X.; Fan, J.; Ma, Y.; Zhang, A.; Xie, H.; Tao, D. Perceptual-Sensitive GAN for Generating Adversarial Patches. *Proc. AAAI Conf. Artif. Intell.* **2019**, *33*, 1028–1035. [CrossRef]
33. Castanon, G.; Byrne, J. Visualizing and Quantifying Discriminative Features for Face Recognition. In Proceedings of the 2018 13th IEEE International Conference on Automatic Face & Gesture Recognition (FG 2018), Xi'an, China, 15–19 May 2018.
34. Sharif, M.; Bhagavatula, S.; Bauer, L.; Reiter, M.K. Accessorize to a Crime: Real and Stealthy Attacks on State-of-the-Art Face Recognition. In Proceedings of the 2016 ACM SIGSAC Conference on Computer and Communications Security, Vienna, Austria, 24–28 October 2016.
35. Zhou, Z.; Tang, D.; Wang, X.; Han, W.; Liu, X.; Zhang, K. Invisible Mask: Practical Attacks on Face Recognition with Infrared. Available online: https://arxiv.org/abs/1803.04683 (accessed on 21 September 2022).

6. Komkov, S.; Petiushko, A. AdvHat: Real-World Adversarial Attack on ArcFace Face ID System. In Proceedings of the 2020 25th International Conference on Pattern Recognition (ICPR), Milan, Italy, 10–15 January 2021.
7. Pautov, M.; Melnikov, G.; Kaziakhmedov, E.; Kireev, K.; Petiushko, A. On Adversarial Patches: Real-World Attack on ArcFace-100 Face Recognition System. In Proceedings of the 2019 International Multi-Conference on Engineering, Computer and Information Sciences (SIBIRCON), Novosibirsk, Russia, 21–27 October 2019.
8. Shen, M.; Liao, Z.; Zhu, L.; Xu, K.; Du, X. VLA: A Practical Visible Light-Based Attack on Face Recognition Systems in Physical World. *ACM Interact. Mob. Wearable Ubiquitous Technol.* **2019**, *3*, 1–19. [CrossRef]
9. Nguyen, D.; Arora, S.S.; Wu, Y.; Yang, H. Adversarial Light Projection Attacks on Face Recognition Systems: A Feasibility Study. In Proceedings of the 2020 IEEE/CVF Conference on Computer Vision and Pattern Recognition Workshops (CVPRW), Seattle, WA, USA, 14–19 June 2020.
10. Yin, B.; Wang, W.; Yao, T.; Guo, J.; Kong, Z.; Ding, S.; Li, J.; Liu, C. Adv-Makeup: A New Imperceptible and Transferable Attack on Face Recognition. Available online: https://arxiv.org/abs/2105.03162 (accessed on 21 September 2022).
11. Xiao, C.; Li, B.; Zhu, J.-Y.; He, W.; Liu, M.; Song, D. Generating Adversarial Examples with Adversarial Networks. In Proceedings of the 27th International Joint Conference on Artificial Intelligence, Stockholm, Sweden, 13–19 July 2018.
12. Tong, L.; Chen, Z.; Ni, J.; Cheng, W.; Song, D.; Chen, H.; Vorobeychik, Y. FACESEC: A Fine-grained Robustness Evaluation Framework for Face Recognition Systems. Available online: https://arxiv.org/abs/2104.04107 (accessed on 18 November 2022).
13. Shen, M.; Yu, H.; Zhu, L.; Xu, K.; Li, Q.; Hu, J. Effective and Robust Physical-World Attacks on Deep Learning Face Recognition Systems. *IEEE Trans. Inf. Forensics Secur.* **2021**, *16*, 4063–4077 [CrossRef]
14. Deb, D. AdvFaces: Adversarial Face Synthesis. Available online: https://github.com/ronny3050/AdvFaces (accessed on 19 November 2022).
15. Serengil, S.I. Deepface. Available online: https://github.com/serengil/deepface (accessed on 22 September 2022).
16. Singh, M.; Arora, A.S. Computer Aided Face Liveness Detection with Facial Thermography. *Wirel. Pers. Commun.* **2020**, *111*, 2465–2476. [CrossRef]
17. Theagarajan, R.; Bhanu, B. Defending Black Box Facial Recognition Classifiers Against Adversarial Attacks. In Proceedings of the 2020 IEEE/CVF Conference on Computer Vision and Pattern Recognition Workshops (CVPRW), Seattle, WA, USA, 14–19 June 2020.
18. Chen, P.-Y.; Zhang, H.; Sharma, Y.; Yi, J.; Hsieh, C.-J. ZOO: Zeroth Order Optimization based Black-box Attacks to Deep Neural Networks without Training Substitute Models. Available online: https://arxiv.org/abs/1708.03999 (accessed on 19 November 2022).

Disclaimer/Publisher's Note: The statements, opinions and data contained in all publications are solely those of the individual author(s) and contributor(s) and not of MDPI and/or the editor(s). MDPI and/or the editor(s) disclaim responsibility for any injury to people or property resulting from any ideas, methods, instructions or products referred to in the content.

Article

Enabling Real-Time On-Chip Audio Super Resolution for Bone-Conduction Microphones

Yuang Li [1,2], Yuntao Wang [1,*], Xin Liu [3], Yuanchun Shi [1], Shwetak Patel [3] and Shao-Fu Shih [4]

[1] Key Laboratory of Pervasive Computing, Ministry of Education, Department of Commputer Science and Technology, Tsinghua University, Beijing 100084, China
[2] Department of Engineering, University of Cambridge, Cambridge CB2 1TN, UK
[3] Department of Computer Science and Engineering, Paul G. Allen School of Computer, University of Washington, Seattle, WA 98195, USA
[4] Google Inc., Mountain View, CA 94043, USA
* Correspondence: yuntaowang@tsinghua.edu.cn

Abstract: Voice communication using an air-conduction microphone in noisy environments suffers from the degradation of speech audibility. Bone-conduction microphones (BCM) are robust against ambient noises but suffer from limited effective bandwidth due to their sensing mechanism. Although existing audio super-resolution algorithms can recover the high-frequency loss to achieve high-fidelity audio, they require considerably more computational resources than is available in low-power hearable devices. This paper proposes the first-ever real-time on-chip speech audio super-resolution system for BCM. To accomplish this, we built and compared a series of lightweight audio super-resolution deep-learning models. Among all these models, ATS-UNet was the most cost-efficient because the proposed novel Audio Temporal Shift Module (ATSM) reduces the network's dimensionality while maintaining sufficient temporal features from speech audio. Then, we quantized and deployed the ATS-UNet to low-end ARM micro-controller units for a real-time embedded prototype. The evaluation results show that our system achieved real-time inference speed on Cortex-M7 and higher quality compared with the baseline audio super-resolution method. Finally, we conducted a user study with ten experts and ten amateur listeners to evaluate our method's effectiveness to human ears. Both groups perceived a significantly higher speech quality with our method when compared to the solutions with the original BCM or air-conduction microphone with cutting-edge noise-reduction algorithms.

Keywords: audio super-resolution; bone-conduction microphone; real-time system; convolutional neural network

Citation: Li, Y.; Wang, Y.; Liu, X.; Shi, Y.; Patel, S.; Shih, S.-F. Enabling Real-Time On-Chip Audio Super Resolution for Bone-Conduction Microphones. *Sensors* **2023**, *23*, 35. https://doi.org/10.3390/s23010035

Academic Editors: Shyan-Ming Yuan, Zeng-Wei Hong and Wai-Khuen Cheng

Received: 2 November 2022
Revised: 15 December 2022
Accepted: 15 December 2022
Published: 20 December 2022

Copyright: © 2022 by the authors. Licensee MDPI, Basel, Switzerland. This article is an open access article distributed under the terms and conditions of the Creative Commons Attribution (CC BY) license (https://creativecommons.org/licenses/by/4.0/).

1. Introduction

The most commonly used microphones for voice communication are air-conduction microphones, which pick up sound propagating through the air. Although providing high fidelity capture in quiet scenarios, they are vulnerable to environmental noises. To improve the speech quality of air-conduction microphones in noisy environments, researchers proposed multi-microphone beamforming with noise suppression techniques [1–3] or deep-learning-based speech enhancement methods [4,5]. However, these solutions require a significant amount of additional hardware or computing resources. Further, all these methods fundamentally seek to reduce environmental noises but also inevitably corrupt speech. Moreover, these solutions are still vulnerable to boisterous environments, such as construction sites or strong wind, where extraneous noises overpower speech signals.

Bone-conduction microphones (BCMs) could achieve more robust results against ambient noises due to their physical design and fundamental conduction principles. However, BCMs only have limited frequency response with high-frequency components above 2 kHz significantly attenuated. Reconstructing the high-frequency details can effectively increase the speech audio's quality.

A traditional method of reconstruction is to design a linear phase impulse response filter [6]. However, acoustic paths are different among speakers because their bone structures are unique. Furthermore, it is impossible to ensure uniform BCM placement, which may result in different spectral properties [7]. Therefore, a simple filter is insufficient to accommodate a variety of users.

Audio super resolution [8], also called bandwidth expansion, is the task of increasing the audio sampling rate and restoring the high-frequency components of low-resolution audios. Convolutional Neural Networks have achieved state-of-the-art performance in audio super resolution [9–11]. Additionally, similar neural-network structures have also been proven effective in reconstructing distorted spectrograms [12] and enhancing recordings from low-end microphones [13,14]. Therefore, designing an audio super-resolution model is feasible to reproduce high-fidelity speech from BCMs while maintaining their noise resistance property in multi-speaker settings. However, existing deep-learning-based audio super-resolution methods are commonly computationally intensive, making them unfit for deployment on resource-constrained embedded systems.

This paper proposes the first-ever real-time on-chip speech audio super-resolution system for BCMs. In order to achieve this goal, we first designed and compared a series of lightweight deep-learning models for speech audio super resolution. Among all the models, ATS-UNet is the most cost-efficient. We proposed an audio temporal shift module (ATSM) and introduced this module to ATS-UNet. Therefore, ATS-UNet can reduce the network to one dimension but still learn sufficient features from the temporal information flow in speech audios.

Thus, ATS-UNet can reconstruct high-fidelity speech audios but require minimal computational resources. We further quantized and deployed ATS-UNet and its variants on micro-controllers, including ARM Cortex-M4f and M7 processors, and conducted a full evaluation of our proposed method's performance regarding the audio quality and inference latency. The results show that ATS-UNet outperformed the cutting-edge audio super-resolution method [9] with the perceptual evaluation of speech quality (PESQ) [15] increased by 9% and log-spectral distance (LSD) [16] by 44%.

On the Cortex-M7 processor, our end-to-end latency, comprising model inference, feature extraction, and reconstruction, is 38 ms on average. This is less than the half-frame length (64 ms), meaning that our system can achieve real-time processing with 128 ms frames half-overlapped. To further assess our method's effectiveness in obtaining high quality speech, we recruited 20 participants, including 10 experts and 10 amateur listeners, for the perceptual audio quality evaluation. The results show that our method outperformed the original BCM solution and commodity noise reduction solution with the air-conduction microphone. To the best of our knowledge, our method is the first chip-deployable audio super-resolution solution.

To summarize, our contributions are as follows:

- We propose a lightweight audio super-resolution deep-learning model—ATS-UNet—that utilizes our proposed audio temporal shift module (ATSM) to form a novel one-dimensional UNet architecture. When compared with ATS-UNet's variants without ATSM, ATS-UNet was the most cost-efficient for chip deployment.
- We implement the first-ever real-time on-chip speech audio super-resolution system for the bone-conduction microphone by quantizing and deploying ATS-UNet to popular micro-controllers in commodity hearable devices. We further evaluate its computational complexity on both ARM Cortex-M4f and M7 processors and demonstrated its real-time processing capability.
- We evaluate our system's effectiveness in improving speech quality with a bone-conduction microphone through perceptual audio quality user studies. Audio samples are publicly available (https://sites.google.com/view/audio-sr-for-bcm/home (accessed on 1 November 2022)).

2. Background and Related Work

Voice communication using air-conduction microphones in noisy environments has been a challenging problem. For conventional speech communication, researchers have proposed speech enhancement methods, such as beamforming with a microphone array and blind source separation [17], for background noise removal. These algorithms only remove part of the unwanted noises and introduce the risk of damaging the voice integrity. Beamforming is based on directionality.

Therefore, it is prone to directional noise sources. In other words, when noise and voice sources are on-axis, beamforming will not effectively separate the noise. To reduce on-axis noise, noise suppression algorithms, such as [3,18,19] first estimate the noise with statistical models and then remove the noise from the captured spectrum to recover the original speech. These methods could lead to speech integrity issues due to the noise model estimation accuracy. Moreover, under extreme conditions, such as strong wind noise, air-conduction microphones will not pick up human voices due to saturation.

A bone-conduction microphone, which collects human speech propagated via human bones, naturally suppresses environmental noises with its hardware placement and FSV conduction. However, the speech captured by the BCM has a limited frequency bandwidth which attenuates quickly above 2 kHz [20]. Our motivation is to enhance the BCM's speech sound quality by recovering high-frequency details while keeping its advantage against environmental noise. In this section, we describe existing speech enhancement algorithms for BCM and then give an overview on speech super-resolution techniques.

2.1. Bone-Conduction Microphones

Bone-conduction microphones are commonly used as an accessorial enhancer to air-conduction microphones for capturing human speech. Researchers have proposed speech enhancement methods using BCMs [21–25]. The BCMs can be used for accurate voice activity detection due to their noise suppression characteristics [24]. BCMs can also be incorporated to increase the voice activity detector accuracy and, hence, increase the accuracy for noise model estimation to achieve better denoising results [22].

Further, BCMs can provide additional input for a multi-modal deep-learning network [25]. However, these solutions require multiple microphones that are costly and limited in capability in extreme circumstances, such as strong wind noise. Our work aims to enhance speech quality using a single bone-conduction microphone. In other words, we plan to achieve clean human speech capture while maintaining the microphone's capability against environmental noises.

Similar speech processing techniques based on BCM speech capture with audio super resolution can be found, including the following: speech enhancement approaches for bone-conduction microphones through audio signal processing [20]. Shimamura and Tamiya [6] proposed a reconstruction filter calculated from long-term spectra of human voices from both air- and bone-conduction microphones.

Shimamura et al. [26] further utilized a multi-layer perceptron to model the reconstruction filter more accurately. Rahman and Shimamura [27] excluded the need for the air-conduction microphones by introducing an analysis-synthesis method based on linear prediction. Bouserhal et al. [28] introduced adaptive filtering along with non-linear bandwidth extension method for enhancing the speech sound quality. However, these methods require complex feature engineering and are, thus, difficult to adapt to different users and equipment setups.

Recently, researchers applied deep-learning methods for speech enhancement with BCMs. These methods aim to increase the sound quality of BCMs to be comparable to air-conduction microphones in ideal conditions. For example, Shan et al. [29] proposed an encoder–decoder network with a long short-term memory (LSTM) layer and local attention mechanism, which reconstructs an air conduction log-spectrogram from a bone-conduction log-spectrogram. This method only reconstructs frequency components below 4 kHz and is based on a specific speaker.

Liu et al. [30] introduced Mel-scale features of speech audio from a bone-conduction microphone with a deep denoise auto-encoder for speech enhancement. This work reconstructs high-frequency components up to 8 kHz. This increases the perceptual evaluation of speech quality (PESQ) by 9.38% compared with the original bone-conduction speech; however, the auto-encoder is also trained with a single speaker's speech. Hussain et al. [31] proved that, with only limited training data, the hierarchical extreme-learning machine could outperform the denoise auto-encoder. Zheng et al. [32] adapted structural similarity (SSIM)—a widely used metric in image quality assessment—as the loss function for a Bidirectional LSTM Neural Network. As a result, the model achieved higher PESQ when trained with SSIM loss compared with the standard mean square error (MSE).

Although proven effective, the aforementioned deep-learning methods were not designed to be deployed on real-time embedded systems due to exceeding computation resources and power limits. Moreover, these methods were not evaluated with cross-user validation, which limits their generalizability to adapt to individual users. Some works [29,32] used a sampling rate of 8 kHz, which is not sufficient for the Wideband Speech protocol with required sampling frequency at 16 kHz. Therefore, our work approaches BCM voice capture as a real-time resource-constrained super-resolution problem on embedded systems. Furthermore, to make our solution robust against individual users and various environments, we also introduced transfer learning to make our model generalizable.

2.2. Audio Super-Resolution Techniques

The audio super resolution, also known as bandwidth expansion, aims to increase the sampling rate and restore high-frequency components of the low-resolution audio. Inspired by image inpainting methods, researchers have proposed several frequency domain based deep-learning methods for audio super resolution. These methods can be trained using clean samples of BCMs as input and air-conduction microphones as references. These samples are then converted into snapshots of spectrograms in the frequency domain as snippets of audio features. The learned model restores the missing high-frequency details from BCMs based on pattern recognition at the inference time. Then, the output snippets are reconstructed back into the real-time speech stream as output. Below we describe various audio super-resolution methods, optimization strategies, and on-chip deployment methods.

Audio super-resolution methods either took raw waveforms [8–11] or spectral representations [33–35] as the input. A one-dimensional UNet [8] asymmetrical network with skip connections was the first attempt to use a deep convolutional neural network for end-to-end speech super resolution. To expand the perceptual field, TFiLM [9] utilized bidirectional LSTM as the module to build up a variant 1D-UNet for speech audio super resolution. In another variation of 1D-UNet [10], conventional convolutional layers were replaced by multi-scale convolutional layers to capture information at multiple scales. Mfnet [11] also attempted to facilitate multi-scale information exchange through multi-scale fusion block.

Other deep-learning methods utilized the spectrogram as input. For example, Li and Lee [33] proposed a three-layer fully connected network for speech audio super resolution. UNet [34] was also proven to be effective in performing speech audio super resolution using the power spectrogram. To take advantage of representations in both the time and frequency domain, TFnet [35] incorporated two network branches that operate on both the waveform and the spectrogram, respectively. Although proven effective on speech audio super resolution, the deep-learning models mentioned above have too many parameters, which causes them to exceed the computation and power budgets of micro-controllers by a factor of 100 times.

The optimization strategy for audio super resolution includes the loss function and training optimizations. One of the most commonly used loss functions is the mean square error [8,33]. Although simple to compute, the MSE does not represent human perceptual speech quality. Therefore, perceptually motivated loss [36] that calculates the L1 distance on log mel-spectrogram was proposed. Further, the log spectral distance (LSD) [8,16], which measures the distance between the log–power spectrum of reference and reconstructed signals, was also adopted as one option for the loss function.

For training, WaveNet [37,38], an auto-regressive model, optimized the joint probability of the targeted high-resolution audio. Adversarial learning is another popular training technique. In this technique, a discriminator that works either in the time domain [10,11,39] or frequency domain [34,40,41] guides the generator to predict more realistic high-resolution audio from low-resolution inputs.

Recently, hearable devices, such as TWS earbuds, have become increasingly popular, with 233 million shipments in 2020, while deep-learning-based audio super-resolution methods have been proven effective, deploying such solutions to a resource-limited embedded system has not been fully investigated. Similar to our proposal, several super-resolution deep-learning methods [39,42,43] have proven the feasibility of applying the super-resolution method on a smartphone.

Other state-of-the-art speech super-resolution models require considerable computation resources and cause significant latency, which is not suitable for edge device deployment. This paper proposed a lightweight deep-learning model—ATS-UNet, which can run on power- and space-limited ARM Cortex-M platforms. We expect future hearables embedded with a single BCM will achieve good performance without the need for additional computation resources with the proposed method.

3. Overview

This paper focuses on the uplink portion of the communication system—namely, the capture side of the speech communication protocol. In particular, the capture and recovery of the BCM input as an alternative solution to the conventional air-conduction microphone. Shown in Figure 1, our proof-of-concept prototype is composed of commercially available electronic parts: a pulse density modulation (PDM) bone-conduction microphone (Knowles V2S100D), an analog MEMS air-conduction microphone (InvenSense ICS-40730), and a micro-controller development board (Bestechnic (http://www.bestechnic.com/Home/Index/index/lan_type/2 (accessed on 1 November 2022)) (BES) BES2300YP).

The BES2300YP system on chip (SoC) simultaneously collects audio signals from the Knowles V2S100D and InvenSense ICS-40730, forming a dataset for audio super resolution. Then, we trained ATS-UNet using this dataset on an Nvidia Titan XP GPU (12GB RAM). The floating-point model was further quantized to the 16-bit data format and transformed from a Tensorflow [44] to CMSIS-NN [45] implementation. With the model quantization and optimization, ATS-UNet can then run efficiently on micro-controllers.

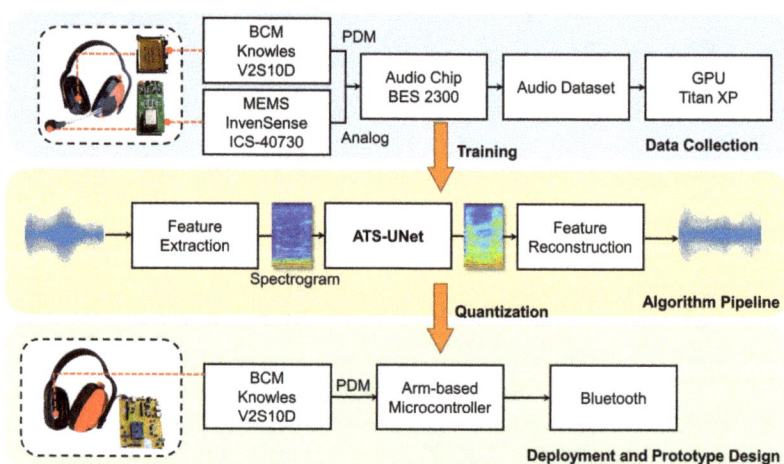

Figure 1. This paper's overview.

In this work, we tested two popular micro-controllers. (1) The BES2300YP with dual ARM-Cortex M4F processors operating at a frequency up to 300 MHz with 992 KB SRAM and 4 MB flash storage. The BES SoC was adopted by many popular TWS earbuds, such as JBL FREE II, Samsung Galaxy Buds Live, and Huawei FreeBuds 2 Pro, for its compact form factor and power efficiency. We only use one single processor in this work since the other processor runs the Bluetooth stack and digital signal processing (DSP)-related algorithms.

Furthermore, the two processors share the SRAM with the storage requirements from the Bluetooth and the operating system taking more than 400 KB. To prevent the memory overflow, we limited the SRAM for the machine-learning model to be below 500 KB. (2) The NXP RT1060 SoC with a single Arm-Cortex M7 processor operating at a frequency up to 600 MHz with 1 MB on-chip SRAM. In this case, only 512 KB general-purpose SRAM can be used to host the machine-learning model. Both micro-controllers support audio applications.

4. Deep-Learning Models for Bone Conduction Speech Audio Super Resolution

In this section, we first describe our general UNet design for bone-conduction speech audio super resolution. We then describe how our models, including our proposed 2D-UNet, Hybrid-UNet, Mixed-UNet, 1D-UNet, and ATS-UNet, were derived from this UNet design. Most importantly, we present the key module called the Audio Temporal Shift Module (ATSM). Finally, we describe the pre-processing and post-processing methods for our deep-learning models.

4.1. UNet Variances for Bone Conduction Speech Audio Super Resolution

The original UNet has a fully convolutional and symmetrical network structure with skip connections to facilitate information flow. Additionally, it can extract temporal and frequency information in the time–frequency domain and reconstruct high-resolution spectrograms. Compared with conventional convolutional and recurrent neural networks, UNet is more efficient, as feature maps are down-sampled, contributing to fewer floating-point operations.

However, UNet's large model size still introduces unfavorable computation for on-chip deployment. Therefore, we reduce the number of channels and network depth. The general UNet architecture (Figure 2) contains five down-sampling blocks (DB) and five up-sampling blocks (UB). Each DB has a max-pooling layer followed by two convolutional layers. The size of the max-pooling layer is 2×1. Therefore, after each DB, the length of the frequency axis of the feature map is halved, while the time dimension remains the same throughout the network.

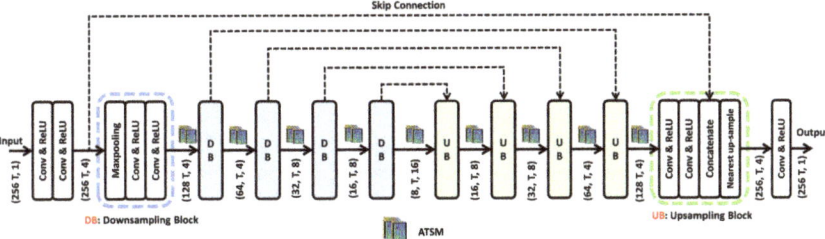

Figure 2. The detailed architecture of our general UNet design. (F, T, C) indicates F for frequency bins, T for temporal frames, and C for channels. When all convolutional layers are 1D and ATSMs are inserted after each DB/UB, it is our ATS-UNet.

In UB, feature maps are up-sampled, concatenated with skip features, and then fed into two convolutional layers. A ReLU activation function is adopted after each convolutional layer except for the last layer. **2D-UNet V1** (Figure 3a) has the same structure and channel numbers as shown in Figure 2; however, it has no ATSM, and every convolutional layer is 2D. We also present **2D-UNet V2**, which has four times the filters of 2D-UNet v1 for comparison.

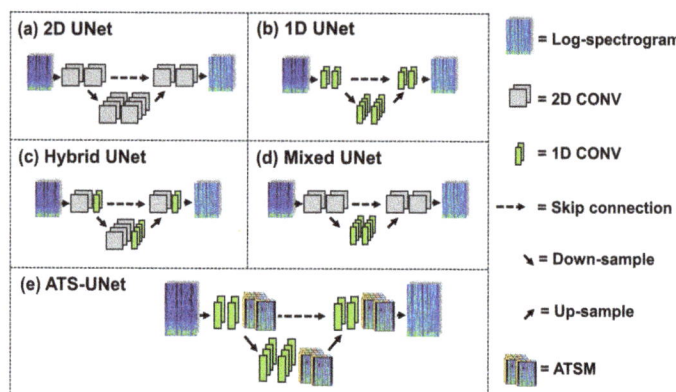

Figure 3. Our proposed series of novel network architectures including a variant 2D-UNet, 1D-UNet, Hybrid-UNet, Mixed-UNet, and ATS-UNet.

Although 2D-UNet V1 is significantly smaller than the original UNet for image segmentation, 2D convolutional layers still introduce unfavorable computation for on-chip deployment. Since low-latency audio super resolution requires a small frame size, the shape of the input spectrogram is narrow (the frequency axis is much longer than the time axis). Therefore, only a few 2D convolutional layers are sufficient to extract information from the full temporal range. Thus, using 2D convolutional layers throughout the network is unnecessary.

Based on the above observation, We present another architecture called **Hybrid-UNet** (Figure 3c) and **Mixed-UNet** (Figure 3d), which replace a portion of the 2D convolutional layers with 1D convolutional layers. 2D convolutional layers enable temporal information flow, while 1D convolutional layers only compute along the frequency dimension to enlarge the perceptual range. Hybrid-UNet adopts 2D and 1D convolutional layers in each DB/UB alternately, which maintains temporal information flow in the whole network. Mixed-UNet replaces 2D convolutional layers with 1D layers in the middle of the network so that temporal information exchange only exits in shallow layers.

Although Hybrid-UNet and Mixed-UNet are more efficient than traditional 2D-UNet, 2D convolutional layers still introduce unfavorable computation for real-time inference on low-end embedded systems. Thus, we replace 2D convolutional layers with 1D convolutional layers completely to obtain a new architecture called **1D-UNet** (Figure 3b), a 1D version of 2D-UNet V1. However, 1D-UNet lacks temporal modeling; therefore, we inserted ATSM after each DB/UB to enable efficient and effective information exchange along the temporal axis. We called 1D-UNet with ATSM **ATS-UNet** (Figure 3e). For all the models, the kernel sizes of 1D and 2D convolutional layers are 3×1 and 3×3, respectively.

4.2. Audio Temporal Shift Module (ATSM)

Conventional deep-learning models for audio processing require massive 2D convolutional operations to extract meaningful features from spectrograms [4] as Figure 4a shows. However, they utilize a large number of computational resources. Therefore, the aforementioned deep-learning models are unlikely to be adopted for on-chip audio super resolution. Instead, we introduced a novel module to accelerate convolutional operations in the time–frequency domain called the Audio Temporal Shift Module (ATSM), as Figure 4b shows. ATSM was inspired by the Temporal Shift Module (TSM) [46], an effective mechanism for video understanding.

This replaces 3D convolutional operations with 2D ones while preserving high-dimensional modeling. This is achieved by shifting the feature maps among video frames to enable temporal information flow. Similarly, ATSM utilizes 1D convolution operations to replace 2D convolution operations for audio processing. In order to utilize information from a longer temporal range for 1D convolutional kernels, ATSM shifts feature maps along the temporal axis of spectrograms. In contrast to the TSM, whose input is four-dimensional feature maps extracted from video frames, the input ATSM is extracted from the 2D log-spectrogram and only has three dimensions: channel, time, and frequency.

More specifically, as illustrated in Figure 4b, feature maps are divided into two chunks along the channel dimension: (1) dynamic and (2) static. The dynamic feature maps are split evenly into two parts, with one part shifted forward (delaying time by one frame) and the other backward (advancing time by one frame). The static feature maps remain unchanged. It is worth noting that ATSM requires no additional computational resources to facilitate information exchange along the temporal dimension in spectrogram computation.

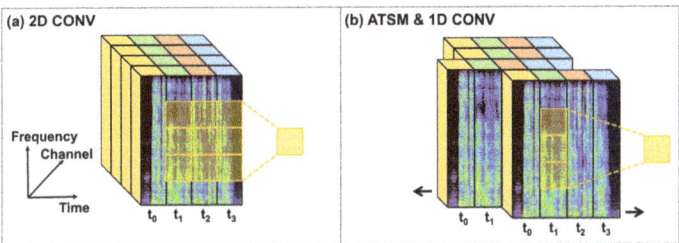

Figure 4. To enable information flow through t_1, t_2, t_3, we can use (**a**) a 2D convolutional layer or (**b**) the proposed ATSM with a 1D convolutional layer. The latter is more lightweight.

4.3. Audio Pre-Processing and Resynthesis

Voice communication's ideal overall latency is below 50 ms, which humans are unable to notice. As latency increases, humans start to notice lip-sync issues; however, communication latency under 150 ms is still considered acceptable. However, a latency that exceeds 400 ms [47] is unacceptable for real-time communication. Therefore, a feasible audio super-resolution system should not add too much latency to the communication process. As a result, our system requires fast computing with an appropriate frame size and short-time Fourier transform (STFT) parameter.

The pre-processing includes the feature extraction from the raw audio signal as shown in the left figure of Figure 5. ATS-UNet processes a single audio frame at a time and outputs audio frames in sequence to resynthesize the audio stream. A large frame provides more information for ATS-UNet but introduces longer latency, since the system has to wait for the time of the entire frame. Therefore, we use a 2048-point (128 ms) frame with half overlap to achieve acceptable latency while maintaining adequate information. These frames are transformed into spectrograms by STFT and fed into ATS-UNet.

The STFT parameter is another major factor in computational intensity. High frequency and time resolution spectrograms can be achieved with a larger fast Fourier transform (FFT) size and overlap between FFT windows, resulting in considerable computation load. Considering the memory and resources on the micro-controller, we adopted a window size of 512 points for the STFT. Further, we also utilized a half overlap strategy to the raw audio data. The detailed trade-off of the STFT parameter is explained in Section 5.

The audio resynthesis, also known as post-processing, converts the reconstructed spectrograms back to the time domain using the inverse short-time Fourier transform (ISTFT). The overlapped frame is then multiplied by the Hanning window (2048-point) and added to the previous frame. We adopted this resynthesis method because the data points in the center of the window are better reconstructed due to richer temporal information. Therefore, the Hanning window function gives the data samples in the center of the window higher weights but weakens the importance of the data samples by the side.

Further, we adopted half-overlapped Hanning window functions so that their summation is a constant value. Thus, it will not distort the signal and can smooth the transition between adjacent frames.

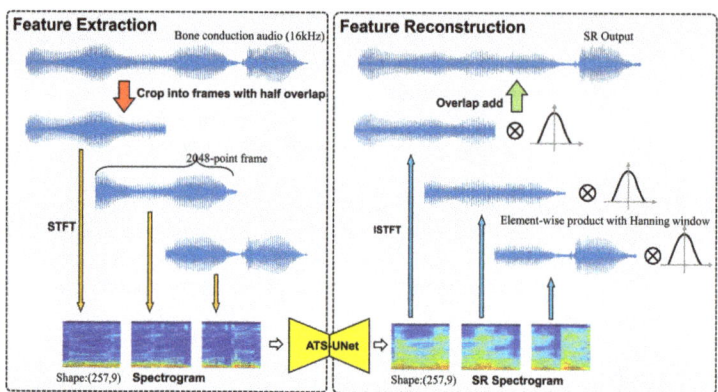

Figure 5. Feature extraction and reconstruction.

5. Model Training and Deployment

In this section, we provide experimental details, including the data collection procedure, training scheme, and quantization procedure.

5.1. Speech Audio Data Collection

We conducted a user experiment to collect an audio dataset using the hardware shown in Figure 6. The MEMS air-conduction microphone was placed near the mouth to collect high-quality ground-truth speech audios. The BCM was secured with an earmuff. Thus, when subjects wore the earmuff, the BCM would be pressed in front of the ear. To prevent reverberation, we placed an acoustic panel in front of the speaker. As Figure 1 indicates, we utilized a BES2300YP micro-controller to simultaneously collect speech audios from both the air- and bone-conduction microphones. The sampling rate and bit depth were set to 44.1 kHz and 16 bits. We recorded the speech audios in a recording studio that was quiet for high speech quality.

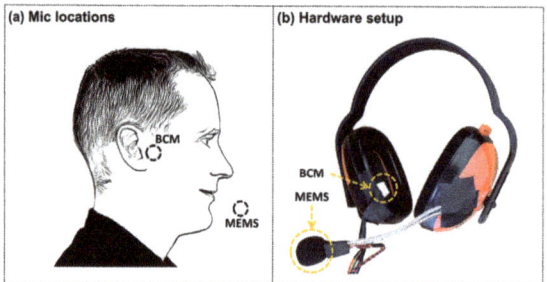

Figure 6. (**a**) The position of the air (MEMS) and the bone (BCM) conduction microphone. (**b**) The headphone prototype for data collection.

We recruited 20 participants (10 males and 10 females). After wearing the headphone, each subject was informed to read six paragraphs of an article, yielding approximately 12 min of speech per subject. We removed the silence clips at the beginning and the end of all audio files and normalized the volume across participants. We then down-sampled each speech audio to 16 kHz, which is sufficient for communication (https://en.wikipedia.org/wiki/Sampling_(signal_processing) (accessed on 1 November 2022)). The processed dataset includes 200 min of speech audios in total. Each participant received a 10 USD gift card after the experiment.

5.2. Implementation and Training Details

Bone-conduction audios were down-sampled to 16 kHz, cropped to 2048-point frames (128 ms), and transformed to time–frequency representations by short-time Fourier transform (STFT) [48] with 512-point Hanning window and half overlap (a stride of 256-point). We adopted the implementation of librosa (https://librosa.org/doc/latest/index.html (accessed on 1 November 2022)) for STFT and ISTFT. The symmetrical component is removed, resulting in 257 Fourier coefficients. Thus, the size of the input spectrogram is 257 × 9.

Then, we converted the power of each coefficient to the log scale and standardize them to normal distribution. We skipped the 0th coefficient (DC component) but fed the remaining 256 coefficients into the network. Lastly, we obtained the enhanced log-spectrogram from the super-resolution model and concatenated it with 0th coefficient; therefore, the output's shape is also 257 × 9. The post-processing audio resynthesis includes denormalization, conversion to linear-scale, and inverse STFT. The model only predicts magnitude, and thus we kept the original phase information from the bone-conduction audio to resynthesize the enhanced speech audio.

All super-resolution models were implemented in Tensorflow [44]. We adopted cross user validation with the training dataset consisting of speech audios from 18 speakers and the remaining audios as the test dataset. We randomly initialized the model and trained it for 100 epochs using the Adam optimizer [49] with a learning rate of 0.0001 and batch size of 64.

5.3. Loss Function

The loss function is given by Equation (1), that consists of two parts: the least absolute deviation (L1) loss and perceptually motivated loss [36]. L1 loss measures the absolute difference between the log-spectrograms of the output speech audio and the ground-truth speech audio—$\log(s(y))$. Perceptually motivated loss is the L1 distance calculated on log-melspectrogram $\log(ms(y))$ considering that the mel-scale is more aligned with human hearing [50]. In Equation (1), $s(y)$ and $s(\hat{y})$ stand for spectrograms of the output and ground-truth audios. $ms(y)$ and $ms(\hat{y})$ represent melspectrograms of the output and ground-truth audios, respectively.

$$Loss = |\log(s(y)) - \log(s(\hat{y}))|_1 + |\log(ms(y)) - \log(ms(\hat{y}))|_1 \quad (1)$$

5.4. Model Quantization

We re-compiled each model using the CMSIS-NN [45] framework for efficient inference on Arm Cortex-M processors. First, we transformed the model from floating-point to fixed-point format. Both weights and activations were quantized to 16-bit integers, given by Equation (2). In practice, quantization is symmetrical around zero with power-of-two scaling; therefore, it can be implemented by bitwise shifts in CMSIS-NN kernels.

$$x_q = \lfloor x \times 2^{15 - \log_2 max|x|} \rfloor, \quad (2)$$

where x represents the weights of a convolutional layer. x_q is the quantized weights.

5.5. Noise Transfer Learning

The primary motivation behind the use of BCM is to enhance the communication quality in noisy environments. Therefore, our system should improve bone-conduction recordings in a quiet environment and in boisterous environments. To this end, we collected voices from BCM in different noisy locations. However, we observed more unwanted noises in the reconstructed speech compared with the quiet laboratory setup. This is because BCM can still pick up some external noises that are further being enhanced by the model.

We adopted transfer learning to fine-tune the model for noise reduction. By collecting pure noises using BCM and adding them to bone-conduction audio in the training dataset, the model can learn to identify the unwanted noises and only recover speech signals. In detail, we collected bone-conduction noises in three locations, including a subway station, a bus stop, and a dining hall. We instructed a participant to wear the prototype without speaking and recorded bone-conduction noises for 20 min in each location.

We then added the noises to the bone-conduction speech recorded in the quiet studio. For each audio clip, the signal-to-noise ratio (SNR) between the bone-conduction speech and the additive noise was randomly sampled from Gaussian distribution with a mean of 18 and a standard deviation of 3.5. Before being deployed in real-world environments, ATS-UNet was fine tuned on the noisy data for another 100 epochs with the same parameters in Section 5.2.

6. Quantitative Speech Quality Evaluation

In this section, we present the quantitative speech quality evaluation regarding the air-conduction microphone as the ground truth. We describe the evaluation metrics, baselines, and results. We then explain the reasons behind the hyper-parameter selection and benchmark the performance of ATS-UNet, UNet variants, and baselines for speech enhancement.

6.1. Evaluation Metrics

We considered the effectiveness, model size, latency, and power consumption to evaluate each model's performance comprehensively. Specifically, the effectiveness includes two metrics: the log spectral distance (LSD) [16], and the perceptual evaluation of speech quality (PESQ) [15]. LSD, given by Equation (3) [8], measures the distance between the log–power spectrum of reference and reconstructed signals. Therefore, a lower value indicates a better performance. PESQ was provided by Recommendation ITU-T P862 [15] for the objective assessment of speech quality. This models the mean opinion score (MOS), which ranges from 1 (bad) to 5 (excellent).

$$LSD(x, \hat{x}) = \frac{1}{T} \sum_{t=1}^{T} \sqrt{\frac{1}{K} \sum_{k=1}^{K} (X(t,k) - \hat{X}(t,k))^2}, \quad (3)$$

where t and k are the frame and frequency index, respectively. X and $\hat{X}$ denote the log–power spectrum of x and $\hat{x}$, which are defined as $X = log|S(x)|^2$. S stands for STFT with 2048-point frames.

6.2. Baselines

Birnbaum et al. [9] inserted temporal feature-wise linear modulation (TFiLM) layers into a time-domain 1D-UNet to expand the receptive field. This improved the performance of audio super resolution compared with the original 1D-UNet [8], achieving cutting-edge audio super resolution performance. Therefore, we adopted TFiLM as the baseline in this paper. We used the open-sourced code of TFiLM implementation (https://github.com/kuleshov/audio-super-res (accessed on 1 November 2022)).

6.3. Effect of the Input Hyper-Parameter

To evaluate the trade-off between frequency resolution and model performance, we first compared two sets of STFT parameters: 1024-point FFT, 256 strides, and Blackman window as well as 512-point FFT, 256 strides, and Hanning window. The experiments were performed on two models. The first model is a lightweight 2D-UNet v1. In the second model, we expanded 2D-UNet v1 by increasing the number of filters in each layer by four times to explore the optimum audio super resolution results without considering computation.

As shown in Table 1, both 2D-UNet v1 and v2 outperformed the baseline method—TFiLM [9] with significantly fewer parameters. This proves the effectiveness of 2D-UNet in speech audio super resolution. The computational intensity of the 1024-point STFT was nearly doubled compared with the 512-point STFT; however, the performances were close. Therefore, we adopted a 512-point window size for the STFT in the following evaluation procedures considering the latency and model size.

Table 1. Performance comparison for different STFT parameters.

Model	Params	Average LSD(dB)/PESQ	
		1024-Point STFT	512-Point STFT
2D-UNet v1	11.8 k	2.028/2.713	2.013/2.790
2D-UNet v2	187.3 k	1.949/2.937	1.961/2.983
TFiLM (baseline) [9]	68,221.2 k	3.646/2.523 (time domain)	

6.4. Model Performance Results and Comparison

To align the loss function with human hearing sensitivity for different frequency ranges, we incorporated perceptually motivated loss [36]. Compared with the L1 loss, this increased the accuracy for every tested architecture (Table 2).

Table 2. Performance comparison for different UNet architectures. Latency is the model inference time on a single 2048-point frame by Arm Cortex-M4f/M7 processor. For 2D-UNet v2 and TFiLM, latency was not provided as they are too large to be deployed on our embedded system.

Our Models	Params	FLOPs	Latency (ms)		Average LSD(dB)/PESQ	
			Cortex-M4f	Cortex-M7	L1	L1 + Perceptual Loss
2D-UNet v2	187.3 k	133.9 M	/	/	1.961/2.983	1.954/3.030
2D-UNet v1	11.8 k	8.6 M	187	44	2.013/2.790	2.004/2.780
Hybrid-UNet	8.4 k	7.0 M	163	38	2.024/2.689	2.015/2.733
Mixed-UNet	6.3 k	6.9 M	166	39	2.026/2.692	2.019/2.743
1D-UNet	4.5 k	4.8 M	129	31	2.063/2.664	2.052/2.717
ATS-UNet	4.5 k	4.8 M	131	32	2.032/2.710	2.032/2.749
TFiLM (baseline) [9]	68,221.2 k	116,420 M	/	/	3.646/2.523	

Although UNet v1 only has about 10 thousand parameters, it still requires a long inference time on a computation restricted platform, and thus we proposed Hybrid-UNet, Mixed-UNet, and ATS-UNet as described in Section 4. Benchmark latencies and accuracies are provided in Table 2 and Figure 7. ATS-UNet and 1D-UNet are the fastest networks, taking 131/32 ms and 129/31 ms, respectively, to inference a 2048-point frame.

Due to the lack of temporal modeling, the accuracy of 1D-UNet is significantly lower than ATS-UNet. ATSM effectively promotes temporal modeling while only adding negligible latency. Although Hybrid-UNet has 2000 more parameters than Mixed-UNet, the two settings achieve nearly the same latency and accuracy because their floating-point operations (FLOPs) are very close. 2D-UNet v1 is the slowest with expensive computation and slightly higher accuracy. Note that 2D-UNet v2 is too large to be run on our embedded system, and thus we leave gaps in the table.

As shown in Figure 7, ATS-UNet is the most cost-efficient model as it is on the upper left of the plot. In addition, spectrum examples in Figure 8 demonstrate that ATS-UNet outperformed TFiLM since it recovered a more accurate high-frequency structure. Considering our embedded system's restricted computational resources and memory, ATS-UNet was the best architecture to enable on-chip audio super resolution for BCM.

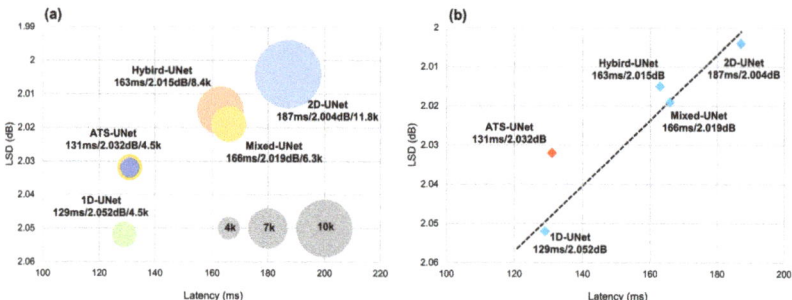

Figure 7. (a) The trade-off between on-chip latency and audio super resolution performance measured by LSD. The number of parameters is represented by circle size. (b) Without ATSM, there is a linear relationship between latency and LSD. ATS-UNet is on the upper left of the dotted line, thereby, proving its superiority.

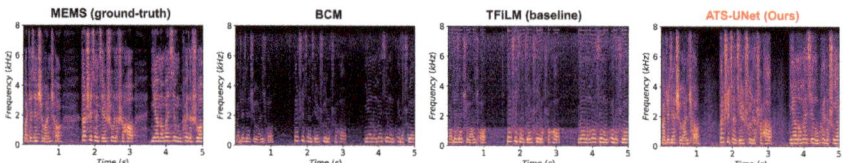

Figure 8. Audio super-resolution results visualized by spectrograms.

6.5. Power Consumption

Since the algorithm pipeline can be run in real-time on Arm Cortex-M7, we measured the power consumption of this microcontroller under two circumstances: (1) during audio super resolution and (2) without audio super resolution. We used a power analyzer (EMK850) to measure the average current within one minute. When our super-resolution module was active, the average consumption was 498 mW (114.4 mA at 4.35 V). After the audio super-resolution module was deactivated, the microcontroller consumed 406mW (93.0 mA at 4.37 V). Therefore, ATS-UNet, feature extraction and reconstruction consumed 92 mW (498 mW–406 mW) on average.

7. Perceptual Speech Quality Evaluation

In this section, we present the perceptual speech quality evaluation of our method under both quiet and noisy environments. We describe the user study design, participants, and results in this section. Specifically, we conducted two user studies. The first was to compare the perceived speech audio quality of different machine-learning models.

The second user study was to evaluate our method's effectiveness against environmental noises. We utilized a within subject user study design. We utilized the Friedman test for non-parametric statistical analysis ($p < 0.05$) and the Wilcoxon signed-rank test for post hoc analysis ($p < 0.05$). We utilized the Mann–Whitney U test to evaluate the difference between user groups for statistical analysis ($p < 0.05$).

7.1. Participants

We recruited 20 participants (14 males and 6 females) with an average age of 33.2 (s.d. = 4.8) separated into two groups. The "Golden Ear" (GE) group had 10 participants (6 males and 4 females) with an average age of 34.2 (s.d. = 5.0). They were specialists who were selected and trained to be able to discern subtle differences in audios. The "Non-Golden Ear" (NGE) group had the other 10 amateur listeners (6 males and 4 females) with an average age of 32.1 (s.d. = 4.6).

The study was conducted in a quiet listening room. During the test, each participant was required to wear headphones (AKG N20 model). A 5 min break was required after

10 trials. The whole study lasted for 60 min. Each participant received a 30 USD gift card for their time and effort.

7.2. User Study 1: Speech Audio Quality in Quiet Environment

This user study included 20 trials. Participants listened to an audio clip from the MEMS air-conduction microphone in each trial, which produced the highest speech audio quality. Then, they listened to and compared three audio clips, including: (1) original speech audio from the BCM (Original); (2) speech audio processed by the 2D-UNet v1; and (3) speech audio processed by the ATS-UNet.

The three audio clips had the same duration, while each set of audio clips lasted between 5 and 10 s with an average duration of 8.2 s. Then, each participant rated the sound quality of these three audio clips by referring to the high-quality audio clip from the MEMS microphone. We utilized a 5-point Likert scale for the rating (5 = very good, 3 = neural, and 1 = very bad). The three audio clips were ordered randomly in each trial before the user study. Participants were allowed to listen to and compare audio clips repeatedly. In total, each participant listened to 80 audio clips.

Results

The results show that both 2D-UNet v1 and ATS-UNet can effectively increase the sound quality of audio from the bone-conduction microphone. Further, 2D-UNet v1 achieved better performance compared with ATS-UNet. As shown in Figure 9a, the mean score of the original audio was 2.09 (s.d. = 0.03), of the ATS-UNet audio was 2.95 (s.d. = 0.03), and of the 2D-UNet v1 audio was 3.03 (s.d. = 0.03). These differences were statistically significant according to a Friedman test ($\chi^2(2, N = 400) = 376.6, p < 0.001$). Post-hoc analysis showed that both the perceived sound quality of audio processed by the ATS-UNet ($Z = -13.6, p < 0.001$) and the 2D-UNet v1 ($Z = -13.9, p < 0.001$) significantly outperformed the original bone-conduction speech audio. Further, 2D-UNet v1 outperformed ATS-UNet ($Z = -2.8, p < 0.01$) significantly.

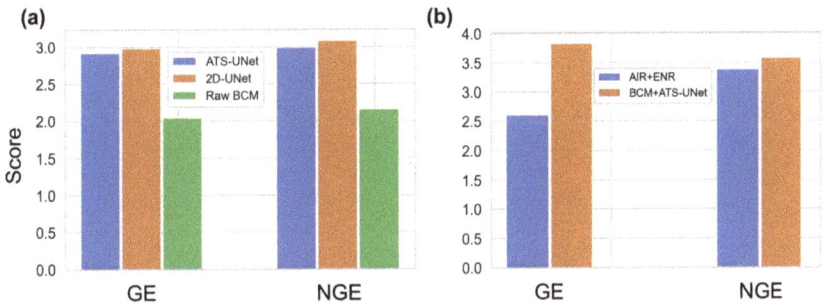

Figure 9. (**a**,**b**) The qualitative speech audio quality evaluation results.

The user group analysis results show that there was significant effect of golden ear status on the perceived sound quality when listening to the original speech audio from the BCM ($Z = -2.1, p = 0.036$) but not under the 2D-UNet v1 ($Z = -0.9, p = 0.37$) or ATS-UNet ($Z = -1.2, p = 0.23$) conditions.

7.3. User Study 2: Effectiveness of ATS-UNet Against Environmental Noises

This user study is to evaluate the effectiveness of our method against environmental noises. We compared our method (BCM + ATS-UNet) to a baseline method—a single air-conduction microphone with environment noise reduction (AIR+ENR). In this evaluation, we adopted Baseus Encok TWS earbuds–WM01 (http://www.baseus.com/product-740?lang=en-us (accessed on 1 November 2022)) as our comparison baseline. It has a built-in signal processing method for environmental noise reduction. During our test with five different brands of earphones with the single speech microphone solution, the Baseus WM01 earbud achieved the best performance in environmental noise reduction.

7.3.1. Speech Audio Data Collection in Noisy Environments

We first recruited five participants to collect speech audios under various environments, including a noisy pedestrian street, a subway station, and a moving car with the window open. They had an average age of 26.5 (s.d. = 2.5). We used the recording hardware presented in Section 5.1 to collect the speech audios from both the air-conduction microphone and the bone-conduction microphone. Further, we streamed the speech audio from the Baseus WM01 earbud to an iPhone 12 for comparison.

We used three hand-clapping events to start each recording and later synchronize the audios. During each data collection session, each participant read the same article that lasts around 2 min. The whole data collection procedure lasted 40 min. Each participant received a 20 USD gift card. As a result, we collected 2 (min) × 5 (participants) × 3 (environments) = 30 min of speech audios with the bone-conduction microphone, the MEMS air-conduction microphone, and the Baseus WM01 earbud. Among these audio recordings, the raw audios from the MEMS microphone contained stronger environmental noises.

7.3.2. Speech Audio Quality Evaluation

This user study included 24 trials. In each trial, participants listened to an audio clip from the MEMS microphone with stronger background noises. Then, they listened to and compared two audio clips, including (1) speech audio from the bone-conduction microphone processed by the ATS-UNet and (2) speech audio from the Baseus WM01 earbud with noise-reduction algorithms. Each pair of audio clips had the same duration and lasted between 8 and 13 s, with an average duration of 10.96 s.

Then, each participant rated the sound quality of these two audio clips by referring to the audio clip from the MEMS microphone. We utilized a 5-point Likert scale for the rating (5 = very good, 3 = neural, and 1 = very bad). The two audio clips were ordered randomly in each trial before the user study. Participants were allowed to listen to and compare the audio clips repeatedly. In total, each participant listened to 60 audio clips.

7.3.3. Results

The results show both golden ear (GE) ($Z = -12.2, p < 0.001$) and non-golden ear (NGE) ($Z = -2.2, p = 0.02$) raters considered the speech audio quality of the BCM + ATS-UNet outperformed the baseline method—AIR + ENR. As shown in Figure 9b, GE raters scored the speech audio quality of BCM + ATS-UNet with an average of 3.83 (s.d. = 0.86) and of the AIR + ENR with an average of 2.61 (s.d. = 0.93). NGE raters scored the speech audio quality of BCM + ATS-UNet and AIR+ENR with average values of 3.58 (s.d. = 1.17) and 3.38 (s.d. = 1.21), respectively.

User group analysis results show that there was a significant effect of the user group on the perceived sound quality of AIR + ENR ($Z = -7.7, p < 0.001$) but not BCM + ATS-UNet ($Z = -1.9, p = 0.053$). These results indicate that GE and NGE raters perceived similar speech audio quality regarding our method. However, GE raters gave the AIR+ENR a significantly lower score, indicating a poorer preference for AIR + ENR.

8. Discussion

In this work, we present the first on-chip audio super-resolution system for BCM. By integrating a novel ATSM into UNet architecture, ATS-UNet makes it possible to recover the missing high-frequency content captured by the BCM on resource-constrained hearable devices. Therefore, model inferences could be run locally on hearable devices without unwanted data transmission and lower latency. In this section, we discuss potential future works and related applications.

8.1. Dual Microphone System and Ambient Awareness

Even though BCM is superior to traditional microphones in noisy environments, and our system significantly improved the BCM's audio quality, air-conduction microphones still provide higher speech quality in low noisy environments. Therefore, a great deal of research [21–25] has focused on using an air-conduction microphone as the primary sensor, accompanied by a BCM for noise reduction. Conversely, low-quality bone-conduction audio is used directly in this research, and thus we hypothesize that there may be an opportunity to apply the audio super-resolution model on bone-conduction speech in conjunction with multi-microphone denoising algorithms.

BCMs and air-conduction microphones are suitable for different scenarios due to their hardware properties. For example, under strong wind noise, BCMs are highly desired, whereas, in a quiet meeting room, BCMs are unnecessary. In this case, the audio super-resolution algorithm leads to unnecessary power consumption. Therefore, another potential future research with a dual-microphone system could be ambient awareness. We anticipate that a dual-microphone system with ambient awareness could achieve the best user experience with optimal power consumption. With the ambient environment information, we could then determine an appropriate microphone and algorithm combination to be utilized at any instance.

8.2. Audio Super Resolution Applications

In this work, our system incorporated a single BCM, which we modeled as an audio super-resolution problem. We have also observed many other potential real-world applications. For example, recently, many people are wearing masks to prevent COVID-19. While these masks prevent the spread of the virus, it also blocks part of the speech signals. Corey et al. [51] showed different masks and microphone placements have different impacts on speech quality.

We believe the audio super-resolution model is a potential solution for recovering the attenuated frequency components from the masks. Increasingly, people pursue high-fidelity music; however, for now, the majority of music on the internet is compressed MP3 files. We anticipate that our model could be used to recover compression losses generated from lossy compression audio codecs. In general, it is encouraged to use our ATS-UNet if audio quality is degraded by frequency loss.

8.3. ATSM for Other Audio Applications

ATSM was designed for processing spectrograms of the audio signal, one of the most widely used input features for audio-related deep neural networks. Therefore, we believe ATSM can be easily adapt to other audio applications, such as speech separation [52] and speech emotion recognition [53]. Researchers can insert ATSM into existing models and reduce the dimension of convolutional layers making the models more lightweight and deployable. Though ATSM was not designed for input features, such as waveforms and MFCC, this provides insight on how to enable information flow and enlarge perceptual range without large convolutional kernels.

8.4. Limitations and Future Work

In this paper, we built a hardware prototype to evaluate the effectiveness of our method to recover high-fidelity speech audio from the bone-conduction microphone. The data collection and performance evaluation procedures were performed on the development board, as Figure 1 shows. We did not develop a wearable hardware solution designed for users to wear it comfortably. Further, during our test and evaluation, the placement of the BCM had a significant effect on the audio quality.

In our implementation, we utilized an earmuff to stabilize the BCM to the user's skin with its location shown in Figure 6. We chose this location for two reasons. (1) It can pick good quality of speech audios during our pilot study (In the pilot study, we compared two locations: in front of the ear and behind the ear). (2) We referred to the cutting-edge design of modern bone-conduction speakers. We expect future work to investigate the optimized location and mounting mechanism. Further, we expect future work to explore sensor fusion approaches to enable better speech audio quality using air- and bone-conduction microphones.

9. Conclusions

In this paper, we proposed a novel real-time embedded audio super-resolution-based speech-capture system with BCM. By integrating a novel ATSM into UNet architecture, ATS-UNet efficiently processed bone-conduction speech audio signals with minimal computational resources among our proposed lightweight audio super-resolution models. Compared with the baseline method (TFiLM), ATS-UNet achieved higher performance in audio quality and reduced the number of parameters by approximately 100 times. Compared to the 2D-UNet v1, ATS-UNet reduced the number of FLOPs by 44% and achieved comparable performance.

With the reduction in computation complexity, our system can achieve real-time processing on a Cortex-M7 with an average power consumption of 92 mW. User studies demonstrated that our system significantly improved the perceptual quality of bone-conduction speech. We propose that our system will promote the usage of BCM in earphones and other deep-learning-based audio-processing applications, particularly those deployed in resource-constrained embedded systems.

Author Contributions: Conceptualization, Y.L., Y.W., X.L., Y.S., S.P., and S.-F.S.; methodology, Y.L., Y.W., X.L., and S.-F.S.; software, Y.L., Y.W., X.L., and S.-F.S.; validation, Y.L., Y.W., X.L., and S.-F.S.; formal analysis, Y.L., Y.W., X.L., Y.S., S.P., and S.-F.S.; investigation, Y.L., Y.W., X.L., Y.S., and S.-F.S.; resources, Y.L., Y.W., X.L., Y.S., S.P., and S.-F.S.; data curation, Y.L., Y.W., X.L., and S.-F.S.; writing—original draft preparation, Y.L., Y.W., X.L., Y.S., S.P., and S.-F.S.; writing—review and editing, Y.L., Y.W., X.L., Y.S., S.P., and S.-F.S.; visualization, Y.L., Y.W., and X.L.; supervision, Y.S., S.P., and S.-F.S.; project administration, Y.S. and S.P.; funding acquisition, Y.S., S.P., and S.-F.S. All authors have read and agreed to the published version of the manuscript.

Funding: This research was funded by the Natural Science Foundation of China under Grant No. 62132010 and No. 62002198, Tsinghua University Initiative Scientific Research Program, Beijing Key Lab of Networked Multimedia, and the Institute for Guo Qiang, Tsinghua University.

Institutional Review Board Statement: The IRB were waived for this study since this study only involved recording users' voices in usual settings such as canteen, sidewalk, or subway.

Informed Consent Statement: Informed consent was obtained from all subjects involved in the study.

Data Availability Statement: Data are available upon request.

Conflicts of Interest: The authors declare no conflict of interest.

References

1. Capon, J. High-resolution frequency-wavenumber spectrum analysis. *Proc. IEEE* **1969**, *57*, 1408–1418. [CrossRef]
2. Pratt, W. Generalized Wiener Filtering Computation Techniques. *IEEE Trans. Comput.* **1972**, *C-21*, 636–641. [CrossRef]
3. Boll, S. Suppression of acoustic noise in speech using spectral subtraction. *IEEE Trans. Acoust. Speech Signal Process.* **1979**, *27*, 113–120. [CrossRef]
4. Park, S.R.; Lee, J.W. A Fully Convolutional Neural Network for Speech Enhancement. In Proceedings of the Interspeech 2017, Stockholm, Sweden, 20–24 August 2017. [CrossRef]
5. Macartney, C.; Weyde, T. Improved speech enhancement with the wave-u-net. *arXiv* **2018**, arXiv:1811.11307.
6. Shimamura, T.; Tamiya, T. A reconstruction filter for bone-conducted speech. In Proceedings of the 48th Midwest Symposium on Circuits and Systems, 2005, Covington, KY, USA, 7–10 August 2005. [CrossRef]
7. McBride, M.; Tran, P.; Letowski, T.; Patrick, R. The effect of bone conduction microphone locations on speech intelligibility and sound quality. *Appl. Ergon.* **2011**, *42*, 495–502. [CrossRef]
8. Kuleshov, V.; Enam, S.Z.; Ermon, S. Audio super-resolution using neural nets. In Proceedings of the ICLR (Workshop Track), Toulon, France, 24–26 April 2017.

9. Birnbaum, S.; Kuleshov, V.; Enam, Z.; Koh, P.W.W.; Ermon, S. Temporal FiLM: Capturing Long-Range Sequence Dependencies with Feature-Wise Modulations. In Proceedings of the Advances in Neural Information Processing Systems, Vancouver, BC, USA, 8–14 December 2019; pp. 10287–10298.
10. Kim, S.; Sathe, V. Bandwidth extension on raw audio via generative adversarial networks. *arXiv* **2019**, arXiv:1903.09027.
11. Hao, X.; Xu, C.; Hou, N.; Xie, L.; Chng, E.S.; Li, H. Time-Domain Neural Network Approach for Speech Bandwidth Extension. In Proceedings of the ICASSP 2020—2020 IEEE International Conference on Acoustics, Speech and Signal Processing (ICASSP), Barcelona, Spain, 4–8 May 2020. [CrossRef]
12. Kegler, M.; Beckmann, P.; Cernak, M. Deep Speech Inpainting of Time-Frequency Masks. In Proceedings of the Interspeech 2020, Shanghai, China, 25–29 October 2020. [CrossRef]
13. Mysore, G.J. Can we Automatically Transform Speech Recorded on Common Consumer Devices in Real-World Environments into Professional Production Quality Speech?—A Dataset, Insights, and Challenges. *IEEE Signal Process. Lett.* **2015**, *22*, 1006–1010. [CrossRef]
14. Su, J.; Jin, Z.; Finkelstein, A. HiFi-GAN: High-Fidelity Denoising and Dereverberation Based on Speech Deep Features in Adversarial Networks. In Proceedings of the Interspeech 2020, Shanghai, China, 25–29 October 2020. [CrossRef]
15. International Telecommunication Union. *Perceptual Evaluation of Speech Quality (PESQ): An Objective Method for End-To-End Speech Quality Assessment of Narrow-Band Telephone Networks and Speech Codecs*; ITU-T Publications: Geneva, Switzerland, 2001.
16. Gray, A.; Markel, J. Distance measures for speech processing. *IEEE Trans. Acoust. Speech Signal Process.* **1976**, *24*, 380–391. [CrossRef]
17. Hidri, A.; Meddeb, S.; Amiri, H. About Multichannel Speech Signal Extraction and Separation Techniques. *J. Signal Inf. Process.* **2012**, *03*, 238–247. [CrossRef]
18. Ephraim, Y.; Malah, D. Speech enhancement using a minimum-mean square error short-time spectral amplitude estimator. *IEEE Trans. Acoust. Speech Signal Process.* **1984**, *32*, 1109–1121. [CrossRef]
19. Scalart, P.; Filho, J. Speech enhancement based on a priori signal to noise estimation. In Proceedings of the 1996 IEEE International Conference on Acoustics, Speech, and Signal Processing Conference Proceedings, Atlanta, GA, USA, 9 May 1996. [CrossRef]
20. Shin, H.S.; Kang, H.G.; Fingscheidt, T. Survey of speech enhancement supported by a bone conduction microphone. In Proceedings of the Speech Communication; 10. ITG Symposium, Braunschweig, Germany, 26–28 September 2012; VDE: Frankfurt, Germany, 2012; pp. 1–4.
21. Liu, Z.; Zhang, Z.; Acero, A.; Droppo, J.; Huang, X. Direct filtering for air- and bone-conductive microphones. In Proceedings of the IEEE sixth Workshop on Multimedia Signal Processing, Siena, Italy, 29 September–1 October 2004. [CrossRef]
22. Lee, C.H.; Rao, B.D.; Garudadri, H. Bone-Conduction Sensor Assisted Noise Estimation for Improved Speech Enhancement. In Proceedings of the Interspeech 2018, Hyderabad, India, 2–6 September 2018. [CrossRef]
23. Takada, M.; Seki, S.; Toda, T. Self-Produced Speech Enhancement and Suppression Method using Air- and Body-Conductive Microphones. In Proceedings of the 2018 Asia-Pacific Signal and Information Processing Association Annual Summit and Conference (APSIPA ASC), Honolulu, HI, USA, 12–15 November 2018. [CrossRef]
24. Zhou, Y.; Chen, Y.; Ma, Y.; Liu, H. A Real-Time Dual-Microphone Speech Enhancement Algorithm Assisted by Bone Conduction Sensor. *Sensors* **2020**, *20*, 5050. [CrossRef] [PubMed]
25. Yu, C.; Hung, K.H.; Wang, S.S.; Tsao, Y.; Hung, J.-w. Time-Domain Multi-Modal Bone/Air Conducted Speech Enhancement. *IEEE Signal Process. Lett.* **2020**, *27*, 1035–1039. [CrossRef]
26. Shimamura, T.; Mamiya, J.; Tamiya, T. Improving Bone-Conducted Speech Quality via Neural Network. In Proceedings of the 2006 IEEE International Symposium on Signal Processing and Information Technology, Vancouver, BC, Canada, 27–30 August 2006. [CrossRef]
27. Rahman, M.S.; Shimamura, T. Intelligibility enhancement of bone conducted speech by an analysis-synthesis method. In Proceedings of the 2011 IEEE 54th International Midwest Symposium on Circuits and Systems (MWSCAS), Seoul, Republic of Korea, 7–10 August 2011. [CrossRef]
28. Bouserhal, R.E.; Falk, T.H.; Voix, J. In-ear microphone speech quality enhancement via adaptive filtering and artificial bandwidth extension. *J. Acoust. Soc. Am.* **2017**, *141*, 1321–1331. [CrossRef] [PubMed]
29. Shan, D.; Zhang, X.; Zhang, C.; Li, L. A Novel Encoder-Decoder Model via NS-LSTM Used for Bone-Conducted Speech Enhancement. *IEEE Access* **2018**, *6*, 62638–62644. [CrossRef]
30. Liu, H.P.; Tsao, Y.; Fuh, C.S. Bone-conducted speech enhancement using deep denoising autoencoder. *Speech Commun.* **2018**, *104*, 106–112. [CrossRef]
31. Hussain, T.; Tsao, Y.; Siniscalchi, S.M.; Wang, J.C.; Wang, H.M.; Liao, W.H. Bone-Conducted Speech Enhancement Using Hierarchical Extreme Learning Machine. In *Lecture Notes in Electrical Engineering*; Springer: Gateway East, Singapore, 2021; pp. 153–162. [CrossRef]
32. Zheng, C.; Yang, J.; Zhang, X.; Sun, M.; Yao, K. Improving the Spectra Recovering of Bone-Conducted Speech via Structural SIMilarity Loss Function. In Proceedings of the 2019 Asia-Pacific Signal and Information Processing Association Annual Summit and Conference (APSIPA ASC), Lanzhou, China, 18–21 November 2019. [CrossRef]
33. Li, K.; Lee, C.H. A deep neural network approach to speech bandwidth expansion. In Proceedings of the 2015 IEEE International Conference on Acoustics, Speech and Signal Processing (ICASSP), South Brisbane, QLD, Australia, 19–24 April 2015. [CrossRef]

34. Eskimez, S.E.; Koishida, K.; Duan, Z. Adversarial Training for Speech Super-Resolution. *IEEE J. Sel. Top. Signal Process.* **2019**, *13*, 347–358. [CrossRef]
35. Lim, T.Y.; Yeh, R.A.; Xu, Y.; Do, M.N.; Hasegawa-Johnson, M. Time-Frequency Networks for Audio Super-Resolution. In Proceedings of the 2018 IEEE International Conference on Acoustics, Speech and Signal Processing (ICASSP), Calgary, AB, Canada, 15–20 April 2018. [CrossRef]
36. Feng, B.; Jin, Z.; Su, J.; Finkelstein, A. Learning Bandwidth Expansion Using Perceptually-motivated Loss. In Proceedings of the ICASSP 2019—2019 IEEE International Conference on Acoustics, Speech and Signal Processing (ICASSP), Brighton, UK, 12–17 May 2019. [CrossRef]
37. Wang, M.; Wu, Z.; Kang, S.; Wu, X.; Jia, J.; Su, D.; Yu, D.; Meng, H. Speech Super-Resolution Using Parallel WaveNet. In Proceedings of the 2018 11th International Symposium on Chinese Spoken Language Processing (ISCSLP), Taipei, Taiwan, 26–29 November 2018. [CrossRef]
38. Gupta, A.; Shillingford, B.; Assael, Y.; Walters, T.C. Speech Bandwidth Extension with Wavenet. In Proceedings of the 2019 IEEE Workshop on Applications of Signal Processing to Audio and Acoustics (WASPAA), New Paltz, NY, USA, 20–23 October 2019. [CrossRef]
39. Li, Y.; Tagliasacchi, M.; Rybakov, O.; Ungureanu, V.; Roblek, D. Real-Time Speech Frequency Bandwidth Extension. In Proceedings of the ICASSP 2021—2021 IEEE International Conference on Acoustics, Speech and Signal Processing (ICASSP), Toronto, ON, Canada, 6–11 June 2021. [CrossRef]
40. Li, X.; Chebiyyam, V.; Kirchhoff, K. Speech Audio Super-Resolution for Speech Recognition. In Proceedings of the Interspeech 2019, Graz, Austria, 15–19 September 2019. [CrossRef]
41. Kumar, R.; Kumar, K.; Anand, V.; Bengio, Y.; Courville, A. NU-GAN: High resolution neural upsampling with GAN. *arXiv* **2020**, arXiv:2010.11362.
42. Liu, X.; Li, Y.; Fromm, J.; Wang, Y.; Jiang, Z.; Mariakakis, A.; Patel, S. SplitSR: An end-to-end approach to super-resolution on mobile devices. *Proc. ACM Interact. Mob. Wearable Ubiquitous Technol.* **2021**, *5*, 25. [CrossRef]
43. Lee, R.; Venieris, S.I.; Dudziak, L.; Bhattacharya, S.; Lane, N.D. MobiSR: Efficient on-device super-resolution through heterogeneous mobile processors. In Proceedings of the The 25th Annual International Conference on Mobile Computing and Networking, Los Cabos, Mexico, 21–25 October 2019. [CrossRef]
44. Abadi, M.; Barham, P.; Chen, J.; Chen, Z.; Davis, A.; Dean, J.; Devin, M.; Ghemawat, S.; Irving, G.; Isard, M.; et al. Tensorflow: A system for large-scale machine learning. In Proceedings of the 12th {USENIX} Symposium on Operating Systems Design and Implementation ({OSDI} 16), Savannah, GA, USA, 2–4 November 2016; pp. 265–283.
45. Lai, L.; Suda, N.; Chandra, V. Cmsis-nn: Efficient neural network kernels for arm cortex-m cpus. *arXiv* **2018**, arXiv:1801.06601.
46. Lin, J.; Gan, C.; Han, S. TSM: Temporal Shift Module for Efficient Video Understanding. In Proceedings of the 2019 IEEE/CVF International Conference on Computer Vision (ICCV), Seoul, Republic of Korea, 27 October–2 November 2019. [CrossRef]
47. Abbas, S.; Mosbah, M.; Zemmari, A. ITU-T Recommendation G. 114, "One way transmission time". In Proceedings of the International Conference on Dynamics in Logistics 2007 (LDIC 2007), Bremen, Germany, August 2007; Lecture Notes in Computer Sciences; Citeseer: Princeton, NJ, USA, 1996. Available online: https://www.itu.int/rec/T-REC-G.114-200305-I/en (accessed on 1 November 2022)
48. Allen, J.; Rabiner, L. A unified approach to short-time Fourier analysis and synthesis. *Proc. IEEE* **1977**, *65*, 1558–1564. [CrossRef]
49. Kingma, D.P.; Ba, J. Adam: A method for stochastic optimization. *arXiv* **2014**, arXiv:1412.6980.
50. Volkmann, J.; Stevens, S.S.; Newman, E.B. A Scale for the Measurement of the Psychological Magnitude Pitch. *J. Acoust. Soc. Am.* **1937**, *8*, 208–208. [CrossRef]
51. Corey, R.M.; Jones, U.; Singer, A.C. Acoustic effects of medical, cloth, and transparent face masks on speech signals. *J. Acoust. Soc. Am.* **2020**, *148*, 2371–2375. [CrossRef] [PubMed]
52. Ochiai, T.; Delcroix, M.; Kinoshita, K.; Ogawa, A.; Nakatani, T. A Unified Framework for Neural Speech Separation and Extraction. In Proceedings of the ICASSP 2019—2019 IEEE International Conference on Acoustics, Speech and Signal Processing (ICASSP), Brighton, UK, 12–17 May 2019. [CrossRef]
53. Drakopoulos, G.; Pikramenos, G.; Spyrou, E.; Perantonis, S. Emotion Recognition from Speech: A Survey. In Proceedings of the 15th International Conference on Web Information Systems and Technologies, Vienna, Austria, 18–20 September 2019; SCITEPRESS—Science and Technology Publications: Setubal, Portugal, 2019. [CrossRef]

Disclaimer/Publisher's Note: The statements, opinions and data contained in all publications are solely those of the individual author(s) and contributor(s) and not of MDPI and/or the editor(s). MDPI and/or the editor(s) disclaim responsibility for any injury to people or property resulting from any ideas, methods, instructions or products referred to in the content.

Deep Learning Anomaly Classification Using Multi-Attention Residual Blocks for Industrial Control Systems

Jehn-Ruey Jiang * and Yan-Ting Lin

Department of Computer Science and Information Engineering, National Central University, Taoyuan City 320317, Taiwan
* Correspondence: jrjiang@csie.ncu.edu.tw

Abstract: This paper proposes a novel method monitoring network packets to classify anomalies in industrial control systems (ICSs). The proposed method combines different mechanisms. It is flow-based as it obtains new features through aggregating packets of the same flow. It then builds a deep neural network (DNN) with multi-attention blocks for spotting core features, and with residual blocks for avoiding the gradient vanishing problem. The DNN is trained with the Ranger (RAdam + Lookahead) optimizer to prevent the training from being stuck in local minima, and with the focal loss to address the data imbalance problem. The Electra Modbus dataset is used to evaluate the performance impacts of different mechanisms on the proposed method. The proposed method is compared with related methods in terms of the precision, recall, and F1-score to show its superiority.

Keywords: anomaly detection; anomaly classification; industrial control system; deep learning; deep neural network; multi-attention block; residual block

1. Introduction

The industrial control system (ICS) integrates information technology (IT) and operational technology (OT) to monitor, control, and manage network-interconnected devices in large-scale industrial production systems or critical infrastructures, such as manufacturing factories, power plants, waterworks, oil refineries, gas pipelines, and public transportation systems [1]. Once cyber attackers intrude into an ICS to launch attacks, its performance degrades and some functions may fail, leading to huge losses. For example, a Taiwan chip-maker was attacked by WannaCrypt malware in 2018. Consequently, many chip-fabrication factories were shut down, leading to a loss of about USD 256 million [2]. For another example, an American oil pipeline system was halted by a ransomware cyberattack, and consequently, a ransom of USD 4.4 million was paid to restore its operations [3].

Anomalies occur before or during major attacks are launched. It is therefore helpful to develop methods to detect and classify anomalies associated with cyberattacks. Alerts are issued once anomalies are detected and/or classified. Traditional anomaly detection and classification methods cannot be directly applied to ICS applications due to differences in protocols and attack types between traditional networks and ICS networks. Several studies proposed ICS anomaly detection and classification methods that inspect network packets of the Modbus and S7Comm protocols. Gomez et al. [4] proposed supervised and unsupervised machine learning methods to detect ICS anomalies. Ning et al. [5] proposed an anomaly detection method based on the generative adversarial network (GAN) model [6] and the deep neural network (DNN) model.

Jiang and Chen [1] proposed an ICS anomaly detection method (abbreviated as the JC-AD method in this paper) and an ICS anomaly classification method (abbreviated as the JC-AC method in this paper). The two methods first utilize the denoising autoencoder (DAE) [7] to reduce data noise and extract core features from packets. Then, the JC-AD method employs the synthetic minority oversampling technique (SMOTE) [8] and the Tomek link (T-Link) [9] mechanism to oversample and undersample data for dealing with imbalance packets, where the majority of class samples (i.e., normal packets) significantly outnumber the minority class samples (i.e., anomalous packets). The SMOTE and the T-LINK mechanisms are for the binary-class samples, so they are not employed by the JC-AC method that addresses multi-class samples. Finally, both methods use extreme gradient boosting (XGBoost) [10] based on the ensemble learning concept to avoid overfitting to achieve good performance.

Among all the above-mentioned ICS anomaly detection methods, the JC-AD method [1] has perfect (i.e., 100%) accuracy, precision, recall, and F1-score. The JC-AC method [1] is the sole ICS anomaly classification method; it has almost perfect (i.e., nearly 100%) precision, recall, and F1-score.

This paper proposes an ICS anomaly classification method integrating difference mechanisms to improve the performance of the JC-AC method. The proposed method is flow-based; that is, it investigates the flow of packets instead of a single packet for classifying anomalies. It builds a DNN with multi-attention blocks [11] for spotting core features, and with residual blocks [12] for avoiding the gradient vanishing problem. The DNN is trained with the Ranger [13] (i.e., RAdam [14] + Lookahead [15]) as the optimizer to prevent the training from being stuck in local minima, and with the focal loss [16] to address the data imbalance problem. The proposed method can be used in conjunction with the JC-AD method. Specifically, it can be used for better anomaly classification after the JC-AD method perfectly detects ICS anomalies. Moreover, the proposed method can also be used for detecting ICS anomalies when viewed as a binary-class (i.e., normal-anomalous) classifier. As will be shown, it has comparably good performance in detecting ICS anomalies.

The Electra Modbus dataset [17] reported in [4] is employed to evaluate the proposed method's performance. As the proposed method integrates several mechanisms, the evaluation also assesses the performance impact of not using a single mechanism. Furthermore, the evaluation results are compared with those of the JC-AC method to show that the proposed method indeed improves the JC-AC method in terms of the precision, recall, and F1-score of anomaly classification. Notably, the proposed method is shown to have comparably good ICS anomaly detection performance when compared with related methods.

The contribution of this paper is three-fold. First, it proposes a novel flow-based method integrating the mechanisms of the muti-attention block, residual block, Ranger optimizer, and focal loss to construct a DNN for monitoring ICS network packets to classify anomalies. Second, extensive experiments using the Electra Modbus dataset are conducted to evaluate the performance impacts of different mechanisms. Third, the performance of the proposed method is compared with those of related methods to show its superiority.

The rest of this paper is organized as follows. Section 2 describes background knowledge. Section 3 elaborates the proposed method. Performance evaluation and comparisons of the proposed method, with related methods, are shown in Section 4. Finally, Section 5 concludes the paper.

2. Background

2.1. Multi-Attention Block

The attention mechanism [11] utilizes the attention block to assign different weights to each part of the DNN input, and extract more critical information for achieving better performance. It is widely used in applications such as machine translation, voice recognition, and image recognition, etc. The structure of the attention block (or the scaled dot-product attention block) is shown in Figure 1a. Based on the input vector, the attention mechanism obtains query vector Q through the Query matrix, key vector K through the Key matrix, and value vector V through the Value matrix. The attention score can be obtained by multiplying Q and K, then scaling and normalizing the product by the SoftMax function to obtain the attention weight, which is in turn multiplied by V to produce the output. Unlike recurrence-based models, such as the recurrent neural network (RNN), which have to sequentially check each input vector one by one, the attention mechanism can check all input vectors simultaneously to determine which input vector has a higher attention score to be paid more attention to. It thus has better performance than its counterparts. If the attention mechanism considers many queries, it is called a multi-attention mechanism and can be realized by multi-attention blocks, as shown in Figure 1b. It can be regarded as running through the attention mechanism multiple times (say, h times) in parallel. Each running of the attention mechanism can pay attention to different parts of input vectors to have an independent output result. All the attention output results are subsequently concatenated and linearly transformed to be the final output.

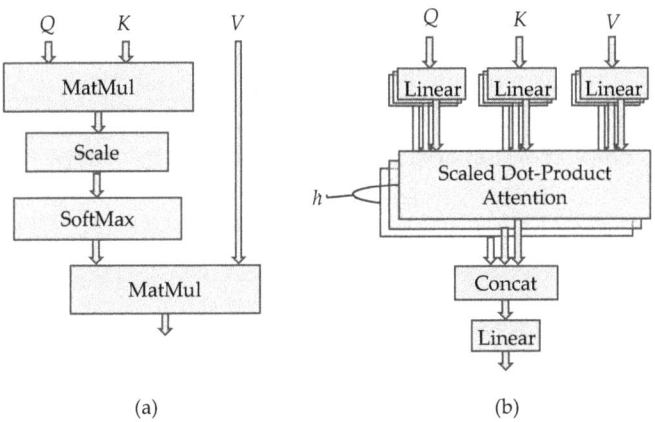

Figure 1. (**a**) The scaled dot-product attention block, and (**b**) the multi-attention block.

2.2. Residual Block

Increasing the depth of a DNN usually improves its performance. However, the depth increase causes the gradient vanishing problem so that the gradient approaches zero; it also causes the gradient exploding problem so that the gradient becomes excessive. The two problems may make the DNN weight update insensitive to output changes; thus, it is sometimes difficult for the DNN training error to converge. Although batch normalization can mitigate the two problems, there still exists the degradation problem that deeper DNNs may have worse performance than shallower DNNs. The residual block (ResBlock) [12], whose structure is shown in Figure 2, can be used to alleviate all the problems, as described below. The ResBlock has the normal dense layer with the ReLU (Rectified Linear Unit) or another activation function. In particular, it has the shortcut connection of identity mapping. With the shortcut, the ResBlock can learn the residual (i.e., difference) between the input and the output of the block. It can thus focus on the residual, which is more significant. As such, it is more sensitive to output changes and can mitigate the gradient vanishing, and the gradient exploding problems, which in turn can alleviate the degradation problem.

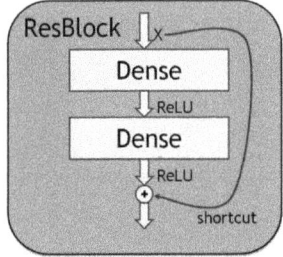

Figure 2. The residual block structure.

2.3. RAdam

RAdam [14] stands for Rectified Adam; it is simply an Adam optimizer [18] with a warmup scheme. Note that Adam stands for the Adaptive Moment Estimation, a well-known gradient descent optimization scheme in the DNN error-backpropagation process. On the one hand, in the early stage of training a DNN with the Adam optimizer, the variance of training errors of all samples is relatively small. However, after several epochs, the variance of training errors of all samples grows large. The reasons are as follows. If the warmup scheme is not employed and a large learning rate is used in the early stage of training a DNN model, the model becomes overfitting to the few samples ever seen. Thus, the training error is quite large for unseen samples. On the other hand, the Adam optimizer with the warmup scheme can reduce the variance of training errors of all data samples, as the learning rate is small in early stages and then grows in later stages of training the DNN. This can prevent the training from getting stuck in local minima for getting better performance.

Figure 3 shows the absolute gradient histogram during training a DNN using Adam without warmup for machine translation on the IWSLT2014 German-English (De-En) dataset [14]. In the histogram, the y-axis is the iteration (epoch), and the x-axis is the absolute gradient value in a logarithmic scale, whose height stands for the frequency. It can be observed from the histogram that using Adam without warmup makes the gradient distribution distorted to have mass centers of relatively small values in the first 10 iterations, which indicates the training may fast converge and be trapped in local minima after the first few iterations. Figure 4 shows the absolute gradient histogram during training a DNN using RAdam (i.e., Adam with warmup) for machine translation on the IWSLT2014 De-En dataset [14]. It can be observed from the histogram that the gradient distribution is not distorted after the first few iterations. This can avoid the bad situation that the training fast converges and is trapped in local minima.

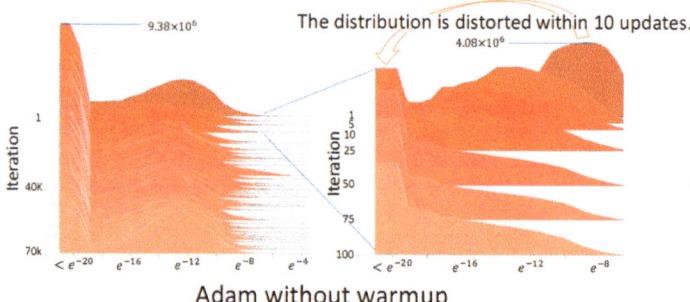

Figure 3. The absolute gradient histogram during training a DNN using Adam without warmup for machine translation on the IWSLT2014 De-En dataset [14].

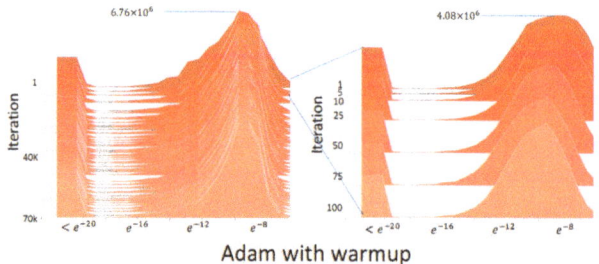

Figure 4. The absolute gradient histogram during training a DNN using RAdam (i.e., Adam with warmup) for machine translation on the IWSLT2014 De-En dataset [14].

2.4. Lookahead

The core concept of Lookahead [15] is to prepare two sets of weights for the neural network, one set of fast weights and one set of slow weights. When the fast weights are updated k times, the slow weights are updated to half the extent of the fast weights. In this way, even if the fast weights get stuck in local minima, the slow weights can still escape local minima to achieve better performance. Note that the Lookahead mechanism and the above-mentioned RAdam mechanism are combined as the Ranger optimizer [13] for training DNNs for improving performance.

Figure 5 [15] is the visualization of Lookahead effects with $k = 10$ for training the ResNet-32 model [12] using stochastic gradient descent (SGD) as an image classifier on the Canadian Institute For Advanced Research-100 (CIFAR-100) dataset [19]. The figure shows the test accuracy surface of the model using fast weights indicated by the blue-dashed path, and slow weights indicated by the purple-line path. It can be observed from Figure 5 that Lookahead can quickly progress to the global minima than SGD.

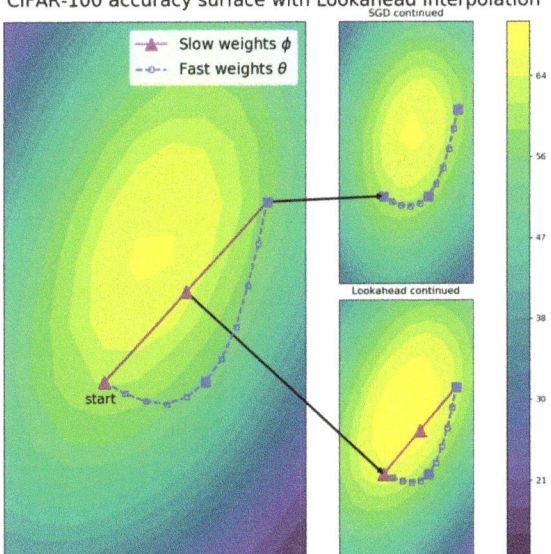

Figure 5. The visualization of Lookahead effects with $k = 10$ through a ResNet-32 test accuracy surface at epoch 100 on the CIFAR-100 dataset [15].

2.5. Focal Loss

The core concept of the focal loss [16] is to set the loss of correctly classified samples (i.e., simple samples) to be small, and to set the loss of misclassified samples (i.e., difficult samples) to be large. Equation (1) is for the simplified cross entropy (CE), where p_t is the probability to correctly predict the input sample to be positive. Equation (2) is for the focal loss (FL), where α_t is the parameter related to data imbalance, and γ is the parameter related to the difficulty of sample detection and classification.

$$\text{CE}(p_t) = -\log p_t \tag{1}$$

$$\text{FL}(p_t) = -\alpha_t (1 - p_t)^\gamma \log p_t \tag{2}$$

Figure 6 shows the CE and the FL with data imbalance parameter $\alpha_t = 1$ (for not considering data imbalance) and $\gamma = 0, 0.5, 1, 2,$ and 5 [16]. Note that the CE is actually the FL with $\gamma = 0$. The y-axis is the loss, and the x-axis is p_t, the probability to correctly predict the input sample to be positive. It can be observed from Figure 6 that the well-classified samples or examples (i.e., those with p_t larger than 0.5 and even close to 1) are associated with the FL that fast approaches 0 when p_t grows. The FL approaches 0 faster than the CE. It can also be observed that the FL sets the loss of well-classified examples to be small, and sets the loss of misclassified examples to be large.

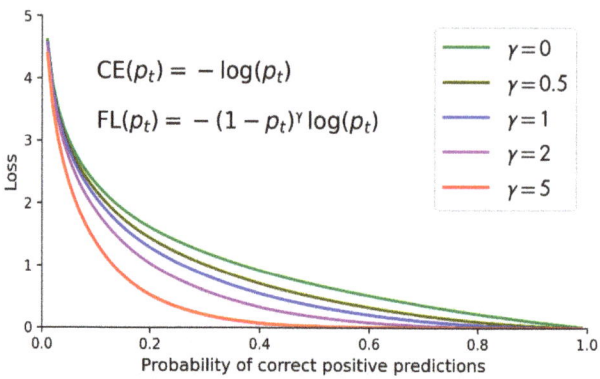

Figure 6. The comparisons of the CE and the FL with different values of γ.

3. Proposed Method

The workflow of the proposed method mainly consists of three steps: data preprocessing, model building, and model training, as shown in Figure 7. Each step is elaborated in a subsection below.

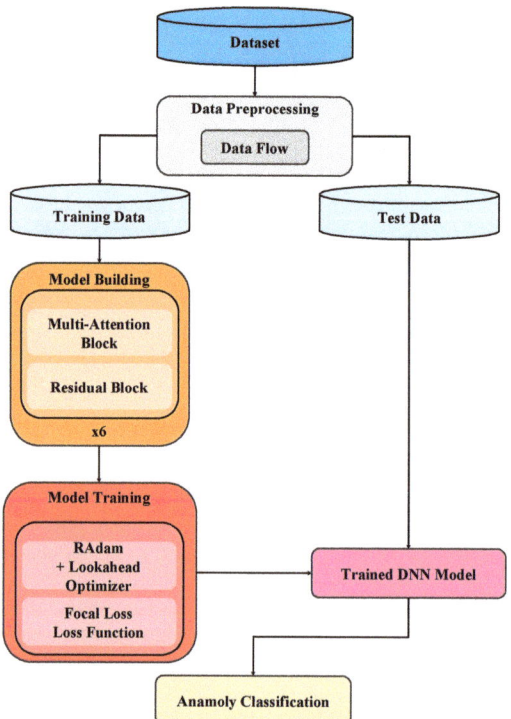

Figure 7. The workflow of the proposed method.

3.1. Data Preprocessing

The data preprocessing contains four sub-steps: label encoding, normalization, data flow processing, and data splitting. The four sub-steps are described below.

1. Label encoding:

In this sub-step, non-numeric features (e.g., categorial features) are converted into integers. If the number of categories of a feature is N, then the feature is converted into an integer in the range of $[0, N-1]$. Using integers to represent non-numeric features may cause the problem that the DNN model mistakes the order relations among integers as the precedence relations among different categories, which in turn may influence the model performance. The one-hot encoding can avoid the problem by inserting N new columns and by assigning 1 in one column, and 0 in the other columns for a category. However, one-hot encoding needs many extra columns, whereas label encoding does not. Since there are likely many categories in the ICS anomaly classification setting, label encoding is adopted to convert non-numerical features.

2. Normalization:

Numerical features may have different ranges, which influence the training of neural networks. The proposed method adopts the min-max normalization to scale values of a feature into the range of $[0, 1]$, while the distribution of feature values remains the same. With normalization, features are transformed to be within a common scale without distorting their distributions, which in turn can improve DNN model training stability and model performance.

3. Data flow processing:

If two packets have an identical pair of the source IP address and the destination IP address, then they are considered to belong to the same flow. Packets of the same flow are sorted by timestamps. As shown in Figure 8, new features are derived form a sliding window of s (say, $s = 4$) consecutive packets of the same flow, and the label of the sth packet is taken as the new label associated with the new features. In this way, the new flow-based data are derived. When an attacker launches an attack, he/she usually sends multiple malicious data packets to the same target device, causing anomalies. Therefore, finding related packets with the same source and destination IP addresses through the data flow processing is very helpful for anomaly classification.

time	sip	dip	...	madd	data	label
1	0	0	...			New Features
2	0	0	...			
3	0	0	...			
4	0	0	...			New Label
5	0	0	...			
6	0	0	...			
7	0	0	...			
8	0	0	...			

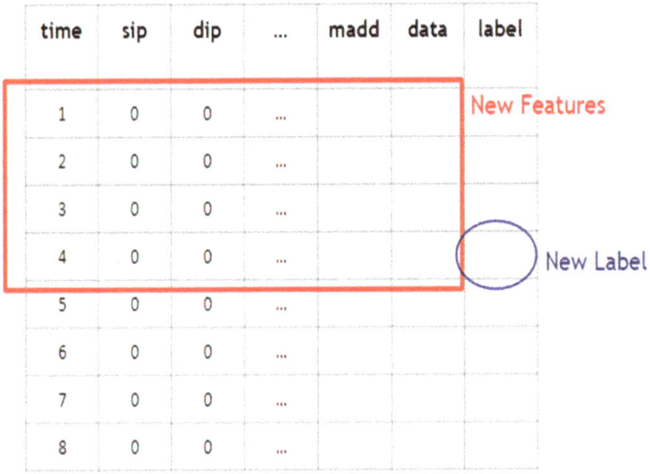

Figure 8. New features and a label are derived from a sliding window of $s = 4$ consecutive packets of a flow.

4. Data splitting:

New flow-based features and associated labels of data are split into the training dataset, the validation dataset, and the test dataset according to the ratios of 0.6, 0.2, and 0.2. The datasets are used to train, validate, and test data. Specifically, the training dataset is used to train a DNN model to have a small error, and the validation dataset is applied to the trained model to check (or validate) if the error is still small to prevent the model from overfitting to the training dataset. The above-mentioned actions constitute an iteration or an epoch of the training process. The training process stops when the error keeps decreasing and stays very small or when the maximum number of iterations is reached. Afterwards, the best model with the smallest error ever encountered is output as the final model. At last, the test dataset is applied to the final DNN model for assessing the model performance.

3.2. Model Building

The proposed method builds the DNN that combines multiple attention blocks and residual blocks. Figure 9a shows the DNN without multi-attention blocks or residual blocks, whereas Figure 9b shows the DNN with multi-attention blocks and residual blocks adopted by the proposed method. In the DNN shown in Figure 9b, the first layer is the input layer. Then, there are t (say $t = 6$) copies of the dense layer, the multi-attention block layer, and the residual block layer, with each layer having eight neurons. The flatten layer and the dense layer then follow. The last is the output layer. In total, the DNN has 51 layers, with the last layer using the SoftMax, and the other layers using Swish [20] as the activation function. It is shown that the Swish function can help mitigate the gradient vanishing problem [20]. The He normal initializer [21] is employed to initialize neural weights. The initializer draws samples from a truncated normal distribution and is shown to have good performance.

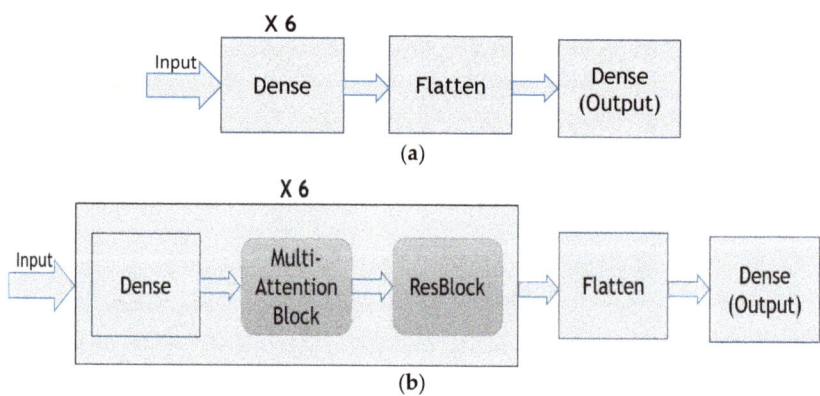

Figure 9. (a) The DNN without multi-attention blocks or residual blocks, (b) the DNN with multi-attention blocks and residual blocks adopted by the proposed method.

3.3. Model Training

The parameters for training the DNN model are described as follows. The focal loss is used as the loss function with the parameter α set as 0.25, and the parameter γ set as 2, which are suggested by [16]. Ranger is used as the optimizer, and the minimum and the maximum learning rates are set as 0.000001 and 0.001, respectively. The early stopping scheme with the parameter patience set as 32 is employed to early stop the model training when the training loss and the validation loss no longer decrease. This can prevent the DNN model from overfitting.

4. Performance Evaluation

The Electra Modbus dataset [17] is applied for evaluating the performance of the proposed method in terms of some metrics. This section first introduces the dataset,

4.1. The Electra Modbus Dataset

Electra Modbus dataset [17] reported in [4] collects Modbus TCP protocol packet data generated during normal and abnormal operation of an electric traction substation used in the railway industry. It uses an ICS testbed to gather data. The testbed is composed of a supervisory control and data acquisition (SCADA) system, a firewall, a switch, a programmable logic controller (PLC) master, four PLC slaves, some ICS devices, and a special device that can launch man-in-the-middle (MitM) attacks for generating anomalous network packets. The special device launches attacks, records the features of all packets, and labels packets as normal or anomalous for 12 hours for generating the dataset. Recorded features of the Electra Modbus dataset are the timestamp, source MAC address, destination MAC address, source IP address, destination IP address, request, function code, error, memory address, and data sent between the PLC master and slaves, as shown in Table 1.

Table 1. Descriptions of the Electra Modbus dataset [4].

Feature	Description	Data Type
time	Timestamp	String
smac	Source MAC address	String
dmac	Destination MAC address	String
sip	Source IP address	String
dip	Destination IP address	String
request	Indicates whether the packet is a request (packet from master to slave)	Boolean
fc	Function code	Integer
error	Indicates whether there has been an error in reading/writing operation	Boolean
madd	Memory address to perform read/write operation	Integer
data	In the case of a read operation, it indicates the data that the slave sends to the master. In the case of a write operation, it indicates the data that the master sends back to the slave	Integer
label	Label for attacks and normal samples	String

Packets generated during the period of normal operations are labelled as normal, whereas packets generated during the period of attacks are regarded as anomalous. There are in total 5.2% of anomalous packets associated with three categories of attacks, namely, reconnaissance, false data injection, and replay attacks. As shown in Table 2, the three categories of attacks are further classified into seven classes, as described below. The "function code recognition attack" is launched by generating malicious packets to scan all possible function codes of the attacked PLC. Attackers inject fake packets for performing the "read attack" or "write attack" on the PLC. The "response modification attack" and the "force error in response attack" are launched by modifying the response of a slave device. The "command modification attack" is launched by modifying command packets of a master device. The "replay attack" is launched by retransmitting packets ever sent by the master or slave devices. The ratios of various attack classes are also shown in Table 2.

Table 2. Attack classes of the Electra Modbus dataset [4].

Classes	Percentage of Samples
Normal	94.8%
Function code recognition attack	0.19%
Response modification attack	0.1%
Force error in response attack	0.007%
Read attack	4.83%
Write attack	0.06%
Replay attack	0.006%

4.2. Performance Metrics

The performance of the proposed method is evaluated in the following metrics: the precision, recall, and F1-score, as defined in the following equations. Note that in the equations, TP (True Positive) stands for the number of classifying anomalous packets as anomalous ones; TN (True Negative) stands for the number of classifying normal packets as normal ones; FP (False Positive) stands for the number of classifying normal packets as anomalous one, and FN (False Negative) stands for the number of classifying anomalous packets as normal ones.

This paper evaluates the performance of the proposed method in terms of only the precision, recall, and F1-score. This is because many existing ICS anomaly classification methods, such as those proposed in [1], also adopt the three metrics to evaluate their performance. In order to compare with related methods properly, the proposed method also adopts the three metrics for its performance evaluation.

$$\text{Precision} = \frac{\text{TP}}{\text{TP} + \text{FP}} \qquad (3)$$

$$\text{Recall} = \frac{\text{TP}}{\text{TP} + \text{FN}} \qquad (4)$$

$$\text{F1} - \text{score} = 2 \times \frac{\text{Precision} \times \text{Recall}}{\text{Precision} + \text{Recall}} \qquad (5)$$

4.3. Performance Evaluation and Comparison

The proposed method utilizes many mechanisms. First, it uses the flow-based data investigation mechanism. It also uses the DNN with the mechanisms of the multi-attention block, the residual block, the Ranger optimizer, and the focal loss. The performance impact of not adopting a single mechanism is assessed. Figure 10 shows the performance evaluation of six different cases of mechanism combinations. In case 1, flow-based data investigation mechanism is not used; instead, per-packet investigation mechanism is used. Case 2 omits the muti attention block mechanism, whereas case 3 omits the residual block mechanism. Case 4 uses the Adam optimizer to replace the Ranger optimizer. The cross-entropy, instead of the focal loss, is used in case 5. All mechanisms are used in case 6. It can be observed that the combination of all mechanisms has the best performance. Furthermore, the residual block mechanism has the most impact on performance, as not using it leads to the worst performance. As to other mechanisms, they have less and similar impacts on performance than the residual block mechanism.

The confusion matrix of the proposed ICS anomaly classification method using all mechanisms is shown in Figure 11. It can be observed that most anomalies can be classified correctly. However, some anomalies, especially those associated with the replay attack and the read attack, are misclassified. This is probably due to the fact that most replay attacks behave similarly to read attacks; that is to say, most replay attacks read data illegally [4].

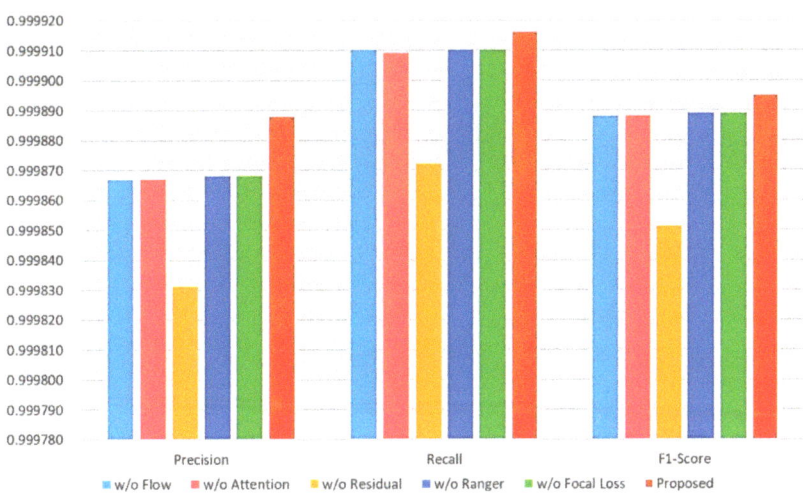

Figure 10. Performance impact of various schemes adopted by the proposed method.

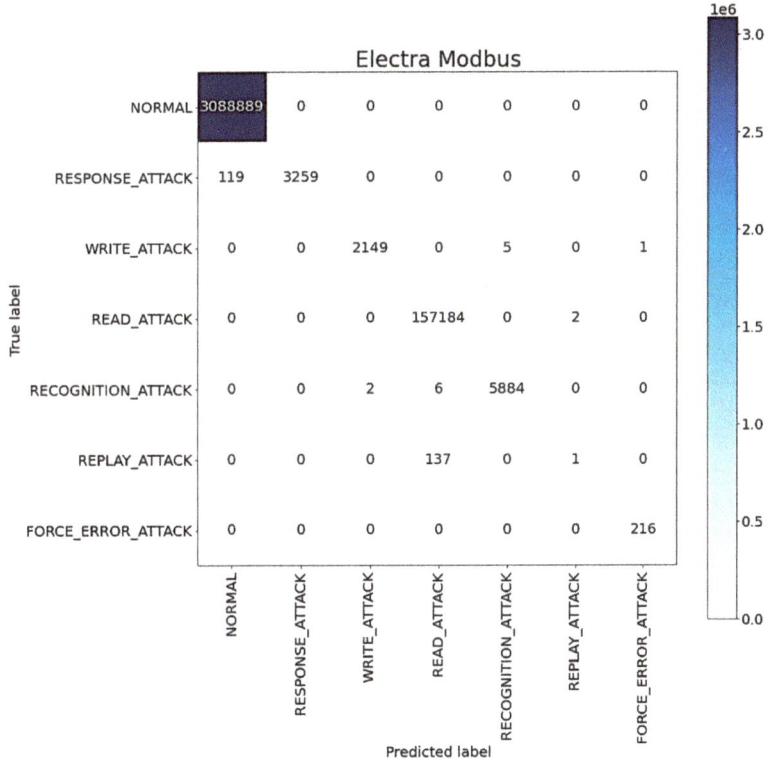

Figure 11. The confusion matrix of the proposed ICS anomaly classification method.

The performance evaluation results of the proposed method are compared with those of the related ICS anomaly classification methods for the multi-class case, as shown in Figure 12 and Table 3. Based on Figure 12 and Table 3, it can be observed that the proposed method is better than the JC-AC method in terms of the precision, recall, and F1-score.

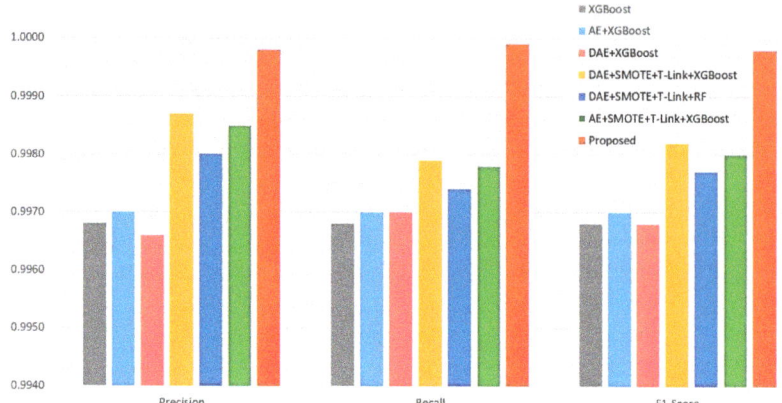

Figure 12. Performance comparisons of the proposed method and related ICS anomaly classification methods for the multi-class case.

Table 3. Performance comparisons of the proposed method and the JC-AC method for the multi-class case.

Method	Precision	Recall	F1-Score
JC-AC [1]	0.9985	0.9978	0.9980
Proposed Method	0.9998	0.9999	0.9998

As mentioned earlier, the proposed method can also be used for detecting ICS anomalies when viewed as a binary-class (i.e., normal-anomalous) classifier. For the binary class case, the precision, recall, and F1-score of the proposed method are all 0.9999, which is quite high. Table 4 shows the comparison results of the proposed method with related ICS anomaly detection methods for the binary-class case. The methods for comparison are the JC-AD method [1] and the methods based on the support vector machine (SVM) [4], the one class support vector machine (OCSVM) [4], the random forest (RF) [4], the isolation forest (IF) [4], the DNN [4], and the integration of GAN and DNN (GAN + DNN) [5]. The proposed method is inferior to the JC-AD method, which has the perfect performance; however, it almost outperforms all other related methods in all performance metrics like the precision, recall, and F1-score.

Table 4. Performance comparisons of the proposed method and related ICS anomaly detection methods for the binary-class (i.e., normal-anomalous) case.

Method	Precision	Recall	F1-Score
JC-AD [1]	1.0000	1.0000	1.0000
SVM [4]	0.9756	1.0000	0.9876
OCSVM [4]	0.9862	0.9856	0.9859
RF [4]	0.9877	0.9871	0.9874
IF [4]	0.8739	1.0000	0.9327
DNN [4]	0.9692	1.0000	0.9843
GAN + DNN [5]	-	0.98	-
Proposed Method	0.9999	0.9999	0.9999

5. Conclusions

This paper proposes a flow-based ICS anomaly classification method. The method first obtains new features based on the flow of packets. It then employs multi-attention blocks for spotting core features, and uses residual blocks for alleviating the gradient vanishing problem. Furthermore, it adopts the Ranger optimizer to avoid the overfitting problem and to accelerate the convergence of the DNN training. The focal loss is finally used as the loss function to deal with the data imbalance problem.

The Electra Modbus dataset is used to evaluate the performance of the proposed method. It is observed that the residual block has the most impact on the performance of the proposed method. The proposed method is shown to outperform the JC-AC method in terms of the precision, recall, and F1-score for ICS anomaly classification. When viewed as a binary-class (i.e., normal-anomalous) classifier, the proposed method can also be used for detecting ICS anomalies. For the binary class case, the proposed method also has comparably high performance of 0.9999 in the precision, recall, and F1-score metrics.

In the future, we plan to apply the proposed method to other ICS environments, such as those under different types of attacks, such as distributed denial-of-service (DDoS), and those using various protocols such as S7 Communication (S7Comm), and Distributed Network Protocol 3 (DPN3), etc. Furthermore, we also plan to employ new techniques, such as graph neural networks, to improve the ICS anomaly detection and classification performance.

Author Contributions: Conceptualization, J.-R.J. and Y.-T.L.; funding acquisition, J.-R.J.; investigation, J.-R.J. and Y.-T.L.; methodology, J.-R.J. and Y.-T.L.; software, Y.-T.L.; supervision, J.-R.J.; validation, J.-R.J. and Y.-T.L.; writing—original draft, J.-R.J. and Y.-T.L.; writing—review and editing, J.-R.J. All authors have read and agreed to the published version of the manuscript.

Funding: This research was funded by the National Science and Technology Council (NSTC), Taiwan, under the grant number 109-2622-E-008-028.

Institutional Review Board Statement: Not applicable.

Informed Consent Statement: Not applicable.

Data Availability Statement: Not applicable.

Acknowledgments: We would like to thank the anonymous reviewers for their valuable comments.

Conflicts of Interest: The authors declare no conflict of interest.

References

1. Jiang, J.-R.; Chen, Y.-T. Industrial control system anomaly detection and classification based on network traffic. *IEEE Access* **2022**, *10*, 41874–41888. [CrossRef]
2. Zhou, C.; Hu, B.; Shi, Y.; Tian, Y.-C.; Li, X.; Zhao, Y. A unified architectural approach for cyberattack-resilient industrial control systems. *Proc. IEEE* **2020**, *109*, 517–541. [CrossRef]
3. Eaton, C.; Volz, D. U.S. Pipeline Cyberattack Forces Closure. Wall Street J. Available online: https://www.wsj.com/articles/cyberattack-forces-closure-of-largest-u-s-refined-fuel-pipeline-11620479737 (accessed on 15 November 2022).
4. Gómez, Á.L.P.; Maimó, L.F.; Celdrán, A.H.; Clemente, F.J.G.; Sarmiento, C.C.; Masa, C.J.D.C.; Nistal, R.M. On the generation of anomaly detection datasets in industrial control systems. *IEEE Access* **2019**, *7*, 177460–177473. [CrossRef]
5. Ning, B.; Qiu, S.; Zhao, T.; Li, Y. Power IoT attack samples generation and detection using generative adversarial networks. In Proceedings of the 2020 IEEE 4th Conference on Energy Internet and Energy System Integration (EI2), Wuhan, China, 30 October–1 November 2020; pp. 3721–3724.
6. Goodfellow, I.; Pouget-Abadie, J.; Mirza, M.; Xu, B.; Warde-Farley, D.; Ozair, S.; Courville, A.; Bengio, Y. Generative adversarial networks. *Commun. ACM* **2020**, *63*, 139–144. [CrossRef]
7. Vincent, P.; Larochelle, H.; Bengio, Y.; Manzagol, P.-A. Extracting and composing robust features with denoising autoencoders. In Proceedings of the 25th International Conference on Machine Learning, Helsinki, Finland, 5–9 July 2008; pp. 1096–1103.
8. Chawla, N.V.; Bowyer, K.W.; Hall, L.O.; Kegelmeyer, W.P. SMOTE: Synthetic minority over-sampling technique. *J. Artif. Intell. Res.* **2002**, *16*, 321–357. [CrossRef]
9. Tomek, I. Two modifications of CNN. *IEEE Trans. Syst. Man Cybern.* **1976**, *6*, 769–772.
10. Chen, T.; Guestrin, C. XGBoost: A scalable tree boosting system. In Proceedings of the 22nd Acm Sigkdd International Conference on Knowledge Discovery and Data Mining, San Francisco, CA, USA, 13–17 August 2016; pp. 785–794.
11. Vaswani, A.; Shazeer, N.; Parmar, N.; Uszkoreit, J.; Jones, L.; Gomez, A.N.; Kaiser, L.; Polosukhin, I. Attention is all you need. *Adv. Neural Inf. Process. Syst.* **2017**, *30*, 5998–6008.
12. He, K.; Zhang, X.; Ren, S.; Sun, J. Deep residual learning for image recognition. In Proceedings of the IEEE Conference on Computer Vision and Pattern Recognition, Las Vegas, NV, USA, 27–30 June 2016; pp. 770–778.

13. Wright, L. New Deep Learning Optimizer, Ranger: Synergistic Combination of RAdam + LookAhead for the Best of Both. Available online: https://lessw.medium.com/new-deep-learning-optimizer-ranger-synergistic-combination-of-radam-lookahead-for-the-best-of-2dc83f79a48d (accessed on 15 November 2022).
14. Liu, L.; Jiang, H.; He, P.; Chen, W.; Liu, X.; Gao, J.; Han, J. On the variance of the adaptive learning rate and beyond. *arXiv* **2017**, arXiv:1908.03265.
15. Zhang, M.; Lucas, J.; Ba, J.; Hinton, G.E. Lookahead optimizer: K steps forward, 1 step back. In Proceedings of the 33rd Conference on Neural Information Processing Systems (NeurIPS 2019), Vancouver, BC, Canada, 8–14 December 2019.
16. Lin, T.Y.; Goyal, P.; Girshick, R.; He, K.; Dollár, P. Focal loss for dense object detection. In Proceedings of the IEEE International Conference on Computer Vision, Venice, Italy, 22–29 October 2017; pp. 2980–2988.
17. Dataset for Cybersecurity Research in Industrial Control Systems. Available online: http://perception.inf.um.es/ICS-datasets/ (accessed on 15 November 2022).
18. Kingma, D.P.; Ba, J. Adam: A method for stochastic optimization. *arXiv* **2014**, arXiv:1412.6980.
19. Krizhevsky, A. *Learning Multiple Layers of Features from Tiny Images*; Technical Report; University of Toronto: Toronto, ON, Canada, 2009.
20. Ramachandran, P.; Zoph, B.; Le, Q.V. Searching for activation functions. *arXiv* **2017**, arXiv:1710.05941.
21. He, K.; Zhang, X.; Ren, S.; Sun, J. Delving deep into rectifiers: Surpassing human-level performance on imagenet classification. In Proceedings of the IEEE International Conference on Computer Vision, Santiago, Chile, 7–13 December 2015; pp. 1026–1034.

Article

Convolutional Long-Short Term Memory Network with Multi-Head Attention Mechanism for Traffic Flow Prediction

Yupeng Wei * and Hongrui Liu

Department of Industrial and Systems Engineering, San Jose State University, San Jose, CA 95192, USA
* Correspondence: yupeng.wei@sjsu.edu

Abstract: Accurate predictive modeling of traffic flow is critically important as it allows transportation users to make wise decisions to circumvent traffic congestion regions. The advanced development of sensing technology makes big data more affordable and accessible, meaning that data-driven methods have been increasingly adopted for traffic flow prediction. Although numerous data-driven methods have been introduced for traffic flow predictions, existing data-driven methods cannot consider the correlation of the extracted high-dimensional features and cannot use the most relevant part of the traffic flow data to make predictions. To address these issues, this work proposes a decoder convolutional LSTM network, where the convolutional operation is used to consider the correlation of the high-dimensional features, and the LSTM network is used to consider the temporal correlation of traffic flow data. Moreover, the multi-head attention mechanism is introduced to use the most relevant portion of the traffic data to make predictions so that the prediction performance can be improved. A traffic flow dataset collected from the Caltrans Performance Measurement System (PeMS) database is used to demonstrate the effectiveness of the proposed method.

Keywords: traffic flow prediction; deep learning; convolutional LSTM; attention mechanism

Citation: Wei, Y.; Liu, H. Convolutional Long-Short Term Memory Network with Multi-Head Attention Mechanism for Traffic Flow Prediction. *Sensors* **2022**, *22*, 7994. https://doi.org/10.3390/s22207994

Academic Editors: Shyan-Ming Yuan, Zeng-Wei Hong and Wai-Khuen Cheng

Received: 9 October 2022
Accepted: 18 October 2022
Published: 20 October 2022

Copyright: © 2022 by the authors. Licensee MDPI, Basel, Switzerland. This article is an open access article distributed under the terms and conditions of the Creative Commons Attribution (CC BY) license (https://creativecommons.org/licenses/by/4.0/).

1. Introduction

Traffic congestion results in reduced efficiency of transportation infrastructure, increased traveling time, and a waste of energy fuel [1–3]. According to a report by Nationwide, 1.9 billion gallons of fuel are wasted every year as a result of traffic congestion [4]. Traffic congestion could be induced by numerous factors, such as bottlenecks, traffic accidents, and severe weather conditions. To address the issue of traffic congestion, traffic flow prediction has gained much attention in the recent decade. Accurate predictive modeling of traffic flow is critically important as it allows transportation users to make wise decisions to circumvent traffic congestion regions [5]. Therefore, commuter and shipment activities could be effectively scheduled to increase moving efficiency. Moreover, accurate predictive modeling of traffic flow can also assist in reducing carbon emissions and traffic incident possibilities.

The advanced development of sensing technology makes big data more affordable and accessible, and thus, data-driven methods have been increasingly adopted for the predictive modeling of traffic flow. Data-driven methods can be classified into two categories: machine learning methods and deep learning methods [6–10]. In comparison with machine learning methods, deep learning methods have gained more attention from both academia and industry in traffic flow predictions due to their extraordinary prediction fidelity and robustness. Among these deep learning methods, artificial neural networks (ANNs) and autoencoder-based methods have been widely used for traffic flow predictions as these methods are capable of decomposing the original traffic flow data into features located at a higher dimensional feature space, and these high-dimensional features can reveal the latent information in the traffic flow data. However, there are two primary issues for ANNs and autoencoders: (1) they can not take the temporal correlation of traffic flow data into account;

(2) they can not consider the correlation of the extracted high-dimensional features. To consider the temporal correlation of traffic flow data, deep learning methods with recurrent characteristics are adopted, such as long short-term memory (LSTM), recurrent neural network (RNN), and gated recurrent unit (GRU). While these deep learning methods with recurrent characteristics are promising, they are not able to use the most relevant part of the traffic flow data to make predictions, which leads to a higher prediction time and a worse prediction accuracy. To address these issues, this work introduces a novel deep learning-based framework to consider the temporal correlation of traffic flow data, the correlation of the extracted high-dimensional features, and the most relevant part of the traffic flow data to make predictions in a unified manner. More specifically, a decoder network is firstly proposed to decompose the traffic flow data into high-dimensional features. Second, a convolutional LSTM network is introduced to simultaneously consider the correlation of the decomposed high-dimensional features and the temporal correlation of traffic flow data, where the convolutional operation is used to consider the correlation of the high-dimensional features, and the LSTM network is used to consider the temporal correlation of traffic flow data. Next, the multi-head attention mechanism is introduced to use the most relevant portion of the traffic data to make predictions so that the prediction performance can be improved. The primary contribution of this work can be summarized as follows:

- A decoder network is introduced to decompose the original traffic flow data into features located at a higher-dimensional feature space.
- A convolutional LSTM network is introduced to consider the correlation of the high dimensional features and the temporal correlation of traffic flow data.
- A multi-head attention mechanism is introduced to use the most relevant portion of the traffic data to make predictions so that the prediction performance can be improved.

The remainder of this paper is organized as follows. Section 2 reviews data-driven methods reported in the literature for traffic flow predictions. Section 3 introduces the proposed deep learning model. Section 4 demonstrates the effectiveness of the proposed method utilizing the traffic flow data from the Caltrans Performance Measurement System (PeMS) database. Section 5 concludes this research work and directs future work.

2. Literature Review

In the context of traffic flow predictions, data-driven methods can be classified into two categories: machine learning [11–13] and deep learning methods [14,15]. These machine learning methods include support vector regression [16], random forest [17], Gaussian process [18], Bayesian models [19], and so on. For example, Tang et al. [20] combined the support vector machine method with multiple denoising mechanisms to predict the traffic flow. A dataset collected by the real-time detectors located in the city of Minneapolis was used to evaluate the performance of the proposed methods. The simulation results have shown that the denoising mechanisms could boost the performance of the support vector machine. Zhang et al. [21] introduced a hybrid framework based upon support vector regression to predict the traffic flow, where the random forest method was implemented for feature selections, and the genetic algorithm was adopted to determine the model hyperparameters. The simulation results have shown that the proposed methodology enables better prediction accuracy. Xu et al. [22] introduced a scalable Gaussian process model for large-scale traffic flow predictions. The proposed model combined the Gaussian process with alternative directional methods for paralleling and optimizing hyperparameters during the training process. Wang et al. [23] presented a vicinity Gaussian process method for short-term traffic flow prediction under the conditions of missing data with measuring errors. In the proposed model, a directed graph was constructed based on the traffic network, a dissimilarity matrix and a proper cost function were selected to boost the prediction performance. Zhu et al. [24] introduced a linear conditional Gaussian process method, where temporal and spatial correlations of traffic flow were taken into account. A simulated traffic dataset was adopted to evaluate the effectiveness of the Gaussian process method, and simulation results have shown that the utilization of both spatial and temporal data can dramatically boost prediction accuracy. Li et al. [25] presented a Bayesian network to tackle the node selection challenge in traffic flow prediction. Experimental results have shown that the proposed directed correlation-based Bayesian network method results in a sparse model and better performance in traffic flow prediction.

With the advanced improvement of computational power, deep learning methods are increasingly adopted in traffic flow prediction due to their extraordinary performance. These deep learning methods include LSTM [26,27], gated recurrent neural network (GRU) [28,29], recurrent neural network (RNN) [30,31], graph neural network (GNN) [32–34], and so on. For instance, Tian et al. [35] introduced LSTM-based predictive modeling of traffic flow, where a smoothing function was implemented to deal with the missing data points, and the LSTM was used to capture the prediction residual. Two traffic flow datasets were used to evaluate the performance of the proposed methodology, and the results have shown that the smoothing function can boost the performance of the predictive model. Dai et al. [36] integrated the spatial-temporal analysis with a GRU network to forecast the traffic flow in a short time interval. In the proposed method, the GRU model was applied to process the spatial-temporal features extracted from the collected traffic data. The simulation results have shown that the GRU outperforms the convolutional neural network (CNN) in both prediction accuracy and robustness. Zhene et al. [37] combined the CNN with RNN for urban traffic flow predictions, where CNN was adopted to extract attributes from traffic flow data and RNN was implemented to make predictions. In comparison with the traditional RNN, the proposed RNN was able to process multiple temporal features simultaneously. The experimental results have demonstrated that online traffic flow prediction could be achieved with high precision by using the proposed methodology. Luo et al. [38] introduced a k-nearest neighbor-based (KNN) LSTM method to extract temporal and spatial correlations, where KNN was utilized to capture spatial correlations and LSTM was adopted to further extract temporal correlations. A dataset provided by the University of Minnesota Duluth Data center was utilized to demonstrate the effectiveness of the proposed methods, and the results have indicated that the proposed method outperforms the auto-regressive integrated moving average and wavelet neural network in terms of prediction accuracy. Zhu et al. [39] integrated the GNN with RNN to extract the spatial and temporal correlations of traffic data. The belief rule-based algorithm was adopted for data fusion, and the fused traffic data were fed into the proposed methodology for traffic flow prediction. Yu et al. [40] presented a novel GNN methodology to predict the traffic flow, in which a weighted undirected graph was utilized to differentiate the density of connected roads. A simulation model was introduced to simulate the traffic propagation, and the simulation results were considered in the GNN model for online traffic flow prediction. The simulation results have shown that the proposed GNN outperforms the traditional GNN in traffic flow predictions. More details about applying GNN for traffic flow predictions can be found in [41].

While numerous data-driven methods have been studied to predict traffic flow under various conditions, some issues still exist with these methods. The existing data-driven methods can not consider the correlation of the extracted high-dimensional features and can not use the most relevant part of the traffic flow data to make predictions, which leads to a higher prediction time and a worse prediction accuracy. To deal with these issues, this work proposes a decoder convolutional LSTM network to simultaneously consider the correlation of the decomposed high-dimensional features and the temporal correlation of traffic flow data, where the convolutional operation is used to consider the correlation of the high-dimensional features, and the LSTM network is used to consider the temporal correlation of traffic flow data. Moreover, a multi-head attention mechanism is introduced to use the most relevant portion of the traffic data to make predictions so that the prediction performance can be improved.

3. Convolutional LSTM with Multi-Head Attention Mechanism

This section introduces the convolutional LSTM with a multi-head attention mechanism. Figure 1 shows the framework of the proposed deep learning approach. First, a moving window with a fixed window size is utilized to split raw traffic flow into historical traffic flow as features and future traffic flow as labels. The historical traffic flow is fed into a decoder network to be decomposed into multiple time-series signals. The decomposed signals are fed into the convolutional LSTM network to consider the correlation of the decomposed high dimensional features and the temporal correlation of traffic flow data. The outputs of the convolutional LSTM are transited to the multi-head attention model for traffic flow prediction. Next, the prediction loss is calculated based on the future traffic flow and predicted traffic flow, and the backpropagation algorithm is adopted to train the proposed method. More details of the proposed deep learning approach are provided in the following subsections.

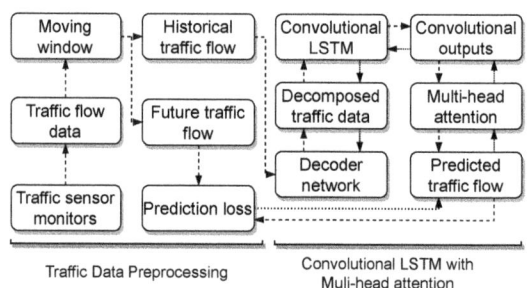

Figure 1. The framework of the convolutional LSTM with a multi-head attention mechanism for traffic flow prediction.

3.1. Decoder Network for Traffic Data Decomposition

The initial step of the proposed method is to decompose the traffic flow so that the most useful latent information can be reflected and the data can be better analyzed. To decompose the traffic flow data, this research uses a decoder network that stacks multiple fully connected layers. The output of the decoder network can be written as Equation (1),

$$\mathbf{D}_{i,L} = f_L \ldots [f_l \ldots [f_2[f_1(\mathbf{X}_i)]]] \tag{1}$$

where $\mathbf{X}_i \in \mathbb{R}^{1 \times T}$ represent the traffic flow data for data sample i; L refers to the total number of stacked fully connected layers in the decoder network; $\mathbf{D}_{i,L} \in \mathbb{R}^{m \times T}$ refers to the output of the decoder network for data sample i; m represents the number of hidden nodes in the fully connected layers of the decoder network; T represents the length of the historical traffic flow; and $f_l(\cdot)$ can be given by Equation (2).

$$f_l(\cdot) := Relu(\mathbf{W}_l \cdot \mathbf{D}_{i,l-1} + \mathbf{b}_l) \tag{2}$$

In Equation (2), $Relu$ represents the rectified linear unit activation function; $\mathbf{W}_l$ refers to the kernel weight matrix at the l-th fully connected layer in the decoder network; $\mathbf{D}_{i,l-1}$ represents the output of the $l-1$-th fully connected layer for data sample i; and $\mathbf{b}_l$ represents the bias weight matrix at the l-th fully connected layer. Next, the output $\mathbf{D}_{i,L}$ of the decoder network is fed into the convolutional LSTM network to consider the correlation of the decomposed high-dimensional features and the temporal correlation of traffic flow data.

3.2. Convolutional LSTM Cell

The traditional LSTM is capable of considering the temporal correlation of traffic flow data. However, the traditional LSTM fails to consider the correlation of the decomposed high-dimensional features. To address this issue, this research aims to introduce the convolutional LSTM cell that incorporates a convolutional operation into the traditional LSTM cell so that both the temporal correlation of traffic flow data and the correlation of the decomposed high-dimensional features can be considered in a unified manner [42]. Figure 2 shows the framework of the convolutional LSTM cell. In the convolutional LSTM cell, the output vector $\mathbf{d}_{i,L}^{(t)}$ of the decoder network at time t and the hidden state $\mathbf{h}_{i,t-1}$ of the one-dimensional convolutional LSTM cell at the prior time point $t-1$ are fed into the one-dimensional convolutional LSTM cell to perform the weighted convolutional operations. Such convolutional operations can consider the correlation of the decomposed high dimensional features $\mathbf{D}_{i,L}$. The recurrent usage of the convolutional LSTM cell can extract temporal correlations, and the output of this cell can be written as Equation (3),

$$\begin{aligned}
\mathbf{f}_{i,t} &= \sigma(C_{i,f} + \mathbb{W}_{f,c} \circ \mathbf{c}_{i,t-1} + \mathbf{b}_f) \\
\mathbf{a}_{i,t} &= \sigma(C_{i,a} + \mathbb{W}_{a,c} \circ \mathbf{c}_{i,t-1} + \mathbf{b}_a) \\
\mathbf{c}_{i,t} &= \mathbf{f}_{i,t} \circ \mathbf{c}_{i,t-1} + \mathbf{a}_{i,t} \circ Tanh(C_{i,c} + \mathbf{b}_c) \\
\mathbf{o}_{i,t} &= \sigma(C_{i,o} + \mathbb{W}_{o,c} \circ \mathbf{c}_{i,t} + \mathbf{b}_o) \\
\mathbf{h}_{i,t} &= \mathbf{o}_{i,t} \circ \sigma(\mathbf{c}_{i,t})
\end{aligned} \qquad (3)$$

where $\mathbf{f}_{i,t}, \mathbf{a}_{i,t}, \mathbf{c}_{i,t}, \mathbf{o}_{i,t}$, respectively, refer to the outputs of the forget gate, input gate, memory cell, and output gate; $\mathbb{W}_{f,c}, \mathbb{W}_{a,c}, \mathbb{W}_{o,c}$ represent the trainable matrices for the forget gate, input gate, and output gate, respectively; $\mathbf{b}_f, \mathbf{b}_a, \mathbf{b}_c, \mathbf{b}_o$ represent the bias vectors for the forget gate, input gate, memory cell, and output gate; σ refers to the sigmoid function; $Tanh$ refers to the hyperbolic tangent function.

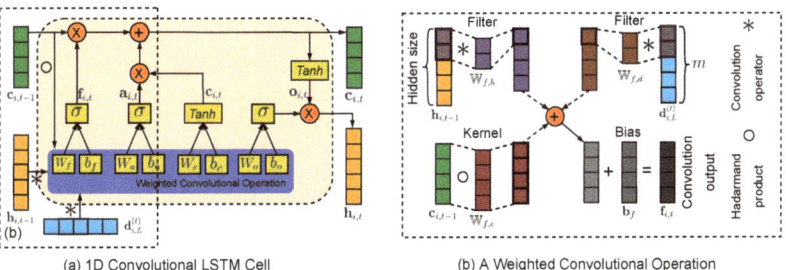

Figure 2. The framework of the one-dimensional convolutional LSTM cell with weighted convolutional operations, where (**a**) is the 1D convolutional LSTM cell and (**b**) gives an example of the weighted convolutional operation.

Moreover, $C_{i,f}, C_{i,a}, C_{i,c}, C_{i,o}$, respectively, refer to the outputs of the convolutional operations at the forget gate, input gate, memory cell, and output gate. These convolutional outputs can be written as Equation (4), where $*$ refers to the convolutional multiplication; $\mathbb{W}_{f,d}$ and $\mathbb{W}_{f,h}$ refer to the kernel matrices of the convolutional operations at the forget gate; $\mathbb{W}_{a,d}$ and $\mathbb{W}_{a,h}$ are the kernel matrices of the convolutional operations at the input gate; $\mathbb{W}_{c,d}$ and $\mathbb{W}_{c,h}$ represent the kernel matrices of the convolutional operations in the memory cell; and $\mathbb{W}_{o,d}$ and $\mathbb{W}_{o,h}$ represent the kernel matrices of the convolutional operations at the output gate.

$$\begin{cases}
C_{i,f} = \mathbb{W}_{f,d} * \mathbf{d}_{i,L}^{(t)} + \mathbb{W}_{f,h} * \mathbf{h}_{i,t-1} \\
C_{i,a} = \mathbb{W}_{a,d} * \mathbf{d}_{i,L}^{(t)} + \mathbb{W}_{a,h} * \mathbf{h}_{i,t-1} \\
C_{i,c} = \mathbb{W}_{c,d} * \mathbf{d}_{i,L}^{(t)} + \mathbb{W}_{c,h} * \mathbf{h}_{i,t-1} \\
C_{i,o} = \mathbb{W}_{o,d} * \mathbf{d}_{i,L}^{(t)} + \mathbb{W}_{o,h} * \mathbf{h}_{i,t-1}
\end{cases} \qquad (4)$$

In summary, the convolutional LSTM cell integrates the convolutional operations with the traditional LSTM cell, where the convolutional operations are adopted to consider the correlation of the decomposed high-dimensional features $D_{i,L}$ and the traditional LSTM cell is utilized to extract the temporal correlations of traffic flow data. The integration of the convolutional operation with the traditional LSTM cell allows the neural network to consider both the correlation of the decomposed high-dimensional features and the temporal correlation of traffic flow data. Next, the hidden outputs, $h_{i,t}$ for all t, of the convolutional LSTM cell are fed into the multi-head attention mechanism for the final prediction.

3.3. Multi-Head Attention Model

In the recent decade, the attention mechanism [43,44] has been introduced to deal with time series as it is capable of using the most relevant proportion of a time series to make predictions. The primary theory of the attention mechanism is simulating the data retrieval process in the data management system. To retrieve data, a query should be inserted into a data management system. If the query is matched with a key, the value associated with the key will be retrieved. Equation (5) shows the construction process of queries Q_i, keys K_i, and values V_i for traffic flow predictions.

$$(\mathbf{W}_Q, \mathbf{W}_K, \mathbf{W}_V) \cdot \mathbf{H}_i = (Q_i, K_i, V_i) \tag{5}$$

In Equation (5), $\mathbf{H}_i$ represents the hidden outputs of the convolutional LSTM network for data sample i, and $\mathbf{H}_i$ can be written as $\mathbf{H}_i = (h_{i,1}, \ldots, h_{i,t}, \ldots, h_{i,T})$; and $\mathbf{W}_Q \in \mathbb{R}^{r \times T}$, $\mathbf{W}_K \in \mathbb{R}^{r \times T}$, $\mathbf{W}_V \in \mathbb{R}^{r \times T}$ are trainable weight matrices. To use the most relevant portion of the values V, the attention vector $\mathbf{a}$ should be obtained by using Equation (6), where $SoftMax$ is the normalized exponential function.

$$\mathbf{a} = SoftMax(Q_i \cdot K_i' / \sqrt{T}) \tag{6}$$

To retrieve the most relevant part of the values V, the attention vector is multiplied by the value matrix, which can be written as $\mathbf{O}_i = \mathbf{a} V_i$.

The multi-head attention mechanism stacks the multiple attention model [45,46]. Figure 3 presents the framework of the multi-head attention model for traffic flow prediction. The attention vector of the multi-head attention mechanism can be written as $\mathbf{a}_h = SoftMax(\mathbf{W}_Q^{(h)} \mathbf{H}_i \cdot (\mathbf{W}_K^{(h)} \mathbf{H}_i)' / \sqrt{T})$, where $\mathbf{W}_Q^{(h)}, \mathbf{W}_K^{(h)}, \mathbf{W}_V^{(h)}$ are trainable weight matrices of the h-th attention model; and $\mathbf{a}_h$ is the attention vector of the h-th attention model. The output of the h-th attention model is written as $\mathbf{O}_{i,h} = \mathbf{a}_h(\mathbf{W}_V^{(h)} \mathbf{H}_i)$.

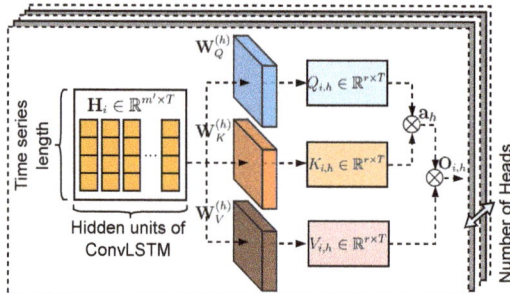

Figure 3. The framework of the multi-head attention model for traffic flow prediction.

Next, the output of all attention models is concatenated, which can be written as Equation (7), where H is the number of attention models and has been stacked in the multi-head attention model.

$$C_i = Concat\{\mathbf{O}_{i,1}, \ldots, \mathbf{O}_{i,h}, \ldots, \mathbf{O}_{i,H}\} \tag{7}$$

Next, the concatenated output **C** is fed into a fully connected layer for final predictions. The training loss of the traffic flow prediction is written as Equation (8), where N refers to the total amount of data samples; $y_{i,j}$ is the true traffic flow for sample i at time j; and $\hat{y}_{i,j}$ is the predicted traffic flow for sample i at time j.

$$L = \frac{1}{N \times T} \sum_{i=1}^{N} \sum_{j=1}^{T} (y_{i,j} - \hat{y}_{i,j})^2 \qquad (8)$$

The backpropagation algorithm is utilized for training the proposed deep learning model. Table 1 presents the training process of the proposed method. First, the weight matrices in the deep learning model are randomly initialized, the traffic flow data and labels are prepared, and the learning rate is initialized. Next, the traffic flow data $\mathbf{X}_i$ for data sample i are fed into the decoder network to decompose the traffic flow data into multiple parts. The output $\mathbf{D}_{i,L}$ of the decoder network is fed into the convolutional LSTM layer to extract temporal and spatial correlations, and the output of this layer is $\mathbf{H}_i$. Next, $\mathbf{H}_i$ is fed into the multi-head attention model to use the most relevant portion of the features extracted by the convolutional LSTM layer. The output of the multi-head attention model $\mathbf{C}_i$ is fed into the fully connected layers for traffic flow predictions, and the trainable weight matrices are updated in each training iteration.

Table 1. The pseudo-code to train the proposed deep learning model for traffic flow predictions.

---
1. Initialize trainable weight matrices
2. Prepare the traffic flow data $\mathbf{X}_i$ and the traffic flow labels $y_{i,j}, \forall i,j$
3. Initialize the learning rate
4. While iteration = $1,\ldots,I$, repeat
   - 4.1. While $l = 1,\ldots,L$, repeat
     - $\mathbf{D}_{i,l} = Relu(\mathbf{W}_l \cdot \mathbf{D}_{i,l-1} + \mathbf{b}_l)$, $\mathbf{D}_{i,l} = \mathbf{X}_i$ if $l=1$
   - 4.2. End iteration
   - 4.3. Feed $\mathbf{D}_{i,L}$ into the convolutional LSTM layer to obtain $\mathbf{H}_i$
   - 4.4. While $h = 1,\ldots,H$, repeat
     - Obtain attention vector $\mathbf{a}_h \leftarrow SoftMax(Q_{i,h} \cdot K'_{i,h}/\sqrt{T})$
     - Obtain attention model's output $\mathbf{O}_{i,h} \leftarrow \mathbf{a}_h \cdot V_{i,h}$
   - 4.5. End iteration
   - 4.6. Obtain $\mathbf{C}_i \leftarrow Concat\{\mathbf{O}_{i,1},\ldots,\mathbf{O}_{i,h},\ldots,\mathbf{O}_{i,H}\}$
   - 4.7. Feed $\mathbf{C}_i$ to FC layers
   - 4.8. Update weight matrices in fully connected layers
   - 4.9. Update weight matrices in the multi-head attention layer
   - 4.10. Update weight matrices in convolutional LSTM layer
   - 4.11. Update weight matrices in the decoder network
5. End iteration
---

4. Case Study

In this section, a real-world traffic flow dataset was used to demonstrate the effectiveness of the proposed deep learning approach. The following subsections provide dataset descriptions, evaluation metrics, model architecture, and prediction results.

4.1. Dataset Description

Traffic flow data collected by the Caltrans Performance Measurement System (PeMS) was utilized to demonstrate the effectiveness of the proposed methodology. The dataset was collected in real-time from over 40,000 unique detectors located on the freeway in the state of California [47]. The collected dataset aggregated hourly traffic flow data obtained from the corresponding detection station. In this study, we used two cases to demonstrate the effectiveness of the proposed method. The first case used the traffic flow data collected from January to March in the year 2022 located at the I5-North freeway, where the post-mile range is from 495.73 to 621.42 in the state of California. The second case used the traffic flow data collected from February to April in the year 2022 located at the I5-North freeway, where the post-mile range is from 495.73 to 621.42 in the state of California. The post-mile refers to the range of routes that move through individual counties in the state of California. For both two cases, the data for the first two months were used to train the proposed deep learning model, and the remaining month was used to test the proposed model. Figure 4 highlights the range of the post-mile 495.73 to 621.42 at the freeway I5-North. To avoid loss of generality, both training and test data were standardized. In this work, we use the data rescaling method to standardize all data to guarantee that both vehicle miles traveled (VMT) and vehicle hours traveled (VHT) are on the same scale. The data rescaling method refers to multiplying each data point by a constant factor, where the factors for VMT and VHT are 10^{-5} and 10^{-3}, respectively.

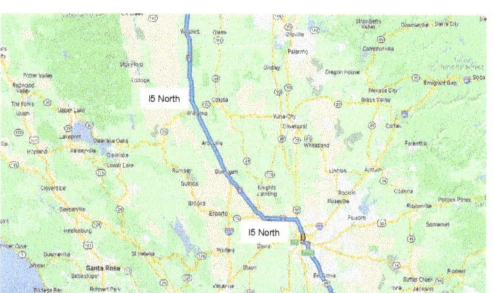

Figure 4. The post-mile ranges from 495.73 to 621.42 located at the freeway I5-North.

4.2. Evaluation Metric

To evaluate the performance of the proposed methodology, this study adopts the root mean squared error (RMSE) and mean absolute error (MAE). The RMSE and MAE can be defined by using Equation (9), where N is the total amount of data samples; $y_{i,j}$ refers to the true traffic flow for the sample i at time j; and $\hat{y}_{i,j}$ represents the predicted traffic flow for the sample i at time j.

$$RMSE = \left(\frac{1}{N \times T} \sum_{i=1}^{N} \sum_{j=1}^{T} (y_{i,j} - \hat{y}_{i,j})^2\right)^{1/2}$$

$$MAE = \frac{1}{N \times T} \sum_{i=1}^{N} \sum_{j=1}^{T} |y_{i,j} - \hat{y}_{i,j}|$$

(9)

4.3. Model Architecture and Hyperparameters

In this case study, we use three tasks to evaluate the prediction performance of the proposed deep learning model for both two cases. These tasks include the next 1st-hour traffic flow prediction (first task), the next 5th-hour traffic flow prediction (second task), and the next 10th-hour traffic flow prediction (third task). The next nth-hour traffic flow prediction refers to using the past 24 h traffic flow data to predict the traffic flow in the $24 + n$ h. Tables 2–4 show the model architecture and hyperparameters used in this case study for three tasks. For these three tasks under two cases, we use the batch size of 100 and utilize the past 24 h traffic flow data to make predictions in each batch. We also use the filter size of 2 in the first task and use the filter size of 10 in the remaining two tasks. Moreover, the number of hidden nodes in the decoder network is 100.

Table 2. The model architecture and hyperparameters used for the next 1st-hour traffic flow prediction task.

No. of Layers	Descriptions	Output Dimensions
1	Input layer	$100 \times 24 \times 1$
2	FC layer	$100 \times 24 \times 100$
3	Convolutional LSTM	$100 \times 24 \times 99$
4	Multi-head attention	$100 \times 24 \times 99$
5	Flatten layer	100×2376
6	Dense layer	100×1

Table 3. The model architecture and hyperparameters used for the next 5th-hour traffic flow prediction task.

No. of Layers	Descriptions	Output Dimensions
1	Input layer	$100 \times 24 \times 1$
2	FC layer	$100 \times 24 \times 100$
3	FC layer	$100 \times 24 \times 100$
4	FC layer	$100 \times 24 \times 100$
5	Convolutional LSTM	$100 \times 24 \times 91$
6	Multi-head attention	$100 \times 24 \times 91$
7	Flatten layer	100×2184
8	Dense layer	100×1

Table 4. The model architecture and hyperparameters used for the next 10th-hour traffic flow prediction task.

No. of Layers	Descriptions	Output Dimensions
1	Input layer	$100 \times 24 \times 1$
2	FC layer	$100 \times 24 \times 100$
3	FC layer	$100 \times 24 \times 100$
4	FC layer	$100 \times 24 \times 100$
5	FC layer	$100 \times 24 \times 100$
6	FC layer	$100 \times 24 \times 100$
7	Convolutional LSTM	$100 \times 24 \times 91$
8	Multi-head attention	$100 \times 24 \times 91$
9	Flatten layer	100×2184
10	Dense layer	100×1

4.4. Traffic Flow Prediction Results for the First Case

Figure 5 shows the traffic flow prediction results for three different tasks under the first case, where VMT refers to vehicle miles traveled, and VHT refers to vehicle hours traveled. From these three figures, we can observe that the proposed methodology can predict the traffic flow with high accuracy, as the true VMT and VHT are close to the predicted VMT and VHT. For example, for the 5th-hour prediction task, the predicted VMT is 1.260 when the true VMT is 1.219. For the 1st-hour prediction task, the predicted VHT is 0.337 when the true VHT is 0.325. To further demonstrate the performance of the proposed method, we compare the proposed method with existing methods reported in the literature, and these methods are listed in Table 5. In this table, the D-ConvoLSTM method refers to the decoder network with the convolutional LSTM network; and the D-Attention method refers to the decoder network with the multi-head attention mechanism; LSTM refers to the long short-term memory network; LASSO refers to the least absolute shrinkage and selection operator; ANN refers to the artificial neural network.

Table 5. Symbols and descriptions of the proposed method and other methods for traffic flow predictions.

Method Symbol	Description
D-ConvLSTM	Decoder with convolutional LSTM
D-Attention	Decoder with multi-head attention
LSTM	Long short term memory network
LASSO	Regression with l1-norm regularization
ANN	Artificial neural network

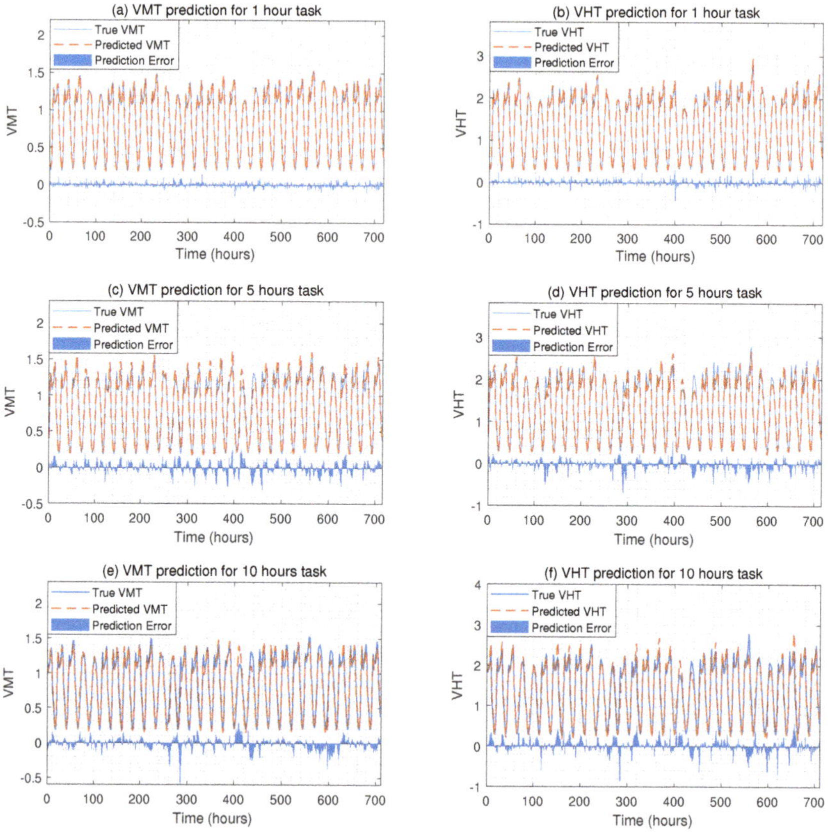

Figure 5. The VMT and VHT prediction results for three different tasks under the first case, where (**a,c,e**) show the VMT predictions for three tasks; and (**b,d,f**) show the VHT predictions for three tasks.

Table 6 compares the traffic flow prediction performance of the proposed method with methods listed in Table 5 in terms of RMSE and MAE. From this table, we can conclude that the proposed method can predict traffic flow with high accuracy and outperforms existing data-driven methods. For example, for the 1st-hour task, the RMSE of the VMT prediction for the proposed method is 0.032, and the RMSE of other data-driven methods ranges from 0.038 to 0.088. For the 5th-hour task, the RMSE of the VHT prediction for the proposed method is 0.128; however, the RMSE of LSTM is 0.145, and the RMSE of ANN is 0.245.

Table 6. The traffic flow prediction errors in terms of RMSE and MAE for the proposed methods and other data-driven methods under the first case.

		1 h Task		5 h Task		10 h Task	
		VMT	VHT	VMT	VHT	VMT	VHT
RMSE	Proposed	0.032	0.066	0.080	0.128	0.084	0.167
	D-ConvLSTM	0.044	0.079	0.099	0.128	0.094	0.157
	D-Attention	0.043	0.086	0.105	0.179	0.113	0.199
	LSTM [30]	0.038	0.064	0.065	0.145	0.104	0.191
	LASSO [48]	0.088	0.141	0.142	0.242	0.141	0.240
	ANN [49]	0.054	0.103	0.137	0.245	0.138	0.241
MAE	Proposed	0.024	0.048	0.059	0.090	0.058	0.116
	D-ConvLSTM	0.034	0.058	0.072	0.097	0.064	0.115
	D-Attention	0.034	0.066	0.076	0.135	0.077	0.138
	LSTM [30]	0.029	0.045	0.046	0.107	0.064	0.130
	LASSO [48]	0.063	0.096	0.099	0.165	0.098	0.163
	ANN [49]	0.039	0.072	0.090	0.172	0.089	0.168

4.5. Traffic Flow Prediction Results for the Second Case

Figure 6 shows the traffic flow prediction results for three different tasks under the second case, where VMT refers to vehicle miles traveled, and VHT refers to vehicle hours traveled. From this figure, we can observe that the proposed methodology can predict the traffic flow with high accuracy as the true VMT and VHT are close to the predicted VMT and VHT. For example, for the 5th-hour prediction task, the predicted VMT is 1.085 when the true VMT is 1.082. For the 1st-hour prediction task, the predicted VHT is 2.110 when the true VHT is 2.138. Table 7 compares the traffic flow prediction performance of the proposed method with methods listed in Table 5 in terms of RMSE and MAE. From this table, we can conclude that the proposed method can predict traffic flow with high accuracy and outperforms existing data-driven methods. For example, for the 1st-hour task, the RMSE of the VMT prediction for the proposed method is 0.053, and the RMSE of other data-driven methods ranges from 0.055 to 0.091. For the 5th-hour task, the MAE of the VHT prediction for the proposed method is 0.093; however, the RMSE of LSTM is 0.129, and the RMSE of ANN is 0.175.

Table 7. The traffic flow prediction errors in terms of RMSE and MAE for the proposed methods and other data-driven methods under the second case.

		1 h Task		5 h Task		10 h Task	
		VMT	VHT	VMT	VHT	VMT	VHT
RMSE	Proposed	0.053	0.100	0.084	0.135	0.100	0.172
	D-ConvLSTM	0.088	0.153	0.094	0.157	0.118	0.184
	D-Attention	0.055	0.087	0.112	0.168	0.141	0.253
	LSTM [30]	0.055	0.106	0.113	0.187	0.119	0.225
	LASSO [48]	0.091	0.145	0.149	0.256	0.145	0.248
	ANN [49]	0.063	0.112	0.143	0.255	0.146	0.256
MAE	Proposed	0.042	0.078	0.062	0.093	0.064	0.109
	D-ConvLSTM	0.060	0.107	0.060	0.104	0.078	0.125
	D-Attention	0.043	0.107	0.075	0.123	0.104	0.187
	LSTM [30]	0.046	0.076	0.077	0.129	0.080	0.153
	LASSO [48]	0.063	0.099	0.102	0.175	0.100	0.168
	ANN [49]	0.044	0.079	0.096	0.175	0.099	0.179

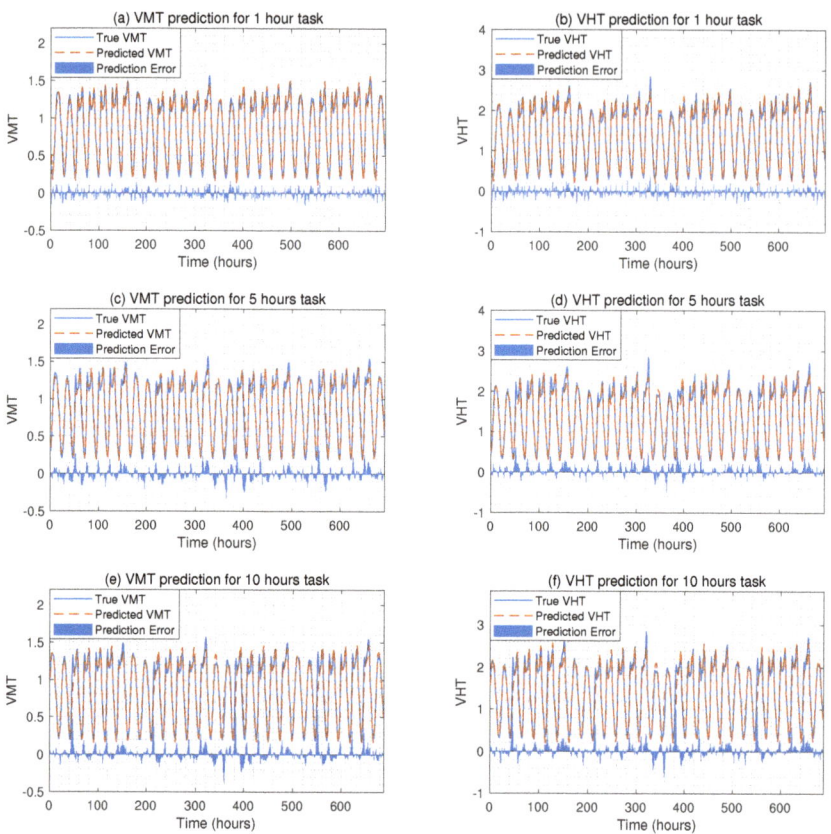

Figure 6. The VMT and VHT prediction results for three different tasks under the second case, where (**a**,**c**,**e**) show the VMT predictions for three tasks; and (**b**,**d**,**f**) show the VHT predictions for three tasks.

5. Conclusions and Future Work

In this study, a deep learning approach was proposed to predict traffic flow. In the proposed deep learning approach, a convolutional long short-term memory network was used to consider the correlation of the extracted high-dimensional features and the temporal correlation of traffic flow data in a unified manner. Moreover, a multi-head attention mechanism was implemented to use the most relevant portion of the traffic flow data to make predictions so that the prediction performance can be improved. A traffic flow dataset collected from the Caltrans Performance Measurement System (PeMS) database was used to demonstrate the effectiveness of the proposed method. Experimental results have shown that the proposed method can accurately predict the traffic flow with a minimum RMSE of 0.032 and outperforms the existing data-driven methods in terms of RMSE and MAE. Future work will be directed to use the convolutional LSTM network to make traffic flow predictions under more complicated environments and conditions.

Author Contributions: Y.W.: conceptualization, methodology, software, investigation, validation, visualization, funding acquisition, writing—original draft preparation. H.L.: conceptualization, writing—review and editing. All authors have read and agreed to the published version of the manuscript.

Funding: This work was supported in part by the Mineta Transportation Institute under Project No. 2211.

Institutional Review Board Statement: Not applicable.

Informed Consent Statement: Not applicable.

Data Availability Statement: The dataset used in this study is available at https://pems.dot.ca.gov/ (accessed on 5 February 2022).

Conflicts of Interest: The authors declare no conflict of interest.

Abbreviations

The following abbreviations are used in this manuscript:

LSTM	Long short-term memory
ANN	artificial neural network
PeMS	Caltrans Performance Measurement System
VMT	vehicle miles traveled
VHT	vehicle hours traveled
RSME	root mean squared error
MAE	mean absolute error
FC	fully connected
ConvLSTM	convolutional long short-term memory

References

1. Bazzan, A.L.; Oliveira, D.d.; Klügl, F.; Nagel, K. To adapt or not to adapt–consequences of adapting driver and traffic light agents. In *Adaptive Agents and Multi-Agent Systems III. Adaptation and Multi-Agent Learning*; Springer: Berlin/Heidelberg, Germany, 2005; pp. 1–14.
2. Ahmad, A.; Arshad, R.; Mahmud, S.A.; Khan, G.M.; Al-Raweshidy, H.S. Earliest-deadline-based scheduling to reduce urban traffic congestion. *IEEE Trans. Intell. Transp. Syst.* **2014**, *15*, 1510–1526. [CrossRef]
3. Zhang, Y.; Li, C.; Luan, T.H.; Fu, Y.; Shi, W.; Zhu, L. A mobility-aware vehicular caching scheme in content centric networks: Model and optimization. *IEEE Trans. Veh. Technol.* **2019**, *68*, 3100–3112. [CrossRef]
4. Falcocchio, J.C.; Levinson, H.S. *Road Traffic Congestion: A Concise Guide*; Springer: Berlin/Heidelberg, Germany, 2015; Volume 7.
5. Wu, Y.; Tan, H.; Qin, L.; Ran, B.; Jiang, Z. A hybrid deep learning based traffic flow prediction method and its understanding. *Transp. Res. Part C Emerg. Technol.* **2018**, *90*, 166–180. [CrossRef]
6. Shi, R.; Du, L. Multi-Section Traffic Flow Prediction Based on MLR-LSTM Neural Network. *Sensors* **2022**, *22*, 7517. [CrossRef] [PubMed]
7. Wang, S.; Zhao, J.; Shao, C.; Dong, C.; Yin, C. Truck traffic flow prediction based on LSTM and GRU methods with sampled GPS data. *IEEE Access* **2020**, *8*, 208158–208169. [CrossRef]
8. Chen, Z.; Wu, B.; Li, B.; Ruan, H. Expressway exit traffic flow prediction for ETC and MTC charging system based on entry traffic flows and LSTM model. *IEEE Access* **2021**, *9*, 54613–54624. [CrossRef]
9. Zhou, Q.; Chen, N.; Lin, S. FASTNN: A Deep Learning Approach for Traffic Flow Prediction Considering Spatiotemporal Features. *Sensors* **2022**, *22*, 6921. [CrossRef]
10. Yu, C.; Chen, J.; Xia, G. Coordinated Control of Intelligent Fuzzy Traffic Signal Based on Edge Computing Distribution. *Sensors* **2022**, *22*, 5953. [CrossRef]
11. Feng, X.; Ling, X.; Zheng, H.; Chen, Z.; Xu, Y. Adaptive multi-kernel SVM with spatial–temporal correlation for short-term traffic flow prediction. *IEEE Trans. Intell. Transp. Syst.* **2018**, *20*, 2001–2013. [CrossRef]
12. Kumar, S.V. Traffic flow prediction using Kalman filtering technique. *Procedia Eng.* **2017**, *187*, 582–587. [CrossRef]
13. Mingheng, Z.; Yaobao, Z.; Ganglong, H.; Gang, C. Accurate multisteps traffic flow prediction based on SVM. *Math. Probl. Eng.* **2013**, *2013*, 418303. [CrossRef]
14. Lv, Y.; Duan, Y.; Kang, W.; Li, Z.; Wang, F.Y. Traffic flow prediction with big data: A deep learning approach. *IEEE Trans. Intell. Transp. Syst.* **2014**, *16*, 865–873. [CrossRef]
15. Miglani, A.; Kumar, N. Deep learning models for traffic flow prediction in autonomous vehicles: A review, solutions, and challenges. *Veh. Commun.* **2019**, *20*, 100184. [CrossRef]
16. Sun, Y.; Leng, B.; Guan, W. A novel wavelet-SVM short-time passenger flow prediction in Beijing subway system. *Neurocomputing* **2015**, *166*, 109–121. [CrossRef]
17. Liu, Y.; Wu, H. Prediction of road traffic congestion based on random forest. In Proceedings of the 2017 10th International Symposium on Computational Intelligence and Design (ISCID), Hangzhou, China, 9–10 December 2017; IEEE: Piscataway, NJ, USA, 2017; Volume 2, pp. 361–364.
18. Sun, S.; Xu, X. Variational inference for infinite mixtures of Gaussian processes with applications to traffic flow prediction. *IEEE Trans. Intell. Transp. Syst.* **2010**, *12*, 466–475. [CrossRef]
19. Pascale, A.; Nicoli, M. Adaptive Bayesian network for traffic flow prediction. In Proceedings of the 2011 IEEE Statistical Signal Processing Workshop (SSP), Nice, France, 28–30 June 2011; IEEE: Piscataway, NJ, USA, 2011; pp. 177–180.

20. Tang, J.; Chen, X.; Hu, Z.; Zong, F.; Han, C.; Li, L. Traffic flow prediction based on combination of support vector machine and data denoising schemes. *Phys. Stat. Mech. Its Appl.* **2019**, *534*, 120642. [CrossRef]
21. Zhang, L.; Alharbe, N.R.; Luo, G.; Yao, Z.; Li, Y. A hybrid forecasting framework based on support vector regression with a modified genetic algorithm and a random forest for traffic flow prediction. *Tsinghua Sci. Technol.* **2018**, *23*, 479–492. [CrossRef]
22. Xu, Y.; Yin, F.; Xu, W.; Lin, J.; Cui, S. Wireless traffic prediction with scalable Gaussian process: Framework, algorithms, and verification. *IEEE J. Sel. Areas Commun.* **2019**, *37*, 1291–1306. [CrossRef]
23. Wang, W.; Zhou, C.; He, H.; Wu, W.; Zhuang, W.; Shen, X. Cellular traffic load prediction with LSTM and Gaussian process regression. In Proceedings of the ICC 2020-2020 IEEE International Conference on Communications (ICC), Dublin, Ireland, 7–11 June 2020; IEEE: Piscatawy, NJ, USA, 2020; pp. 1–6.
24. Zhu, Z.; Peng, B.; Xiong, C.; Zhang, L. Short-term traffic flow prediction with linear conditional Gaussian Bayesian network. *J. Adv. Transp.* **2016**, *50*, 1111–1123. [CrossRef]
25. Li, Z.; Jiang, S.; Li, L.; Li, Y. Building sparse models for traffic flow prediction: An empirical comparison between statistical heuristics and geometric heuristics for Bayesian network approaches. *Transp. Transp. Dyn.* **2017**, *7*, 107–123. [CrossRef]
26. Wei, W.; Wu, H.; Ma, H. An autoencoder and LSTM-based traffic flow prediction method. *Sensors* **2019**, *19*, 2946. [CrossRef]
27. Xiao, Y.; Yin, Y. Hybrid LSTM neural network for short-term traffic flow prediction. *Information* **2019**, *10*, 105. [CrossRef]
28. Fu, R.; Zhang, Z.; Li, L. Using LSTM and GRU neural network methods for traffic flow prediction. In Proceedings of the 2016 31st Youth Academic Annual Conference of Chinese Association of Automation (YAC), Wuhan, China, 11–13 November 2016; IEEE: Piscataway, NJ, USA, 2016; pp. 324–328.
29. Shu, W.; Cai, K.; Xiong, N.N. A short-term traffic flow prediction model based on an improved gate recurrent unit neural network. *IEEE Trans. Intell. Transp. Syst.* **2021**, *23*, 16654–16665. [CrossRef]
30. Yang, B.; Sun, S.; Li, J.; Lin, X.; Tian, Y. Traffic flow prediction using LSTM with feature enhancement. *Neurocomputing* **2019**, *332*, 320–327. [CrossRef]
31. Xiangxue, W.; Lunhui, X.; Kaixun, C. Data-driven short-term forecasting for urban road network traffic based on data processing and LSTM-RNN. *Arab. J. Sci. Eng.* **2019**, *44*, 3043–3060. [CrossRef]
32. Li, Z.; Xiong, G.; Chen, Y.; Lv, Y.; Hu, B.; Zhu, F.; Wang, F.Y. A hybrid deep learning approach with GCN and LSTM for traffic flow prediction. In Proceedings of the 2019 IEEE Intelligent Transportation Systems Conference (ITSC), Auckland, New Zealan, 27–30 October 2019; IEEE: Piscatway, NJ, USA, 2019; pp. 1929–1933.
33. Chen, J.; Liao, S.; Hou, J.; Wang, K.; Wen, J. GST-GCN: A Geographic-Semantic-Temporal Graph Convolutional Network for Context-aware Traffic Flow Prediction on Graph Sequences. In Proceedings of the 2020 IEEE International Conference on Systems, Man, and Cybernetics (SMC), Melbourne, Australia, 17–20 October 2021; IEEE: Piscataway, NJ, USA, 2020; pp. 1604–1609.
34. Jiang, W.; Luo, J. Graph neural network for traffic forecasting: A survey. *Expert Syst. Appl.* **2022**, *4*, 117921. [CrossRef]
35. Tian, Y.; Zhang, K.; Li, J.; Lin, X.; Yang, B. LSTM-based traffic flow prediction with missing data. *Neurocomputing* **2018**, *318*, 297–305. [CrossRef]
36. Dai, G.; Ma, C.; Xu, X. Short-term traffic flow prediction method for urban road sections based on space–time analysis and GRU. *IEEE Access* **2019**, *7*, 143025–143035. [CrossRef]
37. Zhene, Z.; Hao, P.; Lin, L.; Guixi, X.; Du, B.; Bhuiyan, M.Z.A.; Long, Y.; Li, D. Deep convolutional mesh RNN for urban traffic passenger flows prediction. In Proceedings of the 2018 IEEE SmartWorld, Ubiquitous Intelligence & Computing, Advanced & Trusted Computing, Scalable Computing & Communications, Cloud & Big Data Computing, Internet of People and Smart City Innovation (SmartWorld/SCALCOM/UIC/ATC/CBDCom/IOP/SCI), Guangzhou, China, 8–12 October 2018; IEEE: Piscataway, NJ, USA, 2018; pp. 1305–1310.
38. Luo, X.; Li, D.; Yang, Y.; Zhang, S. Spatiotemporal traffic flow prediction with KNN and LSTM. *J. Adv. Transp.* **2019**, *2019*, 4145353. [CrossRef]
39. Zhu, H.; Xie, Y.; He, W.; Sun, C.; Zhu, K.; Zhou, G.; Ma, N. A novel traffic flow forecasting method based on RNN-GCN and BRB. *J. Adv. Transp.* **2020**, *2020*, 7586154. [CrossRef]
40. Yu, B.; Lee, Y.; Sohn, K. Forecasting road traffic speeds by considering area-wide spatio-temporal dependencies based on a graph convolutional neural network (GCN). *Transp. Res. Part C Emerg. Technol.* **2020**, *114*, 189–204. [CrossRef]
41. Ye, J.; Zhao, J.; Ye, K.; Xu, C. How to build a graph-based deep learning architecture in traffic domain: A survey. *IEEE Trans. Intell. Transp. Syst.* **2020**, *2020*, 7586154. [CrossRef]
42. Shi, X.; Chen, Z.; Wang, H.; Yeung, D.Y.; Wong, W.K.; Woo, W.c. Convolutional LSTM network: A machine learning approach for precipitation nowcasting. *Adv. Neural Inf. Process. Syst.* **2015**, *23*, 3904–3924.
43. Bahdanau, D.; Cho, K.; Bengio, Y. Neural machine translation by jointly learning to align and translate. *arXiv* **2014**, arXiv:1409.0473.
44. Xu, K.; Ba, J.; Kiros, R.; Cho, K.; Courville, A.; Salakhudinov, R.; Zemel, R.; Bengio, Y. Show, attend and tell: Neural image caption generation with visual attention. In Proceedings of the International Conference on Machine Learning, PMLR, Lille, France, 7–9 July 2015; pp. 2048–2057.
45. Vaswani, A.; Shazeer, N.; Parmar, N.; Uszkoreit, J.; Jones, L.; Gomez, A.N.; Kaiser, Ł.; Polosukhin, I. Attention is all you need. In Proceedings of the Advances in Neural Information Processing Systems 30, Long Beach, CA, USA, 4–9 December 2017; Volume 30.
46. Li, J.; Tu, Z.; Yang, B.; Lyu, M.R.; Zhang, T. Multi-head attention with disagreement regularization. *arXiv* **2018**, arXiv:1810.10183.
47. Caltrans. Performance Measurement System (PeMS). Available online: https://pems.dot.ca.gov/ (accessed on 5 February 2022).

48. Sun, S.; Huang, R.; Gao, Y. Network-scale traffic modeling and forecasting with graphical lasso and neural networks. *J. Transp. Eng.* **2012**, *138*, 1358–1367. [CrossRef]
49. Rahman, F.I. Short term traffic flow prediction using machine learning-KNN, SVM and ANN with weather information. *Int. J. Traffic Transp. Eng.* **2020**, *10*, 371–389.

Review

Weed Detection Using Deep Learning: A Systematic Literature Review

Nafeesa Yousuf Murad [1], Tariq Mahmood [1], Abdur Rahim Mohammad Forkan [2], Ahsan Morshed [3], Prem Prakash Jayaraman [2] and Muhammad Shoaib Siddiqui [4,*]

[1] Big Data Analytics Laboratory, Department of Computer Science, School of Mathematics and Computer Science, Institute of Business Administration, Karachi 75270, Pakistan
[2] School of Science, Computing and Engineering Technologies, Swinburne University of Technology, Melbourne 3122, Australia
[3] School of Engineering and Technology, Central Queensland University, Melbourne 3000, Australia
[4] Faculty of Computer and Information Systems, Islamic University of Madinah, Medina 42351, Saudi Arabia
* Correspondence: shoaib@iu.edu.sa

Abstract: Weeds are one of the most harmful agricultural pests that have a significant impact on crops. Weeds are responsible for higher production costs due to crop waste and have a significant impact on the global agricultural economy. The importance of this problem has promoted the research community in exploring the use of technology to support farmers in the early detection of weeds. Artificial intelligence (AI) driven image analysis for weed detection and, in particular, machine learning (ML) and deep learning (DL) using images from crop fields have been widely used in the literature for detecting various types of weeds that grow alongside crops. In this paper, we present a systematic literature review (SLR) on current state-of-the-art DL techniques for weed detection. Our SLR identified a rapid growth in research related to weed detection using DL since 2015 and filtered 52 application papers and 8 survey papers for further analysis. The pooled results from these papers yielded 34 unique weed types detection, 16 image processing techniques, and 11 DL algorithms with 19 different variants of CNNs. Moreover, we include a literature survey on popular vanilla ML techniques (e.g., SVM, random forest) that have been widely used prior to the dominance of DL. Our study presents a detailed thematic analysis of ML/DL algorithms used for detecting the weed/crop and provides a unique contribution to the analysis and assessment of the performance of these ML/DL techniques. Our study also details the use of crops associated with weeds, such as sugar beet, which was one of the most commonly used crops in most papers for detecting various types of weeds. It also discusses the modality where RGB was most frequently used. Crop images were frequently captured using robots, drones, and cell phones. It also discusses algorithm accuracy, such as how SVM outperformed all machine learning algorithms in many cases, with the highest accuracy of 99 percent, and how CNN with its variants also performed well with the highest accuracy of 99 percent, with only VGGNet providing the lowest accuracy of 84 percent. Finally, the study will serve as a starting point for researchers who wish to undertake further research in this area.

Keywords: weed detection; deep learning; machine learning; systematic literature review

Citation: Murad, N.Y.; Mahmoodm, T.; Forkan, A.R.M.; Morshed, A.; Jayaraman, P.P.; Siddiqui, M.S. Weed Detection Using Deep Learning: A Systematic Literature Review. *Sensors* 2023, 23, 3670. https://doi.org/10.3390/s23073670

Academic Editors: Shyan-Ming Yuan, Zeng-Wei Hong, Wai-Khuen Cheng and Francesca Antonucci

Received: 24 February 2023
Revised: 18 March 2023
Accepted: 29 March 2023
Published: 31 March 2023

Copyright: © 2023 by the authors. Licensee MDPI, Basel, Switzerland. This article is an open access article distributed under the terms and conditions of the Creative Commons Attribution (CC BY) license (https://creativecommons.org/licenses/by/4.0/).

1. Introduction

Crop farming is considered a significant agricultural pursuit for the global economy in the modern era, and over a longer time period, it has had a notable impact on countries' GDP. In 2018, it contributed 4% to the global GDP and accounts for more than 25% of the GDP for many developing countries. Moreover, with almost 9% of the world population hungry in 2020, agriculture is a powerful source of food, revenue, and employment and is expected to minimize poverty, raise income levels, and boost prosperity for a projected 9.7 billion population by 2050 [1,2]. However, agricultural growth through crop farming is always at risk due to several reoccurring problems, for example, climate change, greenhouse gas emissions, pollution and waste generation, malnutrition, and food wastage [1]. Another serious problem that plagues crop farming is the growth of weeds which leads to significant crop wastage annually. Hence, weed management and removal practices have been adopted for several decades to control weed growth [3–5].

Weeds are undesired plants that compete against productive crops for space, light, water, and soil nutrients and propagate themselves either through seeding or rhizomes. They are generally poisonous, produce thorns and burrs, and hamper crop management by contaminating crop harvests. Smaller weed seedlings with a slow growth rate are more difficult to detect and manage than larger ones which grow vigorously. Weed management is complicated because the competitive nature of weeds can vary in different conditions and seasons. For instance, the tall and fast-growing fat hen weed is considered dangerous to adjacent crops, but fat hen seedlings that appear in late summer are considerably smaller in size and not potentially dangerous [6]. Similarly, chickweed is smaller and less dangerous during the summer season, but in winter, it can have a high growth rate and can swamp crops such as onions and spring greens [7,8]. Moreover, weeds can co-exist 'peacefully' with the crops earlier on in their growth period but start competing for more natural resources later on. Another difficulty in managing weeds is determining the exact time when a weed actually starts to affect the harvest. Moreover, several weeds, such as couch grass and creeping buttercup, can survive in drought and severe winter weather as they store food in long underground stems. Weeds are also potential hosts for pests and diseases which can easily spread to cultivated crops. For instance, the charlock and shepherd's purse weeds may carry clubroot and eelworm diseases, while chickweed can host the cucumber mosaic virus [7,9,10]. Finally, different weeds have different seeding frequencies, further complicating weed management; for instance, groundsel can produce 1000 seeds per season, while scentless mayweed might produce 30,000 seeds per plant. These seeds might stay in the soil for decades until exposed to light; for instance, the poppy seed can survive even up to 80 years.

For several decades weeds have been managed, detected, and controlled manually [3–5]. The most common method of weed detection is manual surveillance by hiring crop scouts or by tasking crop farmers to do the same, which is expensive, difficult to manage, and infeasible to execute in unfavorable weather conditions Scouts only work on a sample of the field and have to follow a pre-determined randomized pattern (e.g., zigzag). Such a setting does not always ensure that all weeds will be detected and removed. Scouts also carry specialized equipment (e.g., hand-held computers with GPS and geo-tagging), which adds to the expense. They need to repeat the process regularly and fill up a report. All these limitations make crop scouting difficult to manage, and hence, weeds continue to affect crop harvest each year globally.

Motivation and Contribution

In this paper, we focus on smart farming techniques that can detect weeds in crop images through machine learning methods, particularly DL. This can potentially eliminate the need for crop scouts while scanning the entire field for weeds with no management overload. However, ever since the introduction of Graphical Processing Units (GPUs), DL has demonstrated an unparalleled pace of research and superior performance across a wide variety of complicated applications involving images, text, video, and speech datasets [11–15]. DL was considered nascent till 2010 due to a lack of hardware technology to process its complex architectures. One of the initial researches on weed detection found in 1991 [16], which highlighted the limitations of using tractor-mounted weed detectors, and proposed the use of digital image processing (IP) techniques [17,18] to detect weeds from both aerial and previous manually-snapped photographs by crop scouts. Research efforts using pure IP and CV techniques for automated weed detection remained very limited for the next two odd decades [19–22]. This paper demonstrates that such applications are still in their infancy with respect to applications in ML and DL [12,23]. Since 2000, researchers have been using ML sometimes in combination with CV to automatically detect weeds from images [24–29]. Although most of these works detected weeds with reasonable accuracy, they also highlighted the potential of DL for significantly better performances.

Through some initial searches, we determined that DL applications in research for weed detection have increased considerably since 2015, and they primarily use convolutional neural networks (CNNs) and their variants such as SegNet, GoogLeNet, ResNet, DetectNet, and VGGNet [30,31]. Though some survey articles have been published since 2015 on DL applications for weed detection [32,33], they lack proper SLR.

The rapid pace of research in DL and its potential to provide a competitive performance on complicated image-based recognition tasks motivated us to conduct the SLR on DL applications of weed detection. Our general intent is to extract and summarize the relevant and most recent research and provide concrete future directions regarding industrial applications and academic research. As this domain involves working with image data of weeds, we target applications of CNNs, particularly the most recent and standard published research content for the time period between 2015 and January 2023. In our SLR, we answer the following research questions:

RQ1: What is the trend of employing deep learning to address the problem of weed detection in recent years?

RQ2: Which types of weeds and corresponding crops have been detected using deep learning, and what are the characteristics of the corresponding weed datasets?

RQ3: Which deep learning algorithms are best suited for a particular weed/crop combination?

RQ4: What are the tangible future research directions to achieve further benefit from deep learning applications for weed detection?

RQ1 is addressed in Section 4, where we describe our SLR methodology and analyze the trends and other relevant statistics of extracted papers.

RQ2 and RQ3 are addressed in Section 5. For RQ2, we analyze and identify the datasets that have been used in the papers and summarise the relevant information such as weed types, crop types, and characteristics of weed images (e.g., resolution, size).

To address RQ3, we identify the different DL algorithms in the literature and compare their frequency of usage and performance data. Moreover, we categorize each paper by assigning a unique label based on the usage of ML, IP, and DL, along with characteristics such as training time and performance. We then compare the performance of ML and IP algorithms with DL ones. To further strengthen the analysis, we associate the weed types with the algorithm used and its associated evaluation outcome. Finally, we address RQ4 in Section 8 by analyzing our findings and proposing a set of future research directions to motivate and enhance the DL research weed detection domain.

2. Related Surveys

In this section, we discuss in detail the eight articles of literature review which we extracted through our SLR.

In [34], the authors review seven research papers based on deep learning and discuss three previously-used techniques for the classification of weeds such as color-based, threshold-based, and learning-based techniques. The authors review the papers over different parameters such as the type of deep learning used, targeted crops, training setup, the training time of the algorithm, dataset acquisition, dataset strength, and accuracy of the algorithm. Research gaps are also identified, and one of the gaps was the lack of a big dataset which could be a major contribution in this field.

Moreover, in [35], the challenges faced by vision-based plant and weed detection and their solutions have been discussed. Two main challenges of weed detection are the light problems, i.e., the algorithm may work differently due to the presence of light, and discrimination between crop and weed, i.e., sometimes both may look similar. Shading or artificial lighting can be used to control the variation of natural light, or image processing techniques like segmentation of background (and then converting the image into Grayscale) can be used to tackle this problem. For the second problem, different types of IP-based classification techniques were discussed, which were based on shape, texture, height, and DL. The authors discussed the comparison of traditional classification and DL methods. They also highlight the application of online cloud databases as an important future direction to further improve the recognition or detection of weeds and crops.

Furthermore, in [22], the authors summarize different problems and provided solutions to weed classification using IP and DL techniques. Four basic steps of classification, such as pre-processing, image segmentation, feature extraction (biological morphology, spectral feature, visual texture, spatial context), and classification (convolutional machine learning), have been discussed in detail. Some challenges like leaf overlapping, light variation, and stages of plant growth and their solutions were discussed. Semi-supervised learning techniques have been proposed by the authors to improve the current performance of the aforementioned techniques.

In [36], the authors analyze different techniques for weed detection using IoT technology. The authors discuss several DL algorithms employed in the context of IoT and perform their comparative analysis, for example, CNN, SegNet (with a synthetic dataset for achieving higher accuracy), and summarised training set technique with CNet, which is a deep CNN based on image segmentation. The authors also propose an IoT-based architecture where different devices and sensors are connected to one central data server, and users can communicate with the server through the Internet. This model can be controlled by a desktop computer or mobile device. Moreover, in [37], the authors focus on the methods and technologies used in weed detection with particular focus on the requirements of weed detection, its applications, and the system needed for weed detection, such as satellite-based positioning, crop-row following, and multi-spectral images. They have also drawn attention to the limitation of previously constructed detection systems, such as the lack of within-row plant-detection facilities.

In [38], the authors discuss DL techniques and architecture. In the former, they discuss Artificial Neural Networks (ANN), CNN, and Graph Convolutional Networks (GCN), and in the latter, they discuss image classification, object detection, semantic segmentation, and instance segmentation. They also mention the significance of public datasets, specifically carrot-weed, CWF-788, CWF-ID, DeepWeeds, GrassClover, Plant Seedlings, Sugar Beets 2016, Sugar Beet/Weed Dataset, and WeedCorn/Lettuce/Radish, to demonstrate how images were acquired, size of the dataset, pixel-wise annotation and modality. They also discuss data augmentation by mentioning limitations in the size of public datasets to work in varied conditions. They discuss fine-grained learning that overcomes the problem of general deep architectures, which ignores the challenges of similarities between crops and weeds, along with low-rank factorization, quantization, and transferred convolutional filters to solve the resource-consumption problems in analyzing real-time data for weed detection through DL. For the manual collection of datasets for weed, identification could be expensive, so weakly supervised and unsupervised methods can be necessary. For weakly supervised, object detection or segmentation can be used on image-level annotation, and for unsupervised learning, domain adaptation and deep clustering can be used. The existing methods for deep learning cannot deal with new species once a model is trained; to overcome this problem, incremental learning is proposed that is used to extend the existing trained model without retraining it.

Finally, in [39], the importance of reducing the use of herbicides is highlighted, and the authors review current and emerging technologies for this domain in the last 5 years. They classify the discussions into "digital image sensor-based" and "non-digital image sensor-based". In the former, the shapes and morphological features of weeds are used for detection, and in the latter, reflectance spectra are used to detect weeds. A complete workflow example of weed detection of Romaine lettuce has been discussed. This workflow shows the means of automatic weed detection using deep learning based on YOLO. In all the review papers, the authors did not conduct an SLR, and the focus is apparently to review performance over specific tasks rather than conduct a wide-ranging review of DL applications to weed detection.

Perhaps the paper most related to our work is [40], in which the authors review DL approaches to weed detection based on four steps: data acquisition, dataset preparation, weed detection, and localization and classification of weeds in crops. They develop a taxonomy for DL applications specifying the weed and crop type, the DL architecture applied, and the IP technique. In data acquisition, they detail how data or images have been collected, for example, using digital cameras, public datasets, camera moving vehicles, etc. They discuss and classify 19 public datasets according to several standard parameters, such as modality, dataset size, etc. In the data preparation phase, after acquiring images using different sources, images are prepared for training and testing, which includes different techniques, for instance, image processing, image labeling, image augmentation, etc. Weed detection is classified as a plant-based classification or a weed mapping approach. In the former, every plant needs to be localized in an image before detection, and in the latter, the density of the presence of weed in an image is used to detect that weed. In the last step, the authors discuss different algorithms, such as CNN, YOLO, FCN, GCN, and hybrid models, along with learning methods, such as supervised, unsupervised, and semi-supervised.

Several major differences distinguish our paper from [40]. Firstly, our process of review is more standardized because we conduct an SLR and answer concrete research questions (Sections 1 and 4). Secondly, we present a more thorough analysis of the SLR results, specifically through analyzing different combinations of algorithms, binning and analyzing individual algorithmic performance, specifying appropriate thematic labels, and analyzing the literature with respect to these labels (Section 7). Thirdly, we use a table to present a comprehensive association of algorithmic performance across different weed types and their respective crops, which provides a strong guideline to analyze current performance in the literature and to determine directions for future research (Section 6). Fourthly, we specify these directions more thoroughly to provide a type of road map for researchers of this domain (Section 8).

3. Background Knowledge

Before looking deeper into weed detection techniques, it is necessary first to understand weeds. This section discusses weeds, their various types, and weed detection algorithms.

3.1. Weed Types

Weeds can be generally classified as annual, biennial, and perennial [3–5]. Annual weeds germinate, bloom, and die within one year, while biennial weeds have a life cycle of two years, with germination and blooming happening in the first year and dying out in the second year. Perennial includes all weeds which last longer than two years in that they can germinate, bloom, and seed for several years. In our 60 extracted papers, authors have used a total of 34 weeds, of which we identified 26 annual and 8 perennial types. We illustrate these weeds in Figures 1–7 and discuss them in the following two sections.

Figure 1. From left to right: Pigweed, Blackgrass, Bluegrass, Dockleaf, Canadian Thistle (Source: [41]).

Figure 2. From left to right: Chickweed, Cleaver, Cockleblur, Crowfoot, Fat-hen (Source: [41]).

Figure 3. From left to right: Field pansy, Hare's ear mustard, Japanese hop, Jungle rice, Little seed (Source: [41]).

Figure 4. From left to right: Mayweed, Meadow grass, Nutsedge, Paragrass, Shepherd's purse (Source: [41]).

Figure 5. From left to right: Silky-bent, Turnip weed, Dicot, Grass Weed, Velvetleaf (Source: [41]).

Figure 6. From left to right: Benghal dayflower, black nightshade, hedge bindweed, Indian jointvetch, snakeweed (Source: [41]).

Figure 7. From left to right: Fescue grass, Chinee apple, *Lantana camara*, Sedge weed (Source: [41]).

For each weed, we label it by its commonly-used published name and mentioned scientific name in parentheses (wherever applicable). We extracted more detailed information about all these weeds from the Invasive Species Compendium section of the Cab Institute's website [42] and Wikipedia entries [43] along with websites of Garden Organic [44], Crop Protect [45], Gardening Know How [46], Lawn Weeds [47], Farms [48], and the USA Department of Agriculture [49]. Moreover, several weeds are categorized as both annual and perennial, for example, chickweed, but we have considered them as annual weeds for classification purposes.

3.1.1. Annual Weeds

In this section, we list the twenty-six (26) annual weeds used in our extracted papers. (1) Chickweed (*Stellaria media*), (2) Loose silky-bent (*Apera spica-venti*), (3) Velvetleaf (*Abutilon theophrasti*), (4) Shepherd's purse (*Capsella bursa-pastoris*), (5) Cleaver (*Galium aparine*), (6) Black nightshade (*Solanam nigrum*), (7) Blackgrass (*Alopecurus myosuroides*), (8) Littleseed canarygrass (*Phalaris minor*), (9) Crowfoot grass (*Dactyloctenium aegyptium*), (10) Jungle rice (*Echinochloa colona*), (11) Mayweed (*chamomile*), (12) Fat-hen (*Chenopodium album*), (13) Pigweed (*Amaranthus albus*), (14) Chinee apple (*Ziziphus mauritiana*), (15) Snakeweed (*Gutierrezia sarothrae*), (16) Indian jointvetch (*Aeschynomene indica*), (17) Fescue grass, (18) Bluegrass (*Poa annua*), (19) Meadow grass (*Poa trivialis*), (20) Hare's ear mustard (*Conringia orientalis*), (21) Turnip weed (*Rapistrum rugosum*), (22) Cocklebur (*Xanthium strumarium*), (23) Field pansy (*Viola rafinesquii*), (24) Japanese hop (*Humulus scandens*), (25) Dicot, and (26) Grass weed (*Monocot*). Collectively, the annual weeds can attack livestock and diverse cereal and vegetable crops, notably wheat, maize, sugar beet, tomato, cotton, rice, carrots, potato, peanuts, and corn. They also grow in different land types like crop fields, pastures, orchards, home lawns, grasslands, arable lands, roadsides, seashores, and wooded areas and in different soil types (sandy, dirty, wet). They pose severe management challenges: they can weaken the crop by more than 70% in some severe cases and can continue to germinate in the soil for decades, along with causing skin diseases to farmers and affecting the milk taste of livestock [6,50–52].

3.1.2. Perennial Weeds

The following eight (8) perennial weeds have been used in our extracted papers for weed detection: (1) Canadian thistle (*Cirsium arvense*), (2) Paragrass (*urochloa mutica*), (3) Nutsedge (*cyperus rotundus*), (4) Dockleaf (*Rumex obtusifolius*), (5) Benghal dayflower (*tropical spiderwort*), (6) Sedge weed, (7) Lantana (*Lantana camara*) [53], and (8) Hedge bindweed (*morning glory*). Collectively, like the annual weeds, these weeds can cause severe damage to crop fields, gardens, lawns, and other land types and can survive for several decades in diverse soil conditions [54–57]. They can grow in water and on profound soils in non-muddy areas and can attack sugarcane, chrysanthemum, rice, cotton, soybeans, peanuts, and corn crops.

3.2. Deep Learning (DL) Algorithms

Deep Neural Networks (DNNs) are extensions of Artificial Neural Networks (ANNs) in terms of complexity, number of connections, and hidden layers. A CNN is a DNN that assigns learnable weights and biases to various aspects and objects of input images to distinguish and classify objects such as weeds. CNNs do not require manual feature selection; rather, the network learns important features automatically from training data to reveal useful information hidden. CNNs are robust at classifying various objects with different scales, orientations, and levels of occlusion. CNNs capture the spatial and temporal dependencies of the input image through relevant filters autonomously and hence provide better and more efficient image processing with a considerably lesser number of estimable parameters and processing time.

Max pooling is generally preferred as it discards the noise (data values unreliable for machine learning) in the data and performs the de-noising operation. Due to the possibility of saturation in *sigmoid* and *tanh* activation functions, CNNs employ the rectified linear activation unit (ReLU) as the activation function $g(z)$, which outputs the aggregated value z if it is greater than 0, and 0 otherwise, i.e., $g(z) = max\{0, z\}$. Hence, it is a piece-wise linear function. The linearity of ReLU for $z > 0$ allows it to preserve many properties of the input to facilitate stochastic gradient descent [15] and generalize to unseen data. Moreover, batch normalization is performed in CNN, which breaks up the training image data into mini-batches and standardizes the mini-batch input data to each layer. This stabilizes the learning process and considerably reduces the training epochs required to train CNNs [58].

One application of convolution and max pooling with ReLU forms one layer of the CNN pipeline. Typically multiple such layers can be employed. In each layer, we can have parallel processing based on different color channels or feature maps, for example, RGB. The output from the last pooling layer is flattened as a 1-D vector and fed to the fully connected layer (i.e., conventional Multilayer Perceptrons (MLP)) for image classification, e.g., detecting weeds within a given image. MLP outputs the probability of occurrence of each possible object (on which the CNN has been trained) through the softmax activation function $\sigma(\vec{z})_i = e^{z_i} / \sum_{j=1}^{K} e^{z_j}$ where σ is the softmax function for i-th activation input vector $\vec{z}_i$, K represents the number of classes, and e^{z_i} and e^{z_j} represent the standard exponential functions for input and output vectors, respectively.

3.3. Variants of CNNs

The first usable and concrete CNN architecture was LeNet-5, proposed by Yann LeCun in 1998 and developed to recognize handwritten and printed characters [59]. It has a 2-layered architecture with 6 feature maps in the first layer and 16 feature maps in the second layer, followed by two fully-connected layers. A key outcome of this work is that larger image sizes can distinguish more pixels for the stroke end-points for written characters. After LeNet-5, ImageNet [60] has motivated researchers to propose enhancements leading to significant reductions in top-5 error percentages, i.e., the proportion of miss-classified images appearing in the top-5 results sorted in decreasing order of predictive confidence $P(Y_i|X_i)$ (where X_i is the input test data and Y_i is the class label under consideration). For instance, AlexNet [61] is trained on 1.2 million images to achieve the lowest top-5 error rate of 16%, with five convolutional layers, followed by three fully connected dense layers. The authors used ReLU activation in all layers except the last layer, which employed softmax activation. Moreover, VGGNet 16 [62] is a deeper network than AlexNet with a top-5 error rate of 7.5%, with five CNN layers followed by three fully connected layers. VGGNet needs to estimate approximately 140 million parameters for training; however, due to the availability of pre-trained models, VGGNets are still being employed for several image classification tasks.

GoogleNet [63] achieved a top-5 error rate of 6.7% (almost equal to human-level performance) with its *inception modules* merging convolutional operations together rather than implementing them in different layers, and the concatenated output shows results from all convolutional operations. It employs 22 layers containing 9 inception module layers inserted between several pooling, convolutional, and fully connected layers with a drop-out layer used to drop input neurons from processing randomly to prevent over-fitting. GoogleNet achieves a significant reduction in the number of parameters to be estimated (4 million) as compared to AlexNet (60 million) and LeNet-5 (more than 100 million). Inception Module V1 used by GoogleNet was later upgraded to Inception Module V4 and Inception ResNet.

The ground-breaking research in CNN was achieved by ResNet (Residual Network) with a top-5 error rate of 3.6% (better than human performance) and remains unbeatable to date [64]. ResNet is a deep CNN with 152 layers which provides a solution to the vanishing gradient problem, i.e., the gradient becomes very small as it keeps on getting multiplied during backpropagation until it stops influencing any weight updates (learning stops). ResNet assumes that deeper layers should not generate more training errors than the shallower ones. Hence, it employs skip-connections, which transfer the results of a few layers to deeper layers while skipping some layers in between, hence preventing deeper layers from producing higher training errors than shallow layers. The gradient flow through the shortcut connection to the earlier layers, thus, reducing the vanishing gradient problem.

Along with this, SegNet [65] has been used for weed detection through image segmentation. It comprises an encoder network and a decoder network, much similar to Autoencoders [12]. In encoding, convolutions are performed using the 13 convolutional layers from VGGNet, followed by 2×2 max pooling to generate an encoded representation. In decoding, the max pooling indices from the encoding phase are employed to upsample the encoded data; for example, the 2×2 matrix is upsampled to a 4×4 matrix, and convolutions are also applied during the upsampling operations. Finally, the softmax function is applied at the end. In essence, there are no fully connected layers after decoding. Rather 1×1 convolutions are used, which allows outputting of a label for each pixel (a requirement for image segmentation) rather than a label for the whole input image. Such a setup is also called a Fully Convolutional Network (FCN). Another FCN-based algorithm is U-Net [66], which is used for biomedical image segmentation. It does not employ pooling indices during the decoding phase. Rather, the entire feature maps are transferred from encoder to decoder to acquire better segmentation performance but at the cost of time and memory. This makes U-Nets computationally intensive as compared to SegNets.

Finally, Deeplab [67] is a series of image segmentation algorithms invented by Google (Deeplabv1, Deeplabv2, Deeplabv3, Deeplabv3+) in 2018. The iterative application of pooling operations in FCNs reduces the spatial resolution of images. Deeplab uses atrous convolutions to generate much denser decoded feature maps with lesser computational overhead. It also enhances the localization accuracy in FCNs through the use of conditional random fields.

3.4. Machine Learning Algorithms

We now briefly describe the more important Machine Learning (ML) algorithms (for more details, refer to [68–71]). Support Vector Machines (SVMs) estimate an optimal hyperplane between data points to linearly separate two classes by maximizing the margin with respect to the closest points called support vectors. Mathematically, from the equation $y = m*x + c$, we can have $y = a*x + b$ and $a*x + b - y = 0$. Suppose we have vectors $X = (x, y)$ and $W = (a, -1)$, then the vector in hyper-plane become $W*x + b = 0$. Assume n training instances with each instance x of D dimension and belonging to class $y = +1$ or $y = -1$. Then the training would be x_i, y_i where $i = 1 \cdots n$, $y_i \epsilon -1, 1$, and $x \epsilon \mathbb{R}^D$. If D=2, then hyper plane would be described as follows: for $y_i = 1$ as $y_i(W*x + b) >= 1$ and for $y_i = -1$ as $y_i(W*x + b) <= -1$. This leads to equations $h1 : w.x + b = -1$ and $h2 : w.x + b = 1$ for two lines forming the hyper-plane. The distance between $h1$ and the starting point is $(-1-b)/|w|$ and the distance between $h2$ and starting point is $(1-b)/|W|$. The maximum distance between $h1$ and $h2$ is called the margin M: $M = (-1-b)/|w| - (1-b)/|w| = 2/|w|$.

Decision Trees (DTs) model data as a tree whose nodes represent features as decision points, branches as feature values, and leaf nodes as class labels. Different patterns of label classification can then be extracted from the root node to each leaf node. At each decision node, features are selected at each node based on statistical criteria, mostly information gain. Specifically, the entropy of any partition of a dataset D can be expressed as $Entropy(D) = -\sum_i^n p_i * log_2(p_i)$, where p is the probability of occurrence of an instance i in n total instances. The Information Gain $G(D, A)$ represents the change in entropy of D when we consider feature A for decision node: $G(D, A) = E(D) - \sum_i^f (|D_f|/|D|) * E(D)$, where f represents all possible values of F, $|D|$ represents total instances in D, and $|D_f|$ represents the number of rows containing the particular value f. Random Forest (RF) is a well-known DT ensemble algorithm that employs bootstrap aggregation (bagging) to generate m ($m < N$) datasets $D_1, D_2, \cdots, D_m$ by sampling D uniformly and randomly with replacement. A set of m DTs $h_1, h_2, \cdots, h_m$ is generated for each dataset. An unseen instance is then tested on each tree, and the class with the majority vote from amongst all m trees is output as the final predicted value.

In the same vein, Adaptive boosting (Adaboost) uses an ensemble boosting technique to construct a strong learner from a number of weak learners, which are typically DTs. In each iteration, it adapts by finding miss-classified data points from each learner and increases their weights (to learn them with more emphasis in the next iteration) while decreasing the weights of correctly classified points (to learn them with less emphasis in the next iteration). As long as the performance of each learner is better than random guessing, Adaboost is guaranteed to converge to a strong learner. A boosted classifier over T weak classifiers can be represented as $F_T(x) = \sum_{t=1}^{T} f_t(x)$ where each f_t is a weak learner that takes x as input and outputs the class label. A hypothesis $h(x_i)$ for each sample in training data is output by each f_t. At iteration t, each weak learner is assigned a coefficient α_t to minimize the following sum of training error E_t: $E_t = \sum_i E[F_{t-1}(x_i) + \alpha_t h(x_i)]$ where F_{t-1} is the boosted classifier of the previous stage, $E(F)$ represents the error function and $f_t(x) = \alpha h(x)$ is the weak learner under consideration for addition to the Adaboost classifier.

Artificial Neural (ANNs), more commonly known as Multilayer Perceptrons, are ML versions of the CNNs described earlier. They can model complex non-linear stochastic relationships between predictors and label through layers of neurons (processing units). Predictor data are fed to an input layer, processed over one or more hidden layers, and predictions are generated at the output layer. The neurons between each pair of layers are connected to each other through synapses called *weights*. The weight vectors are updated based on numerical values output from a mathematical *activation function* from each neuron in the hidden and output layer, based on the aggregated input at each neuron. A sample output can be represented as $h_i = \sigma(\sum_{j=0}^{N} V_{ij} x_i + T_i^h id)$ where σ is the activation function, N is the number of input neurons to a given neuron, v_{ij} are the weights of these input neurons, x_i shows the input values to input neurons, and T is the threshold for activation.

Moreover, the k-nearest neighbor (KNN) categorizes the input x by its k nearest neighbors. For k, it will observe the adjacent neighbors of hidden data points and assign the data point to a class with the highest number of data points from all classes of k neighbors. It uses Euclidean distance when it calculates the probabilities. KNN gives the input x to the class which has the highest probability: $P(Y = j | X = x) = 1/K \sum_{i \in a} I(y_i = j)$, where a is the set of k nearest neighbors and $I(y_i = j)$ is an indication variable which calculates to 1 if a given neighbor (x_i, y_i) in a is a participant of class j, else it calculates to 0. Finally, K-means is primarily used for cluster analysis. It divides the data into k predefined unique clusters (collection of data points with similar features) where each data point should preferably belong to only a single cluster. It initially sets k centroids randomly and assigns every data point to its nearby cluster. It calculates the centroids for each cluster by averaging all the data points belonging to that cluster. The Euclidean distance between a data point q and centroid p is typically calculated as $d(q, p) = \sqrt{\sum_{i=1}^{n}(q_i - p_i)^2}$, where n is the the number of features.

3.5. Image Processing Techniques

Image Processing (IP) can be used to improve images for further processing with DL or ML algorithms. IP facilitates algorithm tasks by improving image quality and transforming images to meet the needs of the algorithm. Local Binary Patterns (LBP) [72] is a visual descriptor of images in CV based on thresholding. It divides an image into equal-sized cells, for example, with each cell containing 16×16 pixels. Each *center* pixel c is then compared to each of its 8 neighbors n, for example, clockwise starting from top-right, middle-right, then bottom-right, and so on. The thresholding works as follows: if the value of n is greater than the value of c, we set $n = 0$; otherwise, we set $n = 1$, giving us an 8-digit binary number. Then, we compute a histogram over c indicating the frequency of each of the 256 (2^8) combinations of this binary number. Finally, we concatenate the histograms of all cells to form a feature vector for the whole image, which can then be processed in machine learning and deep learning tasks. Mathematically, LBP for pixel c over a neighboring radius

r (set to 8) is estimated as follows: $LBP_{c,r} = \sum_{i}^{P-1} s(v_i - v_c) 2^i$ where P is the number of neighboring pixels, v_i and v_c are values of the neighboring and center pixel respectively, $s(t)$ is thresholding such that $s(t) = 1$ if $t > 0$ and 0 otherwise. The histogram feature vector of size 2^P (256) is then estimated from the obtained LBP code.

Moreover, Simple Linear Iterative Clustering (SLIC) [73] is an image segmentation algorithm that uses k-means clustering to create *superpixels*, which are small-sized clusters of pixels sharing common features. Clustering is done by distance measurement computation in 5D (*labxy*) space, where (*l*, *a*, *b*) is the 3-dimensional color representation of the pixel at coordinate (x, y). The distance measure D_S is then defined as follows: $D_S = d_{lab} + (m/s) * d_{xy}$ where $d_{lab} = \sqrt{(l_k - l_i)^2 + (a_k - a_i)^2 + (b_k - b_i)^2}$ and $d_{xy} = \sqrt{(x_k - x_i)^2 + (y_k - y_i)^2}$ and m controls the density of superpixels proportionally. Moreover, Histogram of Gradients (HoG) [74] is used for feature extraction. It divides an image into a number of *regions*, and for each region, it estimates the gradient (magnitude) and the orientation (direction) of the edges in that region. Then, the histogram of this data (HoG) for each region is generated separately. Suppose that for each pixel (x, y), we define HG as the distance between the adjacent right and adjacent left pixel values and VG as the distance between the adjacent top and bottom pixel values. The gradient magnitude GM is $GM = \sqrt{(HG)^2 + (VG)^2}$ and the gradient angle GA is $GA = tan^{-1}(HG/VG)$ Then, HoG is generated by binning the frequencies of either GM or GA or both together [74].

In addition, Hilbert Transform (HT) [18] is used to separate features of a specific shape within an image, for example, circles, lines, and ellipses. A line is a collection of single points with slope m and intercept c and $y = mx + c$ in the xy plane. In HT, we convert a line from (x, y) plane to (m, c) space, i.e., from $y = mx + c$ to $c = -mx + y$. To avoid unbounded values of m, the well-known Hough space (r, θ) transformation can also be used as follows: $r = x.cos\theta + y.sin\theta$. Moreover, Median filtering [17] is a non-linear IP technique that maintains edges while removing noise. It calculates the median gray-scale value of a pixel's neighborhood. In applying a fixed-size kernel, we sort all pixel values within this kernel based on gray-scale values. Then, the median value of this sorted array will be used, and zeros can be padded in rows and columns to complete the pixel count. Finally, Background Subtraction (BS) [75] is a well-known technique used in IP and CV for detecting moving objects in videos from static cameras for additional processing. It isolates these foreground objects with respect to a reference image by subtracting the current frame from a reference frame called the background model. If the data points are non-linear, then we need to add one more dimension to the data point, which will be $z = x^2 + y^2$.

GANs (generative adversarial networks) [76] are types of generative deep learning algorithms whose purpose is to learn a set of training samples and their probability distributions and then generate data from this distribution. GANs can produce more samples based on the measured probability distribution and are particularly accurate in producing realistic high-resolution images. GANs comprise two different feed-forward artificial neural networks named Generator (Gen) and Discriminator (Dis) that participate in an adversarial game. The input to the generator is Gaussian noise $p_z(x)$ and Gen tries to generate an approximation $p_{model}(x)$ to the probability distribution of the actual data $p_{data}(x)$. Meanwhile, Dis learns to distinguish whether a data point x is sampled from $p_{data}(x)$ or $p_z(x)$, the latter being input to Dis as data sampled from $p_{model}(x)$. The task of Gen is to fool Dis into thinking that data sampled from $p_{model}(x)$ is actually the data sampled from $p_{data}(x)$. Therefore, Dis maximizes the probability of classifying data as $p_{data}(x)$ and minimizes the probability of classifying it as $p_{model}(x)$, while Gen tries to do the exact opposite. In this context, both Dis and Gen participate in a two-player minimax game with the value function $Val(Dis, Gen)$ as $min_{Gen} max_{Dis} Val(Dis, Gen) = E_x \sim p_{data}(x)[log Dis(x)] + E_z \sim p_z(z)[log(1 - Dis(Gen(z))))]$ where Gen maximizes $log Dis(Gen(z))$ rather than minimizing $log(1 - Dis(Gen(z)))$ [12,76].

4. SLR Methodology

In this section, we address RQ1: What is the trend of employing deep learning to address the problem of weed detection in recent years. To answer this, we conduct an SLR by following the standard methodology [77] and dividing our work into three phases: (1) Planning, (2) Execution, and (3) Reporting (see Figure 8).

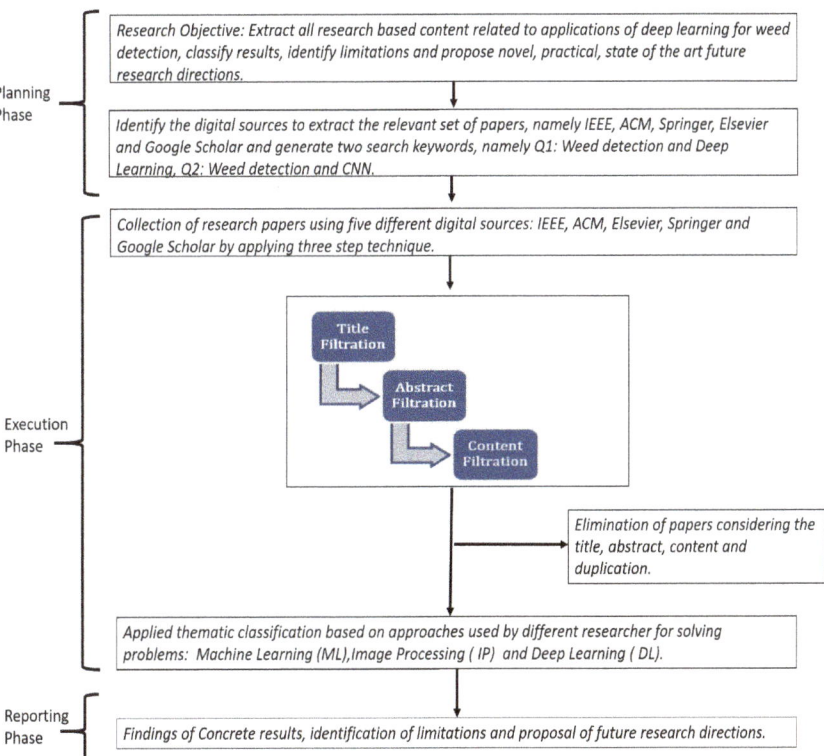

Figure 8. Our SLR process.

In the planning phase, we identify the research objective, research protocols, search keywords, and digital sources for extracting the relevant papers. In the execution phase, we execute our search queries on each of the identified digital sources to acquire the relevant corpus of papers by using a three-step technique (described below) and eliminating duplicates. In the reporting phase, we apply thematic classification to our final list of extracted papers and describe them in detail, identify the limitations of these works, and then propose a concrete set of future work recommendations.

In this SLR, our specific consideration is in the domain of DL applications in weed detection. Our research objective is to extract the state-of-the-art, identify published academic research related to this domain, understand the content of these papers, classify our results using different methods of analyses, identify the gaps or limitations through these classifications, and consequently propose guidelines and directions to motivate and enhance the state-of-the-art research. To achieve this, we adopted the following inclusion and exclusion criteria.

- We targeted original academic research content published in journals, conferences, workshops, and symposiums. We excluded periodicals (magazines and news from newspapers), letters, books, and online content, specifically websites, blogs, and social network feeds.

- We considered papers published in the English language only.
- We selected the following digital sources: IEEE, ACM, Elsevier, Springer, and Google Scholar. Our previous experience [78,79] has shown us that these sources are collectively effective in retrieving required content for data analytics, machine learning, and any computer science domain in general. Moreover, Google Scholar can effectively index published data from other sources, for instance, Taylor and Francis, Wiley, MDPI, and Inderscience.
- Initially, we focused on research published from 2010 and onwards. However, after some preliminary results, we discovered that the content most relevant to the domain of weed detection using DL was published primarily from 2015 onwards. Therefore, we focus our SLR from Jan 2015 till Jan 2023.

The published DL research has seen an exponential rise in different domains in the last several years, for example, after the proposal of generative adversarial networks in 2014 and the discovery of different variants of CNNs, autoencoders, and recurrent neural networks (RNNs). Hence, we decided to focus only on the latest research from 2015 till January 2023. We considered all articles irrespective of the country of the first authors. Moreover, we considered three types of research publications: (1) application papers, i.e., papers that present a novel research idea along with experimental results; (2) literature reviews (both systematic and non-systematic) and (3) frameworks, i.e., papers that present a novel research framework/idea with a concrete design but it has not been validated with experimental work.

We formulated our search queries from our four research questions (mentioned in Section 1). In these questions, we focused on discovering important information on smart farming, particularly deep learning, applications for weed detection, for instance, the different research trends and statistics, types of weeds detected and algorithms used, and performance comparison of algorithms. From our previous experience [78,79], we concluded that all this information could be extracted by using search queries based on different combinations of the following three keywords: (1) weed detection, (2) smart farming, which is used interchangeably with *smart agriculture* and *precision farming*, (3) Weed Classification, and (4) deep learning, in which we particularly targeted CNNs. Based on this, we initially executed the following nine search queries (& = AND): (1) {"weed detection"}, (2) {"precision farming" & "weed detection"}, (3) {"precision agriculture" & "weed detection"}, (4) {"smart farming" & "weed detection"}, (5) {"weed detection" & "deep learning"}, (6) {"precision farming" & "weed detection" & "deep learning"}, (7) {"precision agriculture" & "weed detection" & "deep learning"}, (8) {"smart farming" & "weed detection" & "deep learning"}, (9) {"weed detection" & "CNN"}, (10) {"weed classification"}, (11) {"precision farming" & "weed classification"}, (11) {"precision agriculture" & "weed classification"}, (12) {"smart farming" & "weed classification"}, (13) {"weed classification" & "deep learning"}, (14) {"precision farming" & "weed classifcation" & "deep learning"}, (15) {"weed classification" & "CNN"}

The results from these queries showed that the most relevant papers could be obtained only through the following two queries (which we also used in our SLR): (1) {"weed detection" & "deep learning"} (labeled as Q1), (2) {"weed detection" & "CNN"} (labeled as Q2) and (3) {"weed classification" & "CNN"} (labeled as Q3). Q1 and Q2 also retrieved articles related to applications in ML and IP (without any DL implementation) for weed detection. We considered these papers in our SLR to facilitate a comparison of these algorithms with DL to understand further the strengths and limitations of these approaches (RQ4 in Section 1).

We implemented a three-step procedure to filter out our required subset of research articles (shown in the execution phase in Figure 8). In the first step, we filtered out the articles based on their titles, i.e., we did not consider articles whose titles were not related to the domain of weed detection using DL, for instance, several titles related to smart farming but no research contribution to weed detection. In the second step, we adopted the same approach to filter articles from the first step based on their abstracts, and in the third step, we filtered articles from the second step based on their content, i.e., after reading the article's introduction, methodology, and results section.

Across all the digital sources, we filtered out a total of 129 articles from the title filtration step, of which 25 were duplicates. Thus, we filtered out 92, 81, and 64 articles after the title, abstract and content filtration, respectively. The breakdown of these numbers with respect to each digital source (IEEE, ACM, Elsevier, Springer, Google Scholar) and search query (Q1, Q2, and Q3) is shown in Table 1. Of our 64 articles, 49 (83%) were retrieved by Q1 alone, 10 by Q2, and the remaining 4 by Q3. Across Q1, Q2, and Q3, IEEE retrieved 29 (53%) of these 64 articles, while ACM retrieved 2 (3%) only. The 8 articles were retrieved from Springer, 10 from Elsevier, and 15 from Google Scholar. All the above trends are also applicable for title filtering and abstract filtering data.

Figure 9 shows the frequency distribution of our filtered 64 articles from January 2015 to January 2023. We observed an exponential trend in the number of publications from 2015 onwards. Moreover, Figure 10 demonstrates that out of our 64 articles, 55 were application papers, 8 were literature reviews, and only 1 article introduced a framework. Finally, Figure 11 shows the co-author citation graph for our 64 papers. Out of a total of 221 authors in these papers, the presented 13 authors in Figure 11 have the strongest co-authorship links. The colors red and green represent two clusters of co-authorship links with the author Arnold W. Schumann participating in the red cluster (in the years 2020 and 2021) [38] as well as in the green cluster (in 2019) [80,81].

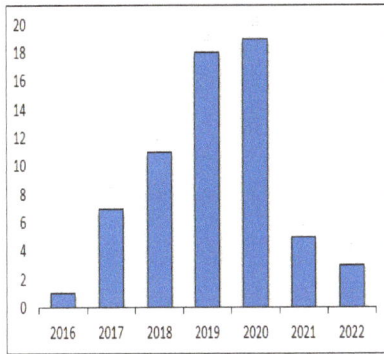

Figure 9. Year Wise Distribution of Articles.

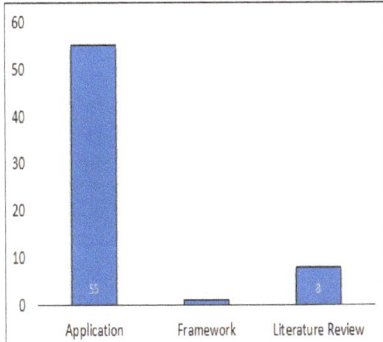

Figure 10. Article types and Frequency.

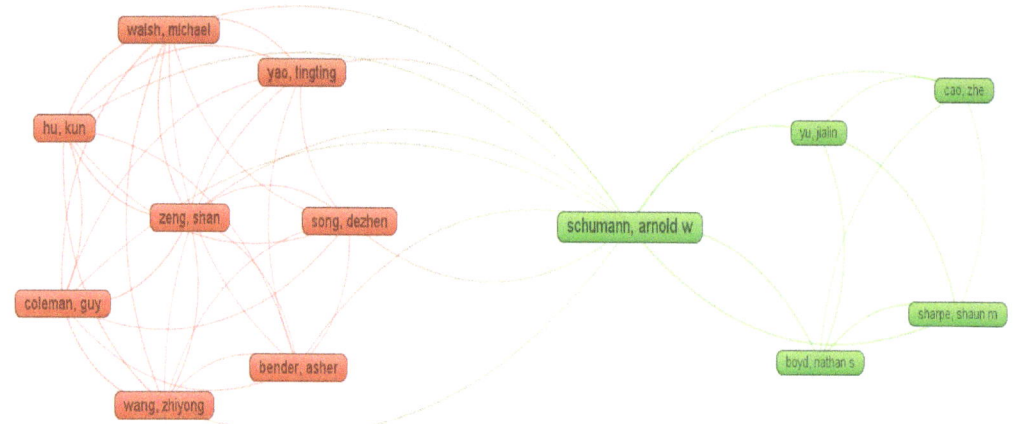

Figure 11. Co-Author Citation Graph.

Table 1. Breakdown of frequency of articles filtered with respect to title, abstract, and content across search queries Q1. Q2 and Q3 and digital sources.

	Q1			Q2			Q3		
Digital Sources	Title	Abstract	Content	Title	Abstract	Content	Title	Abstract	Content
IEEE	50	42	25	4	4	4	0	0	0
ACM	3	2	1	2	2	1	0	0	0
Elsevier	8	7	8	3	3	2	0	0	0
Springer	8	7	6	3	3	2	0	0	0
Google Scholar	10	5	9	9	2	2	11	6	4

5. Review of Weed Detection Algorithms

In this section, we summarize the findings of our surveys on application articles and literature reviews. For convenience, we merged the single paper, which proposed a framework [82] into the application articles (i.e., implementation) category.

5.1. Weed Datasets Available for Deep Learning

In this section, we answer RQ2: Which types of weeds and corresponding crops have been detected using deep learning, and what are the characteristics of the corresponding weed datasets? In this regard, we extracted and classified important characteristics regarding the datasets of weed images used by researchers of the application papers, shown in Table 2. These are (1) corresponds to the reference article, (2) the dataset name or label (Dataset), (3) the size or the number of images in the dataset (Size), (4) the type of crop for which the weed was detected by the authors (Crop), (5) the particular weed type which was detected (Weed), (6) the modality of the dataset (Modality), (7) the data collection technique through which images were acquired (Data Collection), and (8) the resolution of the images (Resolution). We use N/M (Not Mentioned) to indicate any of this information not mentioned by the authors.

Table 2. Information regarding weed datasets used in application papers.

Article	Dataset	Size	Crop	Weed	Modality	Data Collection	Resolution
[32]	Gharo	N/M	Wheat	N/M	N/M	Drone	4000 × 3000
[33]	Sensifly	1500	Wheat, Maize, Sugar beet	Chickweed, Blackgrass	RGB	Referred Dataset	150 × 150
[83]	IARI	60	Rice, Maize	Littleseed canarygrass, Crowfoot, Jungle rice	N/M	Referred Dataset	800 × 450
[84]	Tegucigalpa	N/M	Vegetables	N/M	RGB	Semi-Professional Camera	4512 × 3000
[85]	Carrot Field	N/M	Carrot	N/M	N/M	Camera	N/M
[28]	Tumakuru	2560	Chrysanthemum	Paragrass, Nutsedge	RGB	Digital Camera	250 × 250
[31]	Heide Weed	2907	Carrot	N/M	RGB	Robot(BoniRob)	832 × 832, 416 × 416 and 288 × 288
[86]	WeedMap	11,441	Sugar beet	N/M	RGB, CIR	Drone	480 × 360
[87]	Dundee	6087	Pasture	Dockleaf	RGB, Greyscale	Camera	64 × 64
[88]	Plant Seedling	5539	Sugar beet, Maize, Wheat	Mayweed, Chickweed, Blackgrass, Shepherd's purse, Cleaver, Fat-hen, Loose silky-bent	N/M	Referred Dataset	N/M
[89]	Brimrose	N/M	N/M	N/M	Hyperspectral	Brimrose VA210, JAI BM-141 camera	500 × 500, 250 × 250, and 125 × 125
[90]	Campo Grande	N/M	Soybean	Dockleaf	RGB	Drone	4000 × 3000
[91]	Luye	N/M	N/M	N/M	RGB	Drone	1920 × 1080
[30]	Bosch BoniRob	1300	Sugar beet	Shepherd's purse	RGB + NIR	Robot	1296 × 966
[82]	Cherry Research Farm	N/M	Strawberry	Fescue grass	RGB	Referred Dataset	N/M
[92]	Bosch BoniRob, Stuttgart	N/M	Sugar beet	N/M	RGB + NIR	Robot	N/M
[93]	DeepField, Bosch BoniRob	N/M	Sugar beet	Dicot, grassweed	RGB, RGB + NIR	Drone, Robot	N/M
[94]	Crop Weed	291	N/M	N/M	RGB	Camera	N/M

Table 2. Cont.

Article	Dataset	Size	Crop	Weed	Modality	Data Collection	Resolution
[95]	CWFID	N/M	Carrot	N/M	RGB (Multispectral)	Robot	N/M
[96]	CropDeep	1000	Corn	Bluegrass, Fat-hen, Canadian thistle, Sedge weed	N/M	Referred Dataset	224 × 224
[97]	CWFID	120	N/M	N/M	RGB (Multispectral)	Referred Dataset	N/M
[98]	Bosch BoniRob	N/M	Sugar beet	N/M	RGB + NIR	Robot	1296 × 966
[99]	CWFID	60	Carrot	N/M	RGB (Multispectral)	Referred Dataset	N/M
[100]	CWFID	60	Carrot	Mayweed	RGB (Multispectral)	Referred Dataset	N/M
[81]	Golf Field	36,000	Turfgrass	Dockleaf	N/M	Camera	640 × 360
[101]	Manitoba	906	Canola	N/M	RGB	Camera	1440 × 960
[102]	Campo Grande	400	Soybean	Dockleaf	RGB	Drone	4000 × 3000
[103]	Shiraz University	N/M	Sugar beet	Pigweed, Fat-hen, Hare's-ear mustard, Turnip weed	RGB	Camera	960 × 1280
[104]	ImageNet	N/M	Cereal	Canadian thistle	RGB	Canon PowerShot G15 Camera	200 × 200
[105]	Ecology	3266	Maize, Wheat, Peanut Seedlings	Fat-hen, Japanese hop, Cocklebur	N/M	Drone	227 × 227
[106]	Rumex 100	900	Grass	Dockleaf	N/M	Referred Dataset	200 × 150
[107]	Tobacco Field	76	Tobacco Seedling	N/M	N/M	Camera	65 × 65
[108]	Bosch BoniRob	900	Sugar beet	N/M	RGB + NIR (Multispectral)	Robot	61×61
[109]	Bean, Spinach	5534	Bean, Spinach	N/M	RGB	Drone	64 × 64
[110]	Cereal	N/M	Cereal	N/M	N/M	Camera	1224 × 1024
[80]	Griffin	4550	Turfgrass, Bahigrass	Dockleaf	RGB	Digital Camera	640 × 256
[111]	Heide Feldhof Farm	796	Wheat	Mayweed, Meadowgrass, Chickweed, Field pansy, Pigweed	N/M	Terrestrial Images (Cellphone Camera)	N/M
[112]	DeepLab	N/M	Sugar beet	N/M	RGB, NIR, CIR, NDVI	Drone	480 × 360
[113]	Bosch BoniRob, OilSeed	280	Sugar beet	N/M	RGB, NIR	Robot, Camera	1296 × 966

Table 2. *Cont.*

Article	Dataset	Size	Crop	Weed	Modality	Data Collection	Resolution
[114]	Bok Choy	11,150	Chinese White Cabbage	N/M	RGB	Camera	512 × 512
[115]	Reduit	15,336	Soybean	Dockleaf	N/M	Drone	227 × 227.
[116]	MonoDicot	N/M	Ragi	Grass weed, Dicot	N/M	Camera	N/M
[117]	DeepWeeds	17,509	N/M	Chinee apple, Snake weed, Lantana	RGB	Referred Dataset	256 × 256
[118]	Radish Weed	6000	Corn, Lettuce, Radish	Nutsedge, Fat-hen, Canadian thistle	RGB	Camera	800 × 600
[119]	Bonn, Stuttgart	9070	Sugar beet	N/M	RGB-NIR	Camera	512 × 384
[120]	Greece Farm	504	Tomato, Cotton	Black nightshade, Velvetleaf	RGB	Camera	128 × 128
[121]	FUT Farm	5400	Rice	Indian jointvetch, Benghal dayflower, Jungle rice	N/M	Camera	250 × 250
[122]	Flanders	652	Sugar beet	Hedge Bindweed	RGB	Camera	800 × 1200
[123]	Annotated Imagery Dataset	462	Corn, Soybean	Cocklebur, foxtail, redroot pigweed and giant ragweed	RGB	Sony WX350, Panasonic DMC-ZS50	1200 × 900
[124]	UAV Imagery	N/M	Soybean	N/M	N/M	UAV	N/M
[125]	UAV Imagery	N/M	Spinach, beet, Bean	N/M	N/M	UAV	7360 × 4972
[126]	E.maculata,	N/M	N/M	dandelion, ground ivy, spotted spurge.	N/M	UAV	426 × 240
[127]	Giselsson	5539	N/M	N/M	N/M	Referred Dataset	N/M
[128]	Beni Mellal-Khenifra	1318	Cereal crops.	Monocotyledon and dicotyledon	N/M	Nikon 7000 camera	N/M
[129]	Dataset	9200	Vegetables	N/M	N/M	digital camera	2048 × 1536
[130]	Dataset	1385	Sugar beet	N/M	RGB	digital camera	480 × 640

Table 2 shows that researchers in application papers have used a total of 44 unique weed image datasets. Under 'Dataset,' we present the dataset names as labeled by the corresponding researchers in their papers. Where no specific name was assigned to a dataset by the authors, we have used the city's name wherein the implementation was done as dataset name (e.g., Gharo (Pakistan)). Most researchers have created their personalized weed image datasets by using drones, robots, and a large variety of cameras. However, three available datasets have also been used: (1) Indian Agriculture Research Institute (IARI), (2) Crop Weed Field Image (CWFID), and (3) Plant Seedling (available on Kaggle). The three largest datasets are Gold Field (36,000), WeedMap (11,441), and Bok Choy (11,150), while the three smallest datasets are CWFID (120), Tobacco Field (76), and IARI (60).

A total of 26 unique crops are used in weed detection. Notably, sugar beet is the most common crop (used in 13 papers), followed by carrot and wheat (in 5 papers), maize (4 papers), and rice, soybean, corn, and cereal (in 4 papers). All the remaining crops are used once in our filtered 51 application papers. As far as the distribution of 34 weed types is concerned, Dockleaf is used for weed detection most frequently (in 7 papers), followed by fat-hen (5 papers), Canadian thistle, Chickweed, and Mayweed (in 3 papers), and Blackgrass, Jungle rice, Nutsedge, Shepherd's purse, Fescue grass, Grass weed and Pigweed (in 2 papers). The remaining 22 weeds are used once. Regarding the input modality, RGB is used most frequently in 33 articles, 5 times in combination with Near Infrared (NIR), and 5 times in multi-spectral mode. Moreover, hyperspectral, NIR, grayscale, and NDVI modes are used once, while Color Infra Red (CIR) is employed twice. Regarding data collection, drones (as UAVs) are used to acquire images in 13 papers, cameras are used in 22 papers (with different types such as digital, cellphone, and professional), robots are used in 9 papers, and the remaining are referred to as datasets. Here, we would like to mention about Bonirob, an agricultural robot developed in Germany and used to acquire images in 3 articles. A variety of resolutions are also employed, ranging from a minimum of 61×61 and 64×64 to a maximum of 4512×3000 and 4000×3000. Out of 51 papers, 14 have used a resolution greater than 1000, and the remaining 36 have a resolution less than 1000. Most high-resolution images have RGB Modality, and low resolution have NIR, CIR, Greyscale, hyperspectral, and NIR + RGB. High-resolution images are mostly snapped by drones, while low-resolution images are snapped through semi-professional or cell phone cameras. Only 35 articles have mentioned the size of the dataset, with the GoldField dataset having the most images (36,000) and CWFID being used in two papers with the fewest images (60).

5.2. Algorithms Used for Weed Detection

In this and the next section, we collectively answer RQ3: which deep learning algorithms are best suited for a particular weed/crop combination? In this regard, we initially extract and classify the algorithmic performance of our application papers, shown in Table 3. Here, we present (1) the article reference (Article), (2) the name of the algorithm(s) employed by the authors (Algorithm), (3) the image processing technique used, if any (IPT), (4) performance measure or KPI, such as accuracy or precision/recall (KPI), (5) the maximum value of this KPI achieved by the authors (Result), (6) the training time taken by the algorithm (TR.Time), (7) the split ratio for train and test sets (TR.TS.Split), and (8) a thematic classification label (TCL) which we assigned to each paper based on the algorithm and the weed detection approach used in the paper.

Table 3. Algorithmic Performance and Related Data for our application papers.

Article	Algorithm	IPT	KPI	Result	TR.Time	TR.TS.Split	TCL
[32]	N/M	Background Subtraction	Accuracy	67%	N/M	N/M	IP
[33]	CNN	N/A	Accuracy	97%	N/M	72–27%	DL.CNN
[83]	GoogleNet	N/A	Accuracy	98%	N/M	83–16%	DL.CNN
[84]	N/M	CV (rgb2gray, im2bw, bwlabel, regionprops)	Accuracy	99%	2.98 s	N/M	IP
[85]	CNN	N/A	PR, RC	91.1%, 86.8%	1895 min (CPU), 976 min (GPU)	N/M	DL.CNN
[28]	SVM, ANN, CNN	Median and Gaussian filter	Accuracy	87%, 93%, 98%	N/M	90–10%	ML.DL.CNN
[31]	YOLO, GoogLeNet	N/A	Accuracy	89%, 86%	N/M	90–10%	DL.CNN
[86]	ResNet18, SVM	Hough transform, SLIC	Accuracy	90%	N/M	N/M	ML.DL.CNN
[87]	SVM, KNN, Ensemble Subspace Discriminant, CNN	CV	Accuracy	89%, 84%, 87%, 93.15%	N/M	80–20%	ML.DL.CNN
[88]	Mask R CNN, FCN	N/A	Accuracy	>90%, <90%	N/M	60–40%	DL.CNN
[89]	CNN	HoG	Accuracy	88% (CNN)	N/M	60–40%	DL.CNN
[90]	SVM, ANN	SLIC	Accuracy	95%, 96%	0.0211	60–40%	ML
[91]	CNN	LBP, HoG	Accuracy	96% (CNN)	N/M	N/M	DL.CNN
[30]	SegNet, SegNet-Basic	N/A	Accuracy	84%, 98%	0.14 s, 0.08s	N/M	DL.CNN
[82]	N/M	Feature Extraction, Classification	N/M	N/M	N/M	N/M	IP
[92]	FCN	N/A	PR, RC	97.9%, 87.8%	N/M	N/M	DL.AE
[93]	FCN	N/A	Avg. PR	87.90%	N/M	N/M	DL.AE
[94]	RF	N/A	Accuracy	97%	57.4 ms	75–25%	ML
[95]	N/M	Foreground Extraction, Image Tiling, Moment-Invariant Feature Extraction	N/M	N/M	480s	N/M	IP

Table 3. Cont.

Article	Algorithm	IPT	KPI	Result	TR.Time	TR.TS.Split	TCL
[96]	ResNet-50, YOLO	N/A	Accuracy	99%	N/M	80–20%	DL.CNN
[97]	SegNet256, SegNet512	N/A	Accuracy	96%	11,389 s	75–25%	DL.AE
[98]	RF	Background Subtraction, Masking, Feature Extraction, Markov Random Field	PR, RC	95%, 96%	N/M	N/M	ML
[99]	SVM	Image Segmentation, Feature Extraction	Accuracy	88.99%	N/M	70–30%	ML
[100]	SVM	Image Segmentation (K-Means), Feature Extraction (HoG)	Accuracy	92%	N/M	70–30%	ML
[81]	DetectNet, GoogLeNet VGGNet	N/A	Accuracy	99%, 50%, 90%	N/M	78–22%	DL.CNN
[101]	SegNet, VGGNet, U-NET	Background Subtraction, Image Labeling	Accuracy	99% (SegNet), 96% (VGGNet), 97% (U-Net)	N/M	85–15%	DL.CNN
[102]	CaffeNet, SVM, AdaBoost, RF	N/A	Accuracy	99%, 97%, 96%, 93%	N/M	75–25%	ML.DL.CNN
[103]	SVM, ANN	N/A	Accuracy	95%, 92%	N/M	60–40%	ML
[104]	NBG, DT, KNN, SVM, ANN	Morphological Erosion	Accuracy	97%, 96%, 96%, 96%, 96%, 98% (IP),	N/M	N/M	ML
[105]	AlexNet	N/A	Accuracy	99.89%	468 s with double GPUs	70–30%	DL.CNN
[106]	SVM	Feature Extraction, Codebook Learning (Clustering), Feature Encoding	Accuracy	89%	N/M	50–50%	ML
[107]	ERT, RF	LBP, HOG	Accuracy	52.5%, 52.4%	83 s	68–32%	ML
[108]	CNN	N/A	mAP	95%	3.6–4.5 s	N/M	DL.CNN
[109]	ResNet18, SVM, RF	N/A	AUC	95%, 95%, 97%	N/M	80–20%	ML.DL.CNN
[110]	GoogleNet	N/A	PR, RC	86%, 46%	N/M	N/M	DL.CNN
[80]	VGGNet, GoogLeNet, DetectNet	N/A	Accuracy	99% (VGGNet)	N/M	88–12%	DL.CNN
[111]	VGGNet, CNN	N/A	mAP	84%	N/M	80–20%	DL.CNN

Table 3. Cont.

Article	Algorithm	IPT	KPI	Result	TR.Time	TR.TS.Split	TCL
[112]	SegNet, U-Net, DeepLabV3	N/A	AUC, F1-score	85%, 72%, 92%, 85%, 97%, 92%	N/M	N/M	DL.AE
[113]	DeepLab	Histogram Equalization	MIoU	96%	N/M	70–30%	DL.AE
[114]	CenterNet	Background Subtraction, Image Segmentation	PR	95%	N/M	80%–20%	DL.CNN
[115]	AlexNet, ANN	SLIC	Accuracy	99%, 48.09%	N/M	70–30%	ML.DL.CNN
[116]	SSD, VGGNet	CV	N/M	N/M	N/M	80–20%	DL.CNN
[117]	CNN	Graph Feature Extraction	Accuracy	98%	N/M	80–20%	DL.CNN
[118]	GCN-ResNet101, AlexNet, VGGNet, ResNet101	N/A	Accuracy	98% (GCN-ResNet101)	1.42 s	70–30%	DL.CNN
[119]	ResNet50, U-Net, SegNet, FCN	Image Segmentation (PSPNet, RSS)	mAP	93% (ResNet50)	N/M	N/M	DL.CNN
[120]	Inception-Resnet, VGGNet, MobileNet, DenseNet, Xception, SVM, XGBoost, LR	N/A	F1-score	99% (DenseNet, SVM)	N/M	N/M	ML.DL.CNN
[121]	SSD	N/A	Accuracy	86%	N/M	90–10%	DL.CNN
[122]	YOLO	K-Means	mAP	89%	6.48 ms	85–15%	DL.CNN
[123]	VGG16, ResNet50, Inception30, YOLOv3	N/A	accuracy	98%	354 s	60–40%	DL.ML.CNN
[124]	SSD, Faster RCNN, CNN	N/A	Precision	65%	N/M	N/M	DL.ML.CNN
[125]	CNN	N/A	Precision	93%	N/M	N/M	DL.ML.CNN
[126]	DCNN	N/A	F1-score	92%	N/M	N/M	DL.ML.CNN
[127]	VGG16, ResNet50, DenseNet	N/A	accuracy	91%	N/M	N/M	DL.ML.CNN
[128]	YOLO	N/A	accuracy	83%	N/M	N/M	DL.ML.CNN
[129]	YOLO-v3, CenterNet, and Faster R-CNN	N/A	F1-score	97%	N/M	N/M	DL.ML.CNN
[130]	U-Net, ResNet	N/A	IoU	96%	N/M	N/M	DL.ML.CNN

We found that both ML and DL algorithms are used by researchers for weed detection. In total, 10 different algorithms are used whose frequency distribution is shown in Table 4. For DL, CNN is used in general. Although Autoencoder (AE) is also used, as such, no specific AE has been used separately, for example, denoising AE, stacked AE, or variational AE. Rather, only the AE architecture is used within the algorithmic process of CNNs to improve the CNN performance. Hence, we have not considered AE as separate from CNN in this work. The CNN algorithms are applied 65 (with different variants) times in application papers (details below), followed by ML algorithms, specifically, SVM (12 times), RF (5), ANN (5), KNN (2), Boosting (XGBoost and Adaboost) (2), DT (includes vanilla DT and ERT) (2), NBG (1), and LR (1). This shows that many representative ML algorithms have already been applied for weed detection, although the frequency of applications remains severely limited as compared to CNN.

The 65 CNNs appeared over a total of 19 different variants, whose frequency distribution is shown in Table 5. Specifically, the custom CNN model is used in 12 implementations, and transfer learning is used in 21 implementation articles, including 7 implementations of ResNet, SegNet, and VGGNet. GoogleNet and FCN are used in 6 and 5 articles, respectively. YOLO, U-Net, and AlexNet are each used in three articles while DetectNet, DeepLab (all variants combined), and SSD are used in two articles, and Mask R CNN, CaffeNet, CenterNet, GCN, MobileNet, DenseNet, and Xception are used once. This demonstrates that researchers are not focusing on several CNN models only; rather, there is a trend to explore recently-introduced CNN variants as soon as they are published (as most of the variants are introduced recently and have very limited applications).

Image Processing (IP) techniques are used in 38 application articles, with a total of 16 unique techniques whose frequency distribution is shown in Table 6. There are four (4) articles in which the authors applied only IP techniques for weed detection, i.e., without using any ML or DL algorithm. In these articles, the authors have used techniques such as both background and foreground subtraction (BFS), converting RGB to Grayscale, binarizing and labeling the images (CV), feature extraction (Ftr Ext), classification (Classify), and image tiling techniques. In the remaining 34 articles, IP techniques have been used for pre-processing images for either a separate application of ML or DL or both ML and DL collectively. Table 6 shows that feature extraction, image segmentation, and BFS are used more frequently, along with SLIC, LBP, and HoG, while some less-applied techniques are also used once, for example, morphological erosion and histogram equalization (HE).

Table 4. Frequency Distribution of Algorithms.

Algorithm	Frequency
CNN	65
SVM	12
RF	5
ANN	5
KNN	2
Boosting	2
DT	2
NBG	1
Linear Regression	1

Regarding the use of performance measures, 32 out of the 51 application papers used the accuracy of weed detection, which gives an average of the predictive performance for both classes ('weed detected' and 'no weed detected' respectively). Precision and recall for the 'weed detected' class, which provide a better indicator of the weed detection performance, are employed in only 11 articles. Of the latter, two papers employ the F1-Score, which is estimated from both precision and recall values. Performance measure was not m1entioned in four articles, while Area Under the Curve (AUC), which provides a measure of overall classification performance similar to accuracy, is used in two (2) articles, once in combination with F1-score. In total, 11 papers out of 52 mentioned model training timing, YOLO with K-means took the shortest time to train at 6.8 ms, and CNN took the longest at 1895 min. As there is no standard for dividing data into train and test, most papers used 60–40, 70–30, and 80–20 splits.

Table 5. Frequency Distribution of CNN Variants.

DL Algorithm	Frequency
CNN	11
ResNet	8
SegNet	7
VGGNet	8
GoogleNet	5
FCN	4
YOLO	4
U-Net	3
AlexNet	3
DetectNet	2
DeepLab	2
SSD	3
Mask R CNN	2
CaffeNet	1
CenterNet	1
GCN	1
MobileNet	1
DenseNet	1
Xception	2

Table 6. Frequency Distribution of IP Techniques.

IP Techniques	Frequency
Feature Extraction	6
Background and Foreground Subtraction	5
Image Segmentation	5
HoG	4
CV	3
SLIC	3
LBP	2
Cluster	2
MG Filter	1
Hough Transformation	1
Classify	1
Image Tiling	1
MRF	1
Masking	1
Morphological Erosion	1
Histogram Equalization	1

Here we addressed How does the performance and usage of machine learning and image processing techniques compare with that of deep learning for weed detection? where we categorize each application paper by assigning it a unique thematic classification label (TCL) based on the usage of ML, DL, and IP, along with other information such as training time and KPI. We classify application paper under five TCLs: (1) **DL.AE**, (2) **ML**, (3) **DL.CNN**, (4) **IP**, and (5) **ML.DL.CNN**. In the following, we define these TCLs and present data about their constituent articles. A summary of this is presented in Figure 12. (the details of the articles are provided in Section 6).

- **DL.AE:** This type represents five articles that employ the autoencoder (AE) technology, i.e., the encoder-decoder DNNs for weed detection. In two of these articles, both the encoder and decoder are modeled as an FCN. In the remaining three articles, two articles use the AE-based CNN called SegNet as {SegNet256, SegNet512} and {SegNet, U-Net, DeepLabV3}, while the last article uses only DeepLab. We remark that the U-Net and DeepLab series of CNN variants are also based on AE. Histogram Equalization (HE) is the only technology that is specified for DL.AE category.
- **ML:** This type represents 9 articles that only employ ML techniques for weed detection. In four of these articles, the performance of a set of ML algorithms are compared, while an individual ML algorithm is used in the remaining five articles. The comparison sets include {SVM, ANN} (in two articles), {NBG, DT, KNN, SVM, ANN} (1) and {ERT, RF} (1). The individual applications include RF (2) and SVM (3). In one of the SVM applications, K-means clustering is used to pre-process the image data and then IP techniques such as SLIC, BFS, Masking, Feature Extraction, Markov Random Field, Image Segmentation, and Morphological Erosion. are used with ML
- **DL.CNN:** This type represents the 31 articles that employ the CNN or one/more of its variants for weed detection. Regarding individual applications, CNN is used most frequently in nine articles, GoogleNet in two articles, and YOLO, SSD, AlexNet, and CenterNet in two articles each. Besides this, the following combinations of variants are used: {ResNet, U-Net, SegNet, FCN, GoogleNet}, {GCN, AlexNet, VGGNet, ResNet}, {SSD, VGGNet}, {CNN, VGGNet}, {SegNet, VGGNet, U-Net}, {DetectNet, GoogLeNet VGGNet}, {ResNet-50, YOLO}, {SegNet, SegNet-Basic}, {Mask R CNN, FCN} and {YOLO, GoogleNet}. All these combinations are used once, except {DetectNet, GoogLeNet VGGNet}, which is used twice. The IP techniques employed with DL.CNN papers are as follows: HoG, LBP, BFS (background), image labeling, image segmentation, CV, feature extraction, and clustering.
- **IP:** This type represents four articles that only employ IP techniques for weed detection. As mentioned above, these techniques are image tiling, classification, CV, feature extraction, and background and foreground subtraction.
- **ML.DL.CNN:** This type represents seven articles that employ both CNN (or one/more of its variants) and ML for weed detection. These 7 combinations are as follows: {SVM, ANN, CNN}, {ResNet, SVM}, {SVM, KNN, ESD, CNN}, {CaffeNet, SVM, Adaboost, RF}, {ResNet, SVM, RF}, {AlexNet, ANN} and {ResNet, VGGNet, MobileNet, DenseNet, Xception, SVM, XGBoost, LR} (abbreviations of algorithms are shown for this sequence for brevity). The IP techniques used in ML.DL.CNN includes MGF, HT, SLIC, and CV algorithms.

In this section, we analyze the performance of algorithms with respect to our TLCs. We will first emphasize the training time and train-test split ratio data shown in Table 3. Our purpose in listing the training time was to gauge the delay or speed-up achieved in training the DL algorithms due to the complicated nature of this learning task. In fact, GPU usage has addressed this problem thoroughly. It is considered prudent to conduct a comparative analysis for different tasks and GPU/CPU settings. In our case, no training time was recorded by the authors in 36 (out of 51) articles. In the remaining articles, the maximum recorded time is as follows: (1) 480 s (IP), (2) 83 s (ML), (3) 976 m (DL.CNN), (4) 480 s (ML.DL.CNN), and (5) 11,389 s (DL.AE). Thus, the maximum training time is

976 min (approximately 16 h) for DL.CNN followed by 3.1 h for DL.AE. Where pre-trained DL models were employed, training time is in the order of seconds.

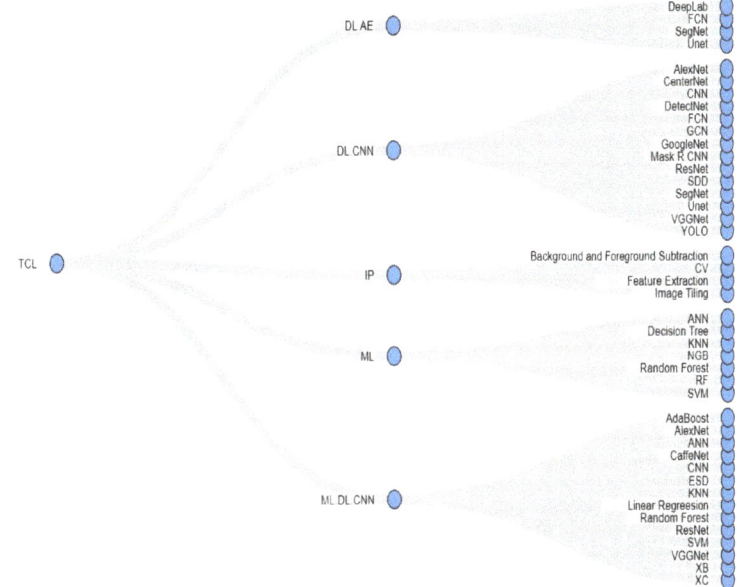

Figure 12. The distribution of different algorithms with respect to five thematic classification labels (TCLs).

The train-test split ratio has been mentioned in 32 articles. After analyzing the ratios, we mapped them into four categories: 60–40, 70–30, 80–20, and 90–10. Out of 31 articles, 11 articles used an 80–20, and 11 others used a 70–30 ratio (these two ratios are the most common ones in ML and DL domains). However, four articles based on DL have used a 90–10 ratio, and five articles (three from ML and two from DL.CNN) have used a 60–40 ratio. Our aim here is to present such information. However, we are not intended to make conclusions about the reason for a ratio selection because these matters are guided by several factors, such as the size of the dataset, previous experience, or trial-and-error experimentation.

To compare the performance of algorithms across our TCLs, we executed the following steps. We decided to analyze all the different performance measures together by focusing on their values. In other words, we do not specifically distinguish between accuracy, precision, recall, etc., but we focus only on their values to compare all application papers together. For this, we analyzed the values in Table 3 and manually created three bins or ranges of values: (1) [45–85) (labeled as low performance), (2) [85–95) (medium), and (3) [95–100] (high). Moreover, we estimated and considered the F1-score in case both precision and recall are measured, and we considered the performance values of each algorithm separately in case multiple algorithms were used in the article. Finally, we considered the F1-score over AUC in articles where both were recorded, and for a recorded performance of '>90' by the authors, we considered it in the range of 85–95. The results are shown in Figure 13 and discussed as follows:

- **ML:** 2 algorithms have a low performance, 4 have medium, while 10 have a high performance (over multiple algorithms).
- **IP:** Two articles mention their results which are low and high, respectively (no multiple algorithms used).

- **DL.CNN:** 5 algorithms have a low performance, 21 have a medium, and 18 have a high performance (over multiple algorithms).
- **ML.DL.CNN:** 2 algorithms have a low performance, 13 have medium while 10 have a high performance (over multiple algorithms), and
- **DL.AE:** One article has a low performance, two have a medium, and the remaining two demonstrate high performance.

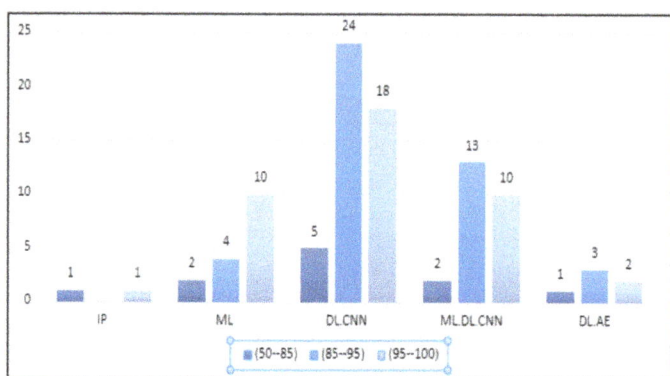

Figure 13. Observed performance evaluation with respect to our defined thematic classification labels.

A notable trend here is that, although ML applications are limited with respect to DL.CNN, the proportion of algorithms that demonstrate a high performance is more (63%) as compared to DL.CNN (44%). In ML.DL.CNN category, of the 10 algorithms demonstrating high performance, 4 of these are ML algorithms while the rest are CNN variants. This clearly shows that traditional ML algorithms, particularly SVM and RF, are demonstrating performances at par with CNN and its variants. Regarding the use of the AE, the performance seems to have remained consistent over the low, medium, and high bands. However, we do acknowledge that a dataset of five articles is limited to making any generalization of this trend.

6. Classification as per Weeds/Crops

To understand the aforementioned trends more thoroughly, we created a table shown in Table 7, which illustrates associations of different weed types and different crop types respectively with their algorithmic applications and demonstrates which algorithm has been used to detect the weed type which belongs to a crop type.

We would like to highlight the following facts before our analysis: (1) We do not present the algorithms for which their corresponding weed/crop types are not mentioned by the authors; (2) In all papers using multiple weed types, the performance achieved is applicable for each type; (3) If algorithm A is used as a feature extractor for algorithm B (e.g., {ResNet-50, YOLO}, {RCNN, SSD} and {VGGNet, SSD})), we have assigned A and B the same performance measure as both contributed to achieving this performance, (4) We do not specifically distinguish between accuracy, precision, recall, etc. as mentioned above for Figure 13, and (5) we show the highest accuracy with which the weed belonging to a crop type is detected using one or more algorithms and in each algorithm. The table assists in understanding the applications and deriving appropriate recommendations.

Table 7. Classification of Weeds and Crops wrt Algorithms.

Weed Type	Crop Type	DL Algorithms	Best Accuracy
Pigweed	Sugar beet, Wheat	CNN, VGGNet, SVM, ANN	95% [SVM]
Blackgrass	Sugar beet, Wheat	CNN, FCN, Mask R CNN	97% [CNN]
Bluegrass	Corn	ResNet, YOLO	99% [ResNet, YOLO]
Dockleaf	Pasture, Soybean, Turfgrass, Bahigrass	CNN, VGGNet, GoogleNet, AlexNet, DetectNet, CaffeNet	99% [AlexNet, CaffeNet]
Canadian Thistle	Corn, Cereal crops	ResNet, YOLO, VGGNet, AlexNet, GCN, SVM, ANN, DT, KNN	99% [YOLO]
Chickweed	Cranesbill, Sugar beet, Maize, Wheat, Charlock	CNN, VGGNet, ,FCN, Mask R CNN	95% [Mask R CNN]
Cleaver	Maize, Sugar beet, Wheat	FCN, Mask R CNN	95% [Mask R CNN]
Cockleblur	Maize, Wheat, Peanut	AlexNet	99% [AlexNet]
Crowfoot	Maize, Rice	GoogleNet	98% [GoogleNet]
Fat-Hen	Wheat, Maize, Peanut, Corn, Sugar beet	ResNet, VGGNet, FCN, YOLO, AlexNet, Mask R CNN, GCN	99% [YOLO]
Field Pansy	Wheat	CNN, VGGNet	84% [CNN, VGGNet]
Hare Ear's Mustard	Sugar beet	SVM, ANN	95% [SVM]
Japanses Hop	Maize, Wheat, Peanut	AlexNet	99% [AlexNet]
Jungle Rice	Rice, Maize	GoogleNet, SSD	98% [GoogleNet]
Little Seed	Rice, Maize	GoogleNet, SSD	98% [GoogleNet]
Mayweed	Wheat, Maize, Carrot, Sugar beet	CNN, VGGNet, , FCN, Mask R CNN	95% [Mask R CNN]
Meadow Grass	Wheat	CNN, VGGNet	84% [CNN, VGGNet]
Nutsedge	Chrysanthemum, Corn	CNN, ResNet, VGGNet, AlexNet, GCN	98% [CNN, GCN]
Paragrass	Chrysanthemum	CNN VGGNet, AlexNet, GCN	98% [CNN, GCN]
Shepherd's Purse	Maize, Sugar beet, Wheat	SegNet, FCN, Mask R CNN	95% [Mask R CNN]
Silky-bent	Cranesbill, Sugar beet, Maize, Wheat, Charlock	FCN, Mask R CNN	95% [Mask R CNN]
Turnip Weed	Sugar beet	SVM, ANN	95% [SVM]
Dicot	Sugar beet, Soybean	FCN	87% [FCN]
Grass Weed	Wheat, Maize, Sugar beet	FCN	87% [FCN]
Velvetleaf	Tomato, Cotton	ResNet, VGGNet, MobileNet, DenseNet, Xception, SVM	99% [SVM, DenseNet]
Benghal Dayflower	Rice	SSD	86% [SSD]

Table 7. Cont.

Weed Type	Crop Type	DL Algorithms	Best Accuracy
Black Nightshade	Tomato, Cotton	ResNet, VGGNet, MobileNet, DenseNet, Xception, SVM	99% [SVM, DenseNet]
Hedge Bindweed	Sugar Beet	YOLO	89% [YOLO]
Indian Jointvetch	Rice, Maize	SSD	86% [SSD]
SnakeWeed	N/M	CNN	98% [CNN]
Chinee Apple	N/M	CNN	98% [CNN]
Lantana Camara	N/M	CNN	98% [CNN]
Cocklebur, foxtail, redroot pigweed and giant ragweed	Corn, Soybean	VGG16, ResNet50, Inception30, YOLOv3	98% [VGG16]
N/M	Soybean	SSD, Faster RCNN, CNN	65% [RCNN]
N/M	Spinach, beet, Bean	CNN	93% [CNN]
Dandelion, ground ivy, spotted spurge.	N/M	DCNN	92% [DCNN]
N/M	N/M	VGG16, ResNet50, DenseNet	91% [ensemble]
Cereal crops.	Monocotyledon and dicotyledon	YOLO	83% [YOLO]
Vegetables	N/M	YOLO-v3, CenterNet, and Faster R-CNN	97% [YOLO-v3]
Sugar beet	N/M	U-Net, ResNet	96% [U-net, ResNet]

The performance of detecting each of the weed types in different crops is high, with a minimum performance of 84% for Field Pansy and Wheat crops and 99% for Bluegrass in Corn, Cockleblur, Dockleaf in Soybean, Pasture, Turfgrass, and bahiagrass, Japanese Hop in Maize, Wheat and Peanut and Black Nightshade and Velveleaf in Tomato and Cotton. The top-4 weeds that are used most frequently by researchers are Canadian thistle, followed by Dockleaf, Fat-hen, and Velvetleaf, while those who are applied only once include Latana Camara, Chinee Apple, SnakeWeed, Indian Jointvetch, Hedge Bind, Benghal Dayflower, Grass Weed, Dicot, Paragrass, Crowfoot and CockleBlur.The top-3 crops that are used most frequently by researchers are Sugar beet, followed by Carrot and Maize, while those that are applied only once include Cranesbill, Charlock, Soybean, Canola, Peanut, White Cabbage, Tobacco Seedling, Lettuce, Tomato, Radish. It is apparent that the applications of the top-4 weed types are spread out over different algorithms, i.e., the novelty in research articles is introduced through the application of novel or other algorithms. From the perspective of algorithms, each algorithm demonstrated high performance in detecting one or more weed types, with a minimum performance of 87% for FCN and SSD and a maximum of 99% for SVM, CNN, VGGNet, and DetectNet. For DL, the top-3 algorithms that are used most frequently are CNN, VGGNet, and FCN, while for ML, the top-3 are SVM, ANN, and KNN. The average performance over all DL algorithms is 94% while the same performance over all the ML algorithms is 90%. In our opinion, this difference is small, and we believe the ML community still has much to offer for weed detection, which could be potentially comparable to DL algorithms. Furthermore, when combined with other algorithms, SVM and Mask R CNN performed equally or better (five times) and only underperformed once (to YOLO). SVM and YOLO have not been used together in other cases. YOLO is the only algorithm that performs every time better or equal to other algorithms when combined. In three cases, CNN alone performed best, and when combined with other algorithms, it performed equally (three times). The following are the findings regarding crop association with algorithms. Sugar beet was used as a crop in 13 of the 51 papers, and different weed types were associated with each paper. Papers with sugar beet crops used a combination of FCN and Mask R CNN six times and FCN, Mask R CNN, and CNN four times. YOLO was only used once to detect sugar beet, and it had the lowest accuracy of 89 percent when compared to FCN, Mask R CNN, and CNN. In nine papers, sugar beet is mostly combined with maize and wheat. Maize and rice were used together four times, and GoogleNet and SSD were frequently used to detect weeds in maize and rice. GoogleNet and SSD are used together twice, and SSD and GoogleNet are used separately once. GoogleNet provided the highest accuracy of 99 percent when combined with SSD and without, but SSD alone provided 98 percent accuracy. Tomato and cotton were used together in two cases where ResNet, VGGNet, MobileNet, DenseNet, Xception, and SVM were used to detect weeds, and both times SVM and DenseNet provided 99 percent accuracy. Furthermore, each crop is used in a unique combination.

7. Summary of Identified Articles in SLR

This section describes a literature survey of our identified 60 articles. We present the application papers and literature review papers in separate sections.

7.1. Summary of Identified Application Papers

In this section, we briefly discuss each application paper according to our TCLs. The sequence in which these papers are discussed is the same as the one found in Table 3.

7.1.1. IP Papers

In [32], the basic idea adopted by the authors is to detect weeds at different stages of growth of the wheat crop, along with detecting the barren land to determine the amount of land used for cultivation. For detection, the authors employ background subtraction techniques in the Hue Saturation Value (HSV) color space, but they can only achieve a maximum weed detection accuracy of 67% with high-resolution images acquired through drones. Moreover, in [84], the authors use CV functions for the classification of weeds and crops, notably, *rgb2gray* for detection of green plants, *im2bw* to convert digital images to binary images, *bwlabel* for labeling binary images, and *regionprops* for measuring feature of images and detection of weed. The classification accuracy obtained with these functions is 99% with a training time of approximately 3 s.

In [82], authors create and implement a framework called the Image Processing Operation (IPO) library for the classification of weeds. IPO stores information about weeds and crops in JSON format which are then automatically converted to MATLAB functions to perform weed discrimination, with the option to add personalized user-defined functions. The authors claim that IPO is partially successful and discuss methods to remove some of its limitations. Finally, in [95], the authors study different features of weed leaves for detection using IP. In their method, the authors propose the execution of several stages in sequences, such as *foreground extraction* with grey-scale images, image tiling, feature extraction, and classification. For classification, authors employ moment-invariant shape features i.e., rotation, scaling, and translation for identifying the weed, with a training time of 480 s.

In [131], the researchers described the importance of large datasets for better weed detection and also emphasized the need for GANs. They also mention a lack of real-world datasets for weeds. To solve this problem, they proposed a model that combines transfer learning or a pre-trained model with GANs. The crop weed dataset at the early growth stage was used, with 202 images of tomato as a crop and 130 images of black nightshade as a weed. To select the best parameters for the model, various combinations of hyperparameter tuning were used. Three pre-trained models were used: Xception, Inception-ResNet, and DenseNet. Xception outperformed with a 99.07% accuracy.

Researchers in [132] study combined Generative Adversarial Networks (GANs) with Deep Convolutional Networks to create a model that detects weed better than existing models. GANs are used to generate synthetic images of weeds, and deep neural networks are used to detect weed images from original and GAN-generated images. They also compared their model to existing models like AlexNet, ResNet, VGG16, and GoogleNet, but their model outperformed with an accuracy of 96.34%.

Researchers focused on a robust image segmentation method in [133], which will be used to distinguish between crops and weeds in real time. They also discussed using annotated images in various studies and stated that annotating images could be time-consuming. However, they used GANs to generate synthetic images to supplement the dataset. Then, for image segmentation, they used CNN variants such as UNet-ResNet, SegNet, and BonNet. UNet-ResNet and SegNet outperformed with 98.3 percent accuracy.

The authors of [134] study developed an algorithm that is used to synthesize real agricultural images. The images were captured with a multi-spectral camera, and Near-Infrared images were collected. They used conditional GAN for segmentation. They also stated that their experiments improved the generalization ability of segmentation and enhanced the model's performance. They used various CNN variants for segmentation, including UNet, SegNet, ResNet, and UNet-ResNet, and UNet outperformed with 97% accuracy in Crop detection and 72% in Weeds.

7.1.2. ML Papers

In [90], the authors attempt to classify soil, soybean, and weed images based on the color indices of these three classes. They compare the performance of SVM and ANN over this task after processing and segregating the datasets through SLIC. The results do not demonstrate any major difference in accuracy between SVM (95%) and ANN (96%). Moreover, in [94], the researchers use ML for the classification of weeds and crops by using RF. The employed dataset is divided into different categories, specifically crop, weed, and irrelevant data. The authors train the model on offline datasets and apply these pre-trained models to real-time images. They have trained their system to give feedback to the flow control system. The RF algorithm gave 97% accuracy with a training time of 57.4 ms.

Moreover, in [98], the authors use RF for crop and weed classification through the following approach. They perform classification using NIR + RGB images, which were captured through a mobile robot. NIR can help distinguish the plant from the soil and background. This process is defined in four steps; firstly, identification of a plant using NIR information, which helps remove unrelated backgrounds so that only relevant regions can be considered for classification. Then masking is computed on pixel location. Secondly, feature selection has been performed on the relevant region. Then in the third step, RF is applied to those computed features, and a binary probability distribution is obtained, which described that the pixel belongs to a crop or a weed. In the fourth and last step, to improve the classification results, the information from the third step is utilized in Markov Random Field (MRF) by computing label assignment independently of the other nearby labels. In this way, authors were able to achieve 97% accuracy with RF.

In [99], the authors focus on identifying weeds from carrot fields to reduce the use of herbicides. During the development of plants, it is very difficult to discriminate between the color of a plant and weeds, which also makes the discrimination process even more difficult when both the plant and weed overlap each other. To address this problem, they proposed a 3-step procedure: (1) image segmentation. In this step, the input images are segregated from weeds using a normalization equation which gives higher weight to the greener part of the plants and removes the other colors from the input image, (2) in the second step, feature extraction is performed from the images got from the first step, and (3) in the third step, weed detection is performed through SVM algorithm. In addition, the overall accuracy obtained by SVM is 88%. In a related paper [100], the authors discussed the problem of overlapping weed and carrots leaves. In the initial stage of plant development, the color of both plant and weed are the same, which makes it more challenging to identify the weed and plant. Therefore, the 3 step procedure has been proposed to improve the detection or identification of plants and weeds. Initially, images are segmented using k-means clustering. Then, features are extracted from these segments by using HoG, which is then fed to SVM to acquire an improved accuracy of 92%.

In [103], the objective of this research is to propose a very accurate identification of weeds against crops using robots. The similarities between the shape of a plant and a weed make it challenging to identify plants precisely from weeds. For that reason, they tried to add different shapes to make a pattern for the individual range of the plants and tried to detect weeds based on these patterns using SVM and ANN to achieve maximum accuracies of 95% and 92%, respectively. Moreover, in [104], the authors compare the performance of several ML algorithms to detect the Canadian Thistle weed, particularly from a limited sample size of 30 images. The intent of the authors is also to demonstrate that, with the use of enhanced IP techniques, it can be possible to attain comparable performance with ML algorithms. Hence, the authors compare the performance of NBG, DT, KNN, SVM, and ANN algorithms with an IP technique in which they initially convert the image to grayscale, remove it from the green channel (in RGB), binarize it, and then perform morphological erosion to detect weed. In fact, this is not a new IP algorithm but rather a sequence of N/M IP techniques. The authors show that this IP method achieves comparable accuracy (98%) to the ML algorithms (97%, 96%, 96%, 96%, and 96%, respectively).

In [106], the authors have focused on developing a system that caters to the effect of using multiple image resolutions in the weed detection process. The authors employ enhancements of feature extraction, codebook learning (a clustering technique), feature encoding, and image classification as IP techniques. Particularly, the system takes an image as an input with 200×50 resolution, then feature extraction is performed by combining fisher encoding with codebook to cater to the limitation of feature extraction by using 2-level image representation. Then the image representation vectors got from feature extraction are given to the SVM algorithm for classification to achieve an overall accuracy of 89%. Finally, in [107], the authors focused on feature engineering, i.e., selecting the best set of features from gray-scale images by using HoG and LBP techniques. The extracted features are fed to two ML algorithms, i.e., ERT and RF, both of which give below-par accuracy of 52.5% and 52.4%, respectively, with a training time of 83 s and a limited customized dataset.

7.1.3. DL.CNN Papers

In [33], the authors introduce the concept of positive (weed present) and negative (weed not present) images. They employ drone-acquired images of 'black-grass' and 'common chickweed' for the positive class and 'wheat,' 'maize,' and 'sugar beet' for the negative class. They pre-process images to avoid overfitting because of a small range of datasets and use the traditional (vanilla) CNN architecture with three combinations of convolution and max pooling layers to extract filters through the former and reduce size through the latter, followed by the one-dimensional fully-connected layer and a single output neuron for classification. The authors achieve an accuracy of 97%. Moreover, in [83], the authors employ transfer learning techniques to reuse the GoogleNet CNN that was previously trained on IARA datasets to classify three types of weeds, namely littleseed canarygrass, crowfoot, and jungle rice. The authors achieve an average accuracy of 98% across these three weeds.

In [85], the authors detect weeds from images of carrot fields to enhance the performance of an existing CNN architecture (with one convolution and max pooling layer only) through the use of GPUs. Although the accuracy remains exactly the same, the authors can attain a maximum speed-up of 2.0× (976 min on GPU as compared to 1895 min on CPU). In another application [31], the authors propose using CNNs to localize and classify weeds simultaneously from carrot field images acquired through robots to replace their current lengthy solution of multi-stage weed detection process through image segmentation. They experiment with both YOLO and GoogleNet to acquire a weed detection accuracy of 89% and 86%, respectively, which is a significant performance improvement over their image segmentation framework.

In [88], the researcher has used Mask R CNN for enhancement of accuracy in weed detection for the following weeds: mayweed, chickweed, blackgrass, shepherd's purse, cleaver, fat-hen, and loose silky-bent. They employ Mask R CNN also for the segmentation of weed images. In both applications, Mask R performs better than FCN through a 100% accuracy in training and greater than 90% in the validation phase. In another application [89], the authors compare the performance of CNN with the HoG image processing method for weed detection. CNN application is conducted on hyperspectral images with four convolutional layers, two fully-connected ones, while RGB images are used with the HoG method. The results show that CNN can extract more discriminative features than HoG and with better accuracy (88%), although the computational processing required by CNN increases with the number of color bands.

Yet another comparison between CNN and IP techniques is done in [91], in which the authors develop a low-cost weed identification system that employs CNN. In the system, the data are initially collected and processed. Then, a relevant set of images is sampled, followed by weed detection through CNN. The authors also employ HOG and LBP approaches and achieve the best accuracy of 96% by initially employing LBP to extract relevant features and then using them as input to CNN. In [30], the authors generate synthetic datasets for weed classification based on real datasets by randomizing different features such as species, soil type, and light conditions. They compare the performance of weed detection over both synthetic and real datasets by using Segnet and Segnet-Basic CNNs and show that there is no performance degradation with synthetic datasets with the accuracy of 84% and 98%, respectively.

In [96], the authors indicate the limitations of detecting weeds with real-life images in that whole image content has to be fed into deep learning architectures, which sometimes makes it difficult to distinguish weeds from their background like soil. Hence, the authors propose using pre-trained deep learning models, particularly ResNet-50 for classification and YOLO for performance speed-up to achieve an accuracy of 99%. The authors create a framework to utilize both these models for weed detection. In a related work [81], the authors experiment with three different deep CNN architectures for weed detection, namely, DetectNet, GoogleNet, and VGGNet. They discovered that, for different types of active turfgrass weeds, VGGNet demonstrated much superior performance as compared to GoogleNet in different surface conditions, mowing heights, and surface densities. Moreover, DetectNet outperformed GoogleNet for dormant turfgrass weeds. The authors also demonstrate that image classification is an easier solution for weed detection as compared to object detection because the latter requires the use of bounding boxes.

In [101], the authors solve the tedious process of manually labeling image data at the pixel level by proposing a 2-step manual labeling process. Here, the first step is the segregation of foreground and background layers using maximum likelihood classification, with manual labeling of segmented pixels of background occurring in the second step. This setting can be used to train segmentation models which can discriminate between crops and other types of vegetation. The authors experiment with this approach using a SegNet model based on ResNet-50 and VGGNet encoder blocks, and UNet. The ResNet-50 SegNet model can demonstrate the best result (99%). Furthermore, in [105], the authors employ the AlexNet CNN architecture for weed classification in the ecological irrigation domain by using three different combinations of weeds and crops as datasets, with both CPU and GPU computing. They demonstrate a maximum accuracy of 99.89%. The authors validate that through their AlexNet application, both multiple and single weeds can be detected simultaneously, hence allowing enhanced irrigation control and management.

In [108], the authors developed intelligent software that is able to perform weed detection on-the-fly on multi-spectral RGB + NIR images acquired from the BOSCH Bonirob farm robot. For this, a lightweight CNN is initially used to extract pixels that represent projections of three-dimensional points belonging to green areas or vegetation. Then, a much deeper CNN uses these pixels to discriminate between crops and weeds. The authors also propose a novel data summarization method that selects relevant subsets of data that are able to approximate the original complete data in an unsupervised manner. The authors are able to achieve a maximum mean average precision (mAP) of 95%. A similar work is done in [110], where the authors use GoogleNet to detect weeds in the presence of a large amount of leaf occlusion. The loss function is guided by the bounding boxes and coverage maps of 17,000 original images collected from a high-speed camera mounted on an all-terrain vehicle. The authors manually annotate these images (which is a time-consuming activity) to achieve a precision of 86%, although the recall performance is poor (46%).

In [80], the author experiments with three CNN architectures, namely VGGNet, GoogLeNet, and DetectNet, for the recognition of broadleaf weeds in turfgrass areas. Through different experiments, the authors show that VGGNet demonstrates the best performance in classifying several different broadleaf weeds, while DetectNet outperformed the others in detecting one particular broadleaf weed. Furthermore, in [111], the authors sought to categorize the weeds in aerial photographs obtained from a height of under ten meters. The photos were taken using a 3024 × 4032 pixel resolution. Images were captured at the Heidfeldhof estate near Stuttgart's Plieningen. Using a mobile, pictures were captured vertically at a height of 50 cm. The captured weed was in its early stages of development, and [135] weed photos were utilized to evaluate the model using pixel-based techniques. They use the CNN model and proposed two approaches, one is object detection, and the second is pixel-wise labeling. The object-based approach was applied to three different datasets, and the highest mAP achieved by this approach was 84.2%, and the pixel-wise approach achieved 77.6% as the highest mean accuracy using FCN.

In [114], the authors combine DL with IP for the classification of crops and weeds. Initially, a previously-trained CenterNet is used for detecting crops and drawing bounding boxes around them. Then, green objects falling outside these boxes are considered to be weeds, and the user can then focus only on crop detection with the reduced number of training images and easier weed detection. Moreover, the authors employ a segmentation-based IP method based on color indexing to facilitate the aforementioned detection of weeds, with the color index being determined through Genetic Algorithm optimization.

This setup achieved a maximum precision of 95% for weed detection in crop/vegetable plantations.

In [116], the authors simply propose a framework for crop and weed classification using deep learning in real-time. They use Dicot and Monocot weeds. Images are being captured using a USB camera and processing of images has been done by using the OpenCV library. For weed classification, SSD objection detection is used, which uses a pre-trained VGG16 for mapping features from images and convolutional filter layers for the detection of weed. For three different settings, i.e., when the weeds and crops are overlapping and the weed size is smaller and larger than the crop size, the authors are able to acquire an average weed detection accuracy of 20% only.

In [117], the authors employ graph-based DL architecture for weed detection from RGB images which are collected from a diverse number of geographical locations, as compared to related works carried out in a controlled environment. Initially, a multi-scale graph is constructed over the weed image with sub-patches of different measures. Then, relevant patch-level patterns are selected by applying a graph pooling layer over the vertices. Finally, RNN architecture is used to predict weeds from a multi-scale graph with a maximum accuracy of 98.1%. In a related work [118], the authors use a feature-based GCN to detect weeds. They construct a GCN graph based on features extracted through CNN and the Euclidean distance between these features. This graph uses both labeled and unlabeled image features for semi-supervised training through information propagation and labeled data for testing. By combining GCN with ResNet-101, the authors were able to acquire accuracies of 97.80%, 99.37%, 98.93%, and 96.51%, respectively, on four different datasets, outperforming the following state-of-the-art methods: AlexNet, VGG16, and ResNet-101, with a reduced running time of 1.42 s.

In [119], the authors propose a semantic segmentation procedure for weed detection with ResNet-50 as the backbone architecture. They employ a particular type of convolution called hybrid dilation for increasing the receptive field and DropBlock for regularization through random dropping of weights. They also optimize RGB-NIR bands into RGB-NIR color indices to make the classification results more robust and employ an attention mechanism to focus the CNN on more correlated regions along with a spatial refinement block for fusing feature maps of differing sizes. The authors test their complicated approach on Bonn and Stuttgart datasets and compare the weed detection performance with UNet, SegNet, and FCN, along with performance over two other semantic segmentation algorithms, i.e., PSPNet and RSS [12]. For both datasets, they achieve better accuracy than the above five algorithms of 75.26% and 72.94%, respectively.

In [121], the authors employ the SSD to detect weeds in rice fields which employs VGG16 to extract features from images. Such a setting gives a maximum accuracy of 86% over different image resolutions, by using multi-scaled feature maps and convolution filters. The authors mention that the accuracy achieved with VGG16 (before re-usage) was 99%.

Finally, in [122], the authors employ the YOLOv3 CNN to discriminate between crops (sugar beet) and weeds (hedge bindweed). They use a combination of synthetic and real images and a K-means algorithm to estimate the anchor box sizes for YOLOv3. A test run on 100 images shows that synthetic images can improve the overall mean average precision (MAP) by more than 7%. The system is also able to demonstrate better performance and trade-off between accuracy and speed as compared to other YOLO variants.

Moreover [123], the researchers compared the performance of pre-trained classification algorithms such as VGG16, ResNet50, and Inceptionv3 for weed classification. Cocklebur, foxtail, redroot pigweed, and gigantic ragweed are four weeds commonly seen in corn and soybean fields in the Midwest of the United States. They also used YOLOv3 object detection to locate and classify weeds in an image dataset. VGG16 outperformed all pre-trained models with an accuracy of 98.90%. They also compare Keras with Pytorch, finding that Pytorch takes less time to train models and has higher accuracy than Keras.

The authors in [124] examined the performance of single shot detector (SSD) and Faster RCNN in terms of weed detection utilizing images of soybean fields recorded with a

UAV in this study. Both the single shot detector and the quicker RCNN object detection algorithms were compared to the patch-based CNN model. According to the authors, Faster RCNN outperformed the SSD Model. Furthermore, faster RCNN outperformed patch-based CNN.

The authors of [125] research proposed a vision-based classification method for weed identification in spinach, beet, and bean. CNN was used for classification. UAV was used to capture the images used in this section. Precision was used to measure model performance, and beet received the highest precision of 93%. Additionally, The researchers in [126] attempted to construct a precision herbicide application using DCNN and its various variations such as VGGNet, DetectNet, GoogleNet, and AlexNet for the detection of various weeds, such as dandelion, ground ivy, and spotted spurge in this work.

To make the algorithms more manageable for hardware with low resources while still retaining accuracy, in this study [127] the authors used ensemble learning approaches, transfer learning, and model compression. The suggested method was carried out in three steps: transfer learning, pruning-based compression, quantization, and Huffman encoding, and model ensembling with a weighted average for improved accuracy. Similarly in [128], researchers presented a method for locating a specific area and applying herbicide based on object detection in real-time as well as crop and weed classification. In this study, two weed types—monocotyledon and dicotyledon—that are typically seen in cereal crops were specifically targeted. They acquired 1318 photos using a Nikon 7000 camera for field recording, trained CNN for classification under various lighting situations, and trained YOLO for object detection. This [129] research study offered a novel deep-learning technique to categorize weeds and vegetable crops. CenterNEt, YOLO-v3, and Faster RCNN were employed in this approach. The YOLO-v3 model was the most effective in identifying weeds in vegetable crops out of the three. For the pixel-by-pixel segmentation of weed, soil, and sugar beet, [130] the author employed ResNet50 and U-Net. For 1385 photos, they employed these models as encoder blocks, and to deal with unbalanced data, they also applied a unique linear loss function. CNN was primarily employed for the classification and spraying of certain areas for herbicide application. The segmentation accuracy in tiny regions was increased by using a bespoke loss function and balanced data.

7.1.4. ML.DL.CNN Papers

In [28], the authors compare the performance of SVM, ANN, and CNN for discriminating between crops and weeds, specifically four different crop types and Paragrass and Nutsedge weed types. They employ median and Gaussian filters for identifying the relevant areas in images and also extract shape features for both crops and weeds. SVM is assessed over two kernel functions, i.e., radial basis and polynomial, while ANN is evaluated with one hidden layer containing six neurons, with the output layer containing two neurons (one each for weed and crop detection). The CNN contains the traditional convolutional and maxpooling layer (with ReLU activation) followed by the fully connected layer. The authors show that, in the best result, ANN is the best classifier for both weed and crop classes, followed by SVM and then CNN.

In [86], the authors use SVM and ResNet-18 classifier to discriminate between weeds and crops from unsupervised (unlabeled) images collected from a UAV. They extract deep features from the images and employ a one-class classification approach with the SVM classifier. Hough transform and SLIC are used to detect the crops' rows and segment the images into superpixels, which are used to train the SVM. It is found that the performance of SVM is comparable with the performance of a ResNet-18 CNN which has been trained through supervised learning (maximum 90%).

In [87], the authors focus on broad-leaf weed detection in pasture fields through an application and comparison of both ML and DL algorithms, namely, SVM (with linear, quadratic, and Gaussian kernel), KNN, Ensemble subspace discriminant, Regression and CNN consisting of six convolutional layers and alternating max-pooling and drop-out layers and three fully connected layers. Local binary pattern histogram (LBPH) is used to extract information from grayscale and RGB images. The authors demonstrate that CNN outperforms all ML variants by giving a maximum accuracy of 96.88%.

In [102], the authors employ CaffeNet (a variant of AlexNet) for grass weed and broadleaf weed detection in soybean crop images captured from the Phantom DJI drone and compare its performance with SVM, Adaboost, and RF algorithms. SLIC was used to extract superpixels for input to all algorithms. Although CaffeNet achieved the best accuracy of 99%, SVM, Adaboost, and RF also achieved similar results with 97%, 96%, and 93% accuracy, respectively.

In [109], the authors address the particular problem of manually annotating and/or segmenting a large number of UAV/drone images for a supervised weed detection task. They propose an automated unsupervised method of weed detection based on CNNs. Initially, they detect crop rows using Hough transform variations and SLIC. The output is a set of lines identifying the center of the crop rows, i.e., around which the crops are growing. Applying a blob-coloring algorithm on these lines to represent the crop regions, anything that falls outside the blob area (crop vegetation) is a potential weed. These weeds are then labeled autonomously and form the dataset for CNN, i.e., ResNet-18. In the data of bean fields, the best accuracy is obtained by ResNet (88.73%), followed by RF (65.4%) and SVM (59.51%), while for the spinach field dataset, RF is the winner with 96.2% accuracy, followed by ResNet-18 (94.34%) and SVM (90.77%).

Moreover, a thorough comparison between ANN and AlexNet CNN has been done by the authors in [115], in which they develop an application to transmit drone-captured images to a machine learning server. The results demonstrate that AlexNet is able to acquire a maximum accuracy of 99.8% while the maximum achieved by ANN is only 48.09%.

In [120], the authors attempt to construct an automated weed detection system that can detect weeds in their different stages of growth and soil conditions. For this, they employ a set of pre-trained CNN architectures, namely Inception-Resnet, VGGNet, MobileNet, DenseNet, and Xception, through transfer learning techniques to extract deep features. Then, each of these feature sets is used for weed classification with a set of traditional ML algorithms, specifically, SVM, XGBoost, and LR. The authors test the system on tomato and cotton fields over black nightshade and velvetleaf weeds. The authors claim that the best F1 score of 99.29% is achieved by Densenet and SVM, while all other CNN-ML combinations give an F1 score greater than 95%.

7.1.5. AE Papers

In [92], the authors focus on the problem of designing an automated weed detection system that can generalize to varying environments and soil conditions, as well as weed and crop types. For this, they propose an autoencoder architecture, embedded within an FCN, which generates two types of features through the downsample-upsample process. First are visual features that are generated for each image through the visual code generated after downsampling, and the second are sequential features that are generated through a sequence code that aggregates data from a batch of images acquired from the Bonirob robot. Both visual and sequence features are combined into a pixel-level label mask that is able to distinguish between both crops and weeds distinctly. In comparison to a baseline method and some previous approaches, the proposed approach demonstrates better precision and recall for both crops and weeds over the Stuttgart and Bonn datasets.

In another paper by the same research group [93], the authors use a similar approach to identify the actual stems of the weeds for mechanical control (e.g., pulling out) and also a surrounding region for effective spraying. For this, they initially generate a visual code for each image, which is then input to two different decoder networks, specifically, one which outputs a pixel map related to weed stem detection and the other for crop detection. This information is used to identify a bounding area around the stems for spraying. The authors show that their system can achieve better average precision for identifying two types of weeds (dicot and grass weed) than a baseline and other related systems.

In [97], the authors employ two variants of the SegNet algorithm (SegNet 512 and SegNet 256) to detect weeds from the CWFID dataset. They also make several architectural changes to the original SegNet architecture to enhance the downsampling performance for both SegNet 512 and SegNet 256, for instance, by adding or removing convolution and batch normalization layers, changing the kernel size and the size of the hidden layers. As the focus of the authors here is on the decoder's performance, we have categorized this paper under the DL.AE label. The validation and test accuracy over SegNet 512 is 92% and 96% respectively, while for SegNet 256, the corresponding accuracy is 92% and 93% respectively. The authors also show that the training, evaluation, and prediction time for SegNet 512 is understandably twice as much for SegNet 256 because the former employs twice as more upsampling and downsampling blocks as compared to the latter.

In [112], the authors conduct a performance comparison of DeepLab V3, U-Net, and SegNet, which are all autoencoder-based CNN variants. Initially, patches or relevant regions are selected from aerial images of sugar beet crops to generate the relevant set of features, which are then input into the three variants. The results show that DeepLab V3 demonstrates the best AUC values for both crop and weed identification, while U-Net performs better than SegNet. However, DeepLab V3 is computationally the most expensive, followed by SegNet and then U-Net. The authors recommend generating smaller patches over a larger training data size with an application of U-Net to achieve a balance of speed and efficiency.

In [113], the authors employ the encoder-decoder architecture for semantic segmentation of weeds and crops. The encoder employs Atrous convolution (similar to DeepLab) over four convolution layers and one pooling layer with an output code of size 1X1. This code is then upsampled in the decoder twice with several low-level features (from the atrous convolution output of the encoder) as input. Different image enhancement techniques were compared and used for improving the quality of images and for making the model to be robust against different lighting conditions. The results demonstrate that when NIR color indices are used with these enhancement techniques, the weed identification performance is significantly improved. However, without NIR indices, pure image enhancement techniques demonstrate an average performance even though they still improve the quality of images under different lighting conditions.

8. Challenges and Future Research Directions

In this section, we answer RQ4: What are the tangible future research directions to achieve further benefit from deep learning applications for weed detection? For this, we identify and divide the directions of future research and challenges in the domain of deep learning applications for weed detection into two parts: domain and technical.

8.1. Domain Challenges

- **Missing integrated image databases:** There is a need to create a general repository of weed image datasets with specified associations to their respective crops, generated with high-speed cameras (either mounted on UAVs/robots or taken manually), of an agreed-upon high resolution, and categorized according to different modality types. This will create proper benchmarks for any future weed detection experiments. For instance, an experiment to detect Canadian thistles in some European countries can employ the standard Canadian Thistle images as the baseline. The need arises from the fact that almost all researchers generate their own datasets using different types of cameras without any baseline images, which makes it difficult to determine the exact impact of their work on the research community.
- **Lack of standards:** The main challenge arising from implementing such a standard repository is that weeds demonstrate significant diversity from each other, as do their associated crops. Both weeds and crops can demonstrate different growth conditions (size, density, etc.) with respect to weather and other external variables, and the effects of shadows and illumination will require further classification (and hence complexity)

of the resulting images. Moreover, manually annotating each image separately over different classifications is a complex task. Catering to all of these requirements in collecting weed and their associated crop images is a challenging task.

- **Environmental Challenges:** The environmental indicators such as soil temperature, soil water potential, exposure to light, fluctuating temperatures, nitrates concentration, soil PH and the gaseous environmental soils impact the composition of weed flora of the cultivated area. Therefore, it is essential to understand the usage of soil profiling and temperate can help predict early weed detection. However, creating a soil profile is a time-consuming task because of the nature of the soil variant.

8.2. Technical-Related

We believe that our analysis of Table 7 provides a clear roadmap for practitioners to derive multiple lines of future research with respect to the selection of algorithm to detect a particular type of weed and/or associated crops or selection of weeds/associated crops for detection. Moreover, from an algorithmic perspective, it is obvious that DL, particularly with the use of CNN and its variants, has the power to generate satisfactory predictive performance for weed detection. As this trend is prevalent and rising, we expect it to continue in the near future. As more variants of CNNs are discovered, there is a high probability that they will be soon applied for weed detection and its related field. In our opinion, the distinction between the performance of ML and DL can only be clarified after thorough experimentation of CNN/variants with more robust ML models, particularly, SVM, Boosting variants (Adaboost, XGBoost, LightGBM), LR, and RF, over different standard weed datasets. Although we have discussed such previous applications, they do not demonstrate clearly that DL has a significant edge over ML applications, and hence, these results cannot be considered comprehensive and generalizable in our opinion.

Furthermore, once we have some standard baseline repository of images as proposed above, we propose an application of CNN on these baseline datasets to provide an actual benchmark performance over different measures, specifically accuracy, precision, recall, F1-score, and AUC score. The reason is that researchers are now focusing on improving CNN's performance further through the use of different variants, notably ResNet, VGGNet, and SegNet. As this trend is increasing rapidly, we expect it to continue. Our proposed baseline performance benchmarks will then provide a standard backbone to compare the performance of any application of CNN variant over any weed type and to position the paper with respect to its comparison with proposed baselines. In doing so, one can also try to address the prevalent problems of natural light variation and weather effects.

Another research direction is to quantify the impact of using standard and well-known IP techniques for both DL and ML algorithm applications, particularly feature selection, BFS, image segmentation, cluster analysis, and different transformations, for the exact problem of distinguishing the weeds from their respective crops in the same image. Moreover, we have seen that very good results have been acquired in both DL and ML applications without the use of any IP technique. So, there is a need to understand, quantify and hence standardize the impact of these techniques for crop-weed discrimination and a generalized perspective.

Moreover, it remains to be explored how ML and DL applications are impacting related fields such as pest and disease detection and the impact of transfer learning of CNN-based models from one domain to the other. Finally, we believe there is a need to design appropriate software architectures for such a weed detection activity which could be generalized for future applications.

9. Conclusions

This paper conducts the first SLR to review deep learning applications in depth for weed detection. We adopt the standard SLR methodology and answer four concrete research questions to thoroughly summarize the state-of-the-art research's impact and articulate domain and technical challenges for future research directions. Furthermore, we created a citation graph to understand the pattern of publications and researchers in this area. We also compare our work with the eight latest literature reviews and demonstrate our approach's superiority and differences with these reviews.

Author Contributions: N.Y.M., T.M., A.R.M.F., A.M., P.P.J. and M.S.S. writers created and designed the study, oversaw the work, and revised the manuscript; N.Y.M. and T.M. authors contributed equally. All authors have read and agreed to the published version of the manuscript.

Funding: This work is funded by the Deanship of Research, Islamic University of Madinah, Kingdom of Saudi Arabia.

Institutional Review Board Statement: Not applicable.

Informed Consent Statement: Not applicable.

Data Availability Statement: Not applicable.

Conflicts of Interest: The authors declare that no conflicts of interest are related to the publication of this paper.

References

1. de Preneuf, F. Agriculture and Food. 2021. Available online: https://www.worldbank.org/en/topic/agriculture/ (accessed on 23 February 2023).
2. Nations, U. Department of Economic and Social Affairs. 2017. Available online: https://www.un.org/development/desa/en/news/population/world-population-prospects-2017.html (accessed on 23 February 2023).
3. Monaco, T.J.; Weller, S.C.; Ashton, F.M. *Weed Science: Principles and Practices*, 4th ed.; Pests Diseases & Weeds; Wiley-Blackwell: Hoboken, NJ, USA, 2002.
4. Fennimore, S.A.; Bell, C. *Principles of Weed Control*, 4th ed.; California Weed Science Society: West Sacramento, CA, USA, 2014.
5. Mishra, A. *Weed Management: (Agriculture Research Associates)*, 2nd ed.; Kalyani: New Delhi, India, 2020.
6. Organic, G. Fat Hen. 2007. Available online: https://www.gardenorganic.org.uk/weeds/fat-hen (accessed on 23 February 2023).
7. Panoff, L. Chickweed: Benefits, Side Effects, Precautions, and Dosage Healthline. 2020. Available online: https://www.healthline.com/nutrition/chickweed-benefits (accessed on 23 February 2023).
8. Organic, G. Common Chickweed. 2021. Available online: https://www.gardenorganic.org.uk/weeds/common-chickweed (accessed on 23 February 2023).
9. Farms. Shepherd's Purse. 2021. Available online: https://www.farms.com/field-guide/weed-management/shepherds-purse.aspx (accessed on 23 February 2023).
10. University, M. Cleavers. 2021. Available online: https://www.massey.ac.nz/massey/learning/colleges/college-of-sciences/clinics-and-services/weeds-database/cleavers.cfm (accessed on 23 February 2023).
11. Pedrycz, W.; Chen, S.M. (Eds.) *Deep Learning: Algorithms and Applications*; Springer International Publishing: Cham, Switzerland, 2020.
12. Goodfellow, I.; Bengio, Y.; Courville, A. *Deep Learning*; MIT Press: Cambridge, MA, USA, 2016. Available online: http://www.deeplearningbook.org (accessed on 23 February 2023).
13. Vargas, R.; Ruiz, L. Deep Learning: Previuos and oresent application. *J. Aware.* **2017**, *2*, 11–20.
14. Hatcher, W.G.; Yu, W. A Survey of Deep Learning: Platforms, Applications and Emerging Research Trends. *IEEE Access* **2018**, *6*, 24411–24432. [CrossRef]
15. Nwankpa, C.; Ijomah, W.; Gachagan, A.; Marshall, S. Activation Functions: Comparison of trends in Practice and Research for Deep Learning. In Proceedings of the 2nd International Conference on Computational Sciences and Technologies, New York, NY, USA, 26–28 October 2018.
16. Thompson, J.; Stafford, J.; Miller, P. Potential for automatic weed detection and selective herbicide application. *Crop. Prot.* **1991**, *10*, 254–259. [CrossRef]
17. Tan, L.; Jiang, J. Chapter 13-Image Processing Basics. In *Digital Signal Processing*, 3rd ed.; Tan, L., Jiang, J., Eds.; Academic Press: Cambridge, MA, USA, 2019; pp. 649–726.
18. Nixon, M.S.; Aguado, A.S. *Feature Extraction and Image Processing for Computer Vision*; Academic Press: Cambridge, MA, USA, 2020.
19. Visser, R.; Timmermans, A.J.M. Weed-It: A new selective weed control system. In *Optics in Agriculture, Forestry, and Biological Processing II*; Meyer, G.E., DeShazer, J.A., Eds.; SPIE: Bellingham, WA, USA, 1996.
20. Hague, T.; Tillett, N.D.; Wheeler, H. Automated Crop and Weed Monitoring in Widely Spaced Cereals. *Precis. Agric.* **2006**, *7*, 21–32. [CrossRef]

21. Siddiqi, M.H.; Ahmad, I.; Sulaiman, S.B. Weed Recognition Based on Erosion and Dilation Segmentation Algorithm. In Proceedings of the 2009 International Conference on Education Technology and Computer, Singapore, 17–20 April 2009; IEEE: Piscataway, NJ, USA, 2009.
22. Wang, A.; Zhang, W.; Wei, X. A review on weed detection using ground-based machine vision and image processing techniques. *Comput. Electron. Agric.* **2019**, *158*, 226–240. [CrossRef]
23. Castrignanò, A.; Buttafuoco, G.; Khosla, R.; Mouazen, A.; Moshou, D.; Naud, O. *Agricultural Internet of Things and Decision Support for Precision Smart Farming*; Academic Press: Cambridge, MA, USA, 2020.
24. Cho, S.; Lee, D.; Jeong, J. AE—Automation and Emerging Technologies: Weed–plant Discrimination by Machine Vision and Artificial Neural Network. *Biosyst. Eng.* **2002**, *83*, 275–280. [CrossRef]
25. Aitkenhead, M.; Dalgetty, I.; Mullins, C.; McDonald, A.; Strachan, N. Weed and crop discrimination using image analysis and artificial intelligence methods. *Comput. Electron. Agric.* **2003**, *39*, 157–171. [CrossRef]
26. Edan, Y.; Han, S.; Kondo, N. Automation in Agriculture. In *Springer Handbook of Automation*; Springer: Berlin/Heidelberg, Germany, 2009; pp. 1095–1128.
27. Pereira, L.A.; Nakamura, R.Y.; de Souza, G.F.; Martins, D.; Papa, J.P. Aquatic weed automatic classification using machine learning techniques. *Comput. Electron. Agric.* **2012**, *87*, 56–63. [CrossRef]
28. Sarvini, T.; Sneha, T.; GS, S.G.; Sushmitha, S.; Kumaraswamy, R. Performance Comparison of Weed Detection Algorithms. In Proceedings of the 2019 International Conference on Communication and Signal Processing (ICCSP), Chennai, India, 4–6 April 2019; IEEE: Piscatawat, NJ, USA, 2019.
29. Etienne, A.; Saraswat, D. Machine learning approaches to automate weed detection by UAV based sensors. In Proceedings of the Autonomous Air and Ground Sensing Systems for Agricultural Optimization and Phenotyping IV, Baltimore, MD, USA, 15–16 April 2019; Thomasson, J.A., McKee, M., Moorhead, R.J., Eds.; International Society for Optics and Photonics, SPIE: Bellingham, WA, USA, 2019; Volume 11008, pp. 202–215.
30. Cicco, M.D.; Potena, C.; Grisetti, G.; Pretto, A. Automatic model based dataset generation for fast and accurate crop and weeds detection. In Proceedings of the 2017 IEEE/RSJ International Conference on Intelligent Robots and Systems (IROS), Vancouver, BC, Canada, 24–28 September 2017; IEEE: Piscataway, NJ, USA, 2017.
31. Czymmek, V.; Harders, L.O.; Knoll, F.J.; Hussmann, S. Vision-Based Deep Learning Approach for Real-Time Detection of Weeds in Organic Farming. In Proceedings of the 2019 IEEE International Instrumentation and Measurement Technology Conference (I2MTC), Auckland, New Zealand, 20–23 May 2019; IEEE: Piscataway, NJ, USA, 2019.
32. Hameed, S.; Amin, I. Detection of Weed and Wheat Using Image Processing. In Proceedings of the 2018 IEEE 5th International Conference on Engineering Technologies and Applied Sciences (ICETAS), Bangkok, Thailand, 22–23 November 2018; IEEE: Piscataway, NJ, USA, 2018.
33. Skacev, H.; Micovic, A.; Gutic, B.; Dotilic, D.; Vesic, A.; Ignjatovic, V.; Lakicevic, S.; Jakovljevic, M.M.; Zivkovic, M. On the Development of the Automatic Weed Detection Tool. In Proceedings of the 2020 Zooming Innovation in Consumer Technologies Conference (ZINC), Novi Sad, Serbia, 26–27 May 2020; IEEE: Piscataway, NJ, USA, 2020.
34. Moazzam, S.I.; Khan, U.S.; Tiwana, M.I.; Iqbal, J.; Qureshi, W.S.; Shah, S.I. A Review of Application of Deep Learning for Weeds and Crops Classification in Agriculture. In Proceedings of the 2019 International Conference on Robotics and Automation in Industry (ICRAI), Rawalpindi, Pakistan, 21–22 October 2019; IEEE: Piscataway, NJ, USA, 2019.
35. Li, N.; Zhang, X.; Zhang, C.; Ge, L.; He, Y.; Wu, X. Review of Machine-Vision-Based Plant Detection Technologies for Robotic Weeding. In Proceedings of the 2019 IEEE International Conference on Robotics and Biomimetics (ROBIO), Dali, China, 6–8 December 2019; IEEE: Piscataway, NJ, USA, 2019.
36. Dankhara, F.; Patel, K.; Doshi, N. Analysis of robust weed detection techniques based on the Internet of Things (IoT). *Procedia Comput. Sci.* **2019**, *160*, 696–701. [CrossRef]
37. Shanmugam, S.; Assunção, E.; Mesquita, R.; Veiros, A.; Gaspar, P.D. Automated Weed Detection Systems: A Review. *KnE Eng.* **2020**, *2020*, 271–284. [CrossRef]
38. Hu, K.; Wang, Z.; Coleman, G.; Bender, A.; Yao, T.; Zeng, S.; Song, D.; Schumann, A.W.; Walsh, M. Deep Learning Techniques for In-Crop Weed Identification: A Review. *arXiv* **2021**, arXiv:2103.14872v1. Available online: http://xxx.lanl.gov/abs/2103.14872 (accessed on 23 February 2023).
39. Liu, B.; Bruch, R. Weed Detection for Selective Spraying: A Review. *Curr. Robot. Rep.* **2020**, *1*, 19–26. [CrossRef]
40. Hasan, A.M.; Sohel, F.; Diepeveen, D.; Laga, H.; Jones, M.G. A survey of deep learning techniques for weed detection from images. *Comput. Electron. Agric.* **2021**, *184*, 106067. [CrossRef]
41. Cabi. Weed Archives, Improving Lives by Solving Problems in Agriculture and the Environment. 2021. Available online: https://www.cabi.org/tag/weed/ (accessed on 23 February 2023).
42. CABI. Invasive Species Compendium: Detailed Coverage of Invasive Species Threatening Livelihoods and the Environment Worldwide. 2021. Available online: https://www.cabi.org/isc/ (accessed on 23 February 2023).
43. wiki. Weed. 2021. Available online: https://en.wikipedia.org/wiki/Weed (accessed on 23 February 2023).
44. Hills, L. Garden Organic:Protecting Rare Heritage Vegetable Varieties through the Heritage Seed Library. 2021. Available online: https://www.gardenorganic.org.uk/ (accessed on 23 February 2023).
45. Protect, C. Intelligent Pest, Weed and Disease Management. 2021. Available online: https://croprotect.com/ (accessed on 23 February 2023).

46. How, G.K. Information About Weeds: Gardening is Easy! Let Us Show You How. 2021. Available online: https://www.gardeningknowhow.com/ (accessed on 23 February 2023).
47. Weeds, L. It's Time For You To Get Rid Of Lawn Weeds: Identify, Remove and Prevent Weeds in Your Garden. 2021. Available online: https://lawnweeds.com/ (accessed on 23 February 2023).
48. Farms. Latest Agricultural Information, Farming News. 2021. Available online: https://www.farms.com/ (accessed on 23 February 2023).
49. USDA. U.S. Department of Agriculture. 2021. Available online: https://www.usda.gov/ (accessed on 23 February 2023).
50. Bugweed. Mayweed Chamomile. 2021. Available online: https://wiki.bugwood.org/HPIPM:Mayweed_chamomile (accessed on 23 February 2023).
51. University, W.S. Mayweed Chamomile. 2021. Available online: http://smallgrains.wsu.edu/weed-resources/common-weed-list/mayweed-chamomile/ (accessed on 23 February 2023).
52. Snakeweed. 2021 Available online: https://www.daf.qld.gov.au/__data/assets/pdf_file/0005/54392/snakeweed.pdf (31 March 2023).
53. Ellis, M.E. Controlling Lantana Weeds: Stopping Lantana Spread In The Garden. 2021. Available online: https://www.gardeningknowhow.com/ornamental/flowers/lantana/controlling-lantana-weeds.htm (accessed on 23 February 2023).
54. Queenland, B. Para Grass. 2021. Available online: https://www.business.qld.gov.au/industries/farms-fishing-forestry/agriculture/land-management/health-pests-weeds-diseases/weeds-diseases/invasive-plants/other/para-grass (accessed on 23 February 2023).
55. Gardening, R. Docks. 2021. Available online: https://www.rhs.org.uk/advice/profile?PID=989 (accessed on 23 February 2023).
56. FactSheet. Fact Sheet-Benghal Dayflower. 2021. Available online: http://www.mdac.ms.gov/wp-content/uploads/bpi_plant_benghal_dayflower.pdf. (accessed on 23 February 20231).
57. County, K. Hedge Bindweed Identification and Control. 2018. Available online: https://www.kingcounty.gov/services/environment/animals-and-plants/noxious-weeds/weed-identification/hedge-bindweed.aspx (accessed on 23 February 2023).
58. Ioffe, S.; Szegedy, C. Batch Normalization: Accelerating Deep Network Training by Reducing Internal Covariate Shift. In *Machine Learning Research, Proceedings of the 32nd International PMLR, Lille France, 6–11 July 2015*; Bach, F., Blei, D., Eds.; Journal of Machine Learning Research (JMLR): Cambridge, MA, USA, 2015; Volume 37, pp. 448–456. Available online: https://arxiv.org/abs/1502.03167 (accessed on 23 February 2023).
59. LeCun, Y.; Bottou, L.; Bengio, Y.; Haffner, P. Gradient-Based Learning Applied to Document Recognition. *Proc. IEEE* **1998**, *86*, 2278–2324. [CrossRef]
60. Lab, S.V. ImageNet Large Scale Visual Recognition Challenge. 2015. Available online: http://image-net.org/challenges/LSVRC/ (accessed on 23 February 2023).
61. Krizhevsky, A.; Sutskever, I.; Hinton, G.E. ImageNet Classification with Deep Convolutional Neural Networks. In Proceedings of the 25th International Conference on Neural Information Processing Systems, Lake Tahoe, NV, USA, 3–6 December 2012; Curran Associates Inc.: Red Hook, NY, USA, 2012; Volume 1, pp. 1097–1105.
62. Simonyan, K.; Zisserman, A. Very Deep Convolutional Networks for Large-Scale Image Recognition. In Proceedings of the 3rd International Conference on Learning Representations, ICLR 2015, San Diego, CA, USA, 7–9 May 2015; Bengio, Y., LeCun, Y., Eds.; ICLR: San Diego, CA, USA, 2015.
63. Szegedy, C.; Liu, W.; Jia, Y.; Sermanet, P.; Reed, S.; Anguelov, D.; Erhan, D.; Vanhoucke, V.; Rabinovich, A. Going Deeper with Convolutions. In Proceedings of the Computer Vision and Pattern Recognition (CVPR), Boston, MA, USA, 7–12 June 2015.
64. He, K.; Zhang, X.; Ren, S.; Sun, J. Deep Residual Learning for Image Recognition. In Proceedings of the 2016 IEEE Conference on Computer Vision and Pattern Recognition (CVPR), Las Vegas, NV, USA, 27–30 June 2016; pp. 770–778.
65. Badrinarayanan, V.; Kendall, A.; Cipolla, R. SegNet: A Deep Convolutional Encoder-Decoder Architecture for Image Segmentation. *arXiv* **2015**, arXiv:1511.00561v3. Available online: http://xxx.lanl.gov/abs/1511.00561 (accessed on 23 February 2023).
66. Ronneberger, O.; Fischer, P.; Brox, T. U-Net: Convolutional Networks for Biomedical Image Segmentation. *arXiv* **2015**, arXiv:1505.04597. Available online: http://xxx.lanl.gov/abs/1505.04597 (accessed on 23 February 2023).
67. Chen, L.C.; Papandreou, G.; Kokkinos, I.; Murphy, K.; Yuille, A.L. DeepLab: Semantic Image Segmentation with Deep Convolutional Nets, Atrous Convolution, and Fully Connected CRFs. *IEEE Trans. Pattern Anal. Mach. Intell.* **2018**, *40*, 834–848. [CrossRef]
68. Bishop, C.M. *Pattern Recognition and Machine Learning (Information Science and Statistics)*; Springer: Berlin/Heidelberg, Germany, 2006.
69. Mitchell, T.M. *Machine Learning*; McGraw-Hill: Singapore, 2006.
70. Tan, P.; Steinbach, M.; Kumar, V. Cluster Analysis: Basic Concepts and Algorithms. *Introduction to Data Mining*; CLANRYE International: New York, NY, USA,2005; pp. 487–568.
71. Zhao, P.W. Some Analysis and Research of the AdaBoost Algorithm. In *Intelligent Computing and Information Science*; Springer: Cham, Switzerland, 2011; Volume 134.
72. Ojala, T.; Pietikäinen, M.; Harwood, D. A comparative study of texture measures with classification based on featured distributions. *Pattern Recognit.* **1996**, *29*, 51–59. [CrossRef]
73. Hsu, C.Y.; Ding, J.J. Efficient image segmentation algorithm using SLIC superpixels and boundary-focused region merging. In Proceedings of the 2013 9th International Conference on Information, Communications & Signal Processing, Tainan, Taiwan, 10–13 December 2013; IEEE: Piscataway, NJ, USA, 2013.

74. Ilas, M.E.; Ilas, C. A New Method of Histogram Computation for Efficient Implementation of the HOG Algorithm. *Computers* **2018**, *7*, 18. [CrossRef]
75. Venetsky, L.; Boczar, R.; Lee-Own, R. Optimization of background subtraction for image enhancement. In Proceedings of the Machine Intelligence and Bio-Inspired Computation: Theory and Applications VII, Baltimore, MD, USA, 2 May 2013.
76. Creswell, A.; White, T.; Dumoulin, V.; Arulkumaran, K.; Sengupta, B.; Bharath, A.A. Generative Adversarial Networks: An Overview. *IEEE Signal Process. Mag.* **2018**, *35*, 53–65. [CrossRef]
77. Sony, M.; Antony, J.; Park, S.; Mutingi, M. Key Criticisms of Six Sigma: A Systematic Literature Review. *IEEE Trans. Eng. Manag.* **2020**, *67*, 950–962. [CrossRef]
78. Zahid, H.; Mahmood, T.; Morshed, A.; Sellis, T. Big data analytics in telecommunications: Literature review and architecture recommendations. *IEEE/CAA J. Autom. Sin.* **2019**, *7*, 18–38. [CrossRef]
79. Imran, S.; Mahmood, T.; Morshed, A.; Sellis, T. Big Data Analytics in Healthcare—A Systematic Literature Review and Roadmap for Practical Implementation. *IEEE/CAA J. Autom. Sin.* **2021**, *8*, 1–22. [CrossRef]
80. Yu, J.; Sharpe, S.M.; Schumann, A.W.; Boyd, N.S. Detection of broadleaf weeds growing in turfgrass with convolutional neural networks. *Pest Manag. Sci.* **2019**, *75*, 2211–2218. [CrossRef]
81. Yu, J.; Sharpe, S.M.; Schumann, A.W.; Boyd, N.S. Deep learning for image-based weed detection in turfgrass. *Eur. J. Agron.* **2019**, *104*, 78–84. [CrossRef]
82. Ashqer, Y.S.; Bikdash, M.; Liang, C.L.K. A Structured Image Processing Operation Library to Automatically Isolate Weeds and Crops. In Proceedings of the 2019 SoutheastCon, Huntsville, AL, USA, 11–14 April 2019; IEEE: Piscataway, NJ, USA, 2019.
83. Tiwari, O.; Goyal, V.; Kumar, P.; Vij, S. An experimental set up for utilizing convolutional neural network in automated weed detection. In Proceedings of the 2019 4th International Conference on Internet of Things: Smart Innovation and Usages (IoT-SIU), Ghaziabad, India, 18–19 April 2019; IEEE: Piscataway, NJ, USA, 2019.
84. Tejeda, A.J.I.; Castro, R.C. Algorithm of Weed Detection in Crops by Computational Vision. In Proceedings of the 2019 International Conference on Electronics, Communications and Computers (CONIELECOMP), Cholula, Mexico, 27 February–1 March 2019; IEEE: Piscataway, NJ, USA, 2019.
85. Umamaheswari, S.; Arjun, R.; Meganathan, D. Weed Detection in Farm Crops using Parallel Image Processing. In Proceedings of the 2018 Conference on Information and Communication Technology (CICT), Jabalpur, India, 26–28 October 2018; IEEE: Piscataway, NJ, USA, 2018.
86. Bah, M.D.; Hafiane, A.; Canals, R.; Emile, B. Deep features and One-class classification with unsupervised data for weed detection in UAV images. In Proceedings of the 2019 Ninth International Conference on Image Processing Theory, Tools and Applications (IPTA), Istanbul, Turkey, 6–9 November 2019; IEEE: Piscataway, NJ, USA, 2019.
87. Zhang, W.; Hansen, M.F.; Volonakis, T.N.; Smith, M.; Smith, L.; Wilson, J.; Ralston, G.; Broadbent, L.; Wright, G. Broad-Leaf Weed Detection in Pasture. In Proceedings of the 2018 IEEE 3rd International Conference on Image, Vision and Computing (ICIVC), Chongqing, China, 27–29 June 2018; IEEE: Piscataway, NJ, USA, 2018.
88. Patidar, S.; Singh, U.; Sharma, S.K.; Himanshu. Weed Seedling Detection Using Mask Regional Convolutional Neural Network. In Proceedings of the 2020 International Conference on Electronics and Sustainable Communication Systems (ICESC), Coimbatore, India, 2–4 July 2020; IEEE: Piscataway, NJ, USA, 2020.
89. Farooq, A.; Hu, J.; Jia, X. Weed Classification in Hyperspectral Remote Sensing Images Via Deep Convolutional Neural Network. In Proceedings of the IGARSS 2018-2018 IEEE International Geoscience and Remote Sensing Symposium, Valencia, Spain, 22–27 July 2018; IEEE: Piscataway, NJ, USA, 2018.
90. Abouzahir, S.; Sadik, M.; Sabir, E. Enhanced Approach for Weeds Species Detection Using Machine Vision. In Proceedings of the 2018 International Conference on Electronics, Control, Optimization and Computer Science (ICECOCS), Kenitra, Morocco, 5–6 December 2018; IEEE: Piscataway, NJ, USA, 2018.
91. Liang, W.C.; Yang, Y.J.; Chao, C.M. Low-Cost Weed Identification System Using Drones. In Proceedings of the 2019 Seventh International Symposium on Computing and Networking Workshops (CANDARW), Nagasaki, Japan, 26–29 November 2019; IEEE: Piscataway, NJ, USA, 2019.
92. Lottes, P.; Behley, J.; Milioto, A.; Stachniss, C. Fully Convolutional Networks With Sequential Information for Robust Crop and Weed Detection in Precision Farming. *IEEE Robot. Autom. Lett.* **2018**, *3*, 2870–2877. [CrossRef]
93. Lottes, P.; Behley, J.; Chebrolu, N.; Milioto, A.; Stachniss, C. Joint Stem Detection and Crop-Weed Classification for Plant-Specific Treatment in Precision Farming. In Proceedings of the 2018 IEEE/RSJ International Conference on Intelligent Robots and Systems (IROS), Madrid, Spain, 1–5 October 2018; IEEE: Piscataway, NJ, USA, 2018.
94. Alam, M.; Alam, M.S.; Roman, M.; Tufail, M.; Khan, M.U.; Khan, M.T. Real-Time Machine-Learning Based Crop/Weed Detection and Classification for Variable-Rate Spraying in Precision Agriculture. In Proceedings of the 2020 7th International Conference on Electrical and Electronics Engineering (ICEEE), Antalya, Turkey, 14–16 April 2020; IEEE: Piscataway, NJ, USA, 2020.
95. Fatma, S.; Dash, P.P. Moment Invariant Based Weed/Crop Discrimination For Smart Farming. In Proceedings of the 2019 International Conference on Computer, Electrical & Communication Engineering (ICCECE), Kolkata, India, 18–19 January 2019; IEEE: Piscataway, NJ, USA, 2019.
96. Abdulsalam, M.; Aouf, N. Deep Weed Detector/Classifier Network for Precision Agriculture. In Proceedings of the 2020 28th Mediterranean Conference on Control and Automation (MED), Saint-Raphael, France, 15–18 September 2020; IEEE: Piscataway, NJ, USA, 2020.

97. Umamaheswari, S.; Jain, A.V. Encoder—Decoder Architecture for Crop-Weed Classification Using Pixel-Wise Labelling. In Proceedings of the 2020 International Conference on Artificial Intelligence and Signal Processing (AISP), Amaravati, India, 10–12 January 2020; IEEE: Piscataway, NJ, USA, 2020.
98. Lottes, P.; Hoeferlin, M.; Sander, S.; Muter, M.; Schulze, P.; Stachniss, L.C. An effective classification system for separating sugar beets and weeds for precision farming applications. In Proceedings of the 2016 IEEE International Conference on Robotics and Automation (ICRA), Stockholm, Sweden, 16–21 May 2016;IEEE: Piscataway, NJ, USA, 2016.
99. Saha, D.; Hanson, A.; Shin, S.Y. Development of Enhanced Weed Detection System with Adaptive Thresholding and Support Vector Machine. In Proceedings of the International Conference on Research in Adaptive and Convergent Systems, Odense, Denmark, 11–14 October 2016; ACM: New York, NY, USA, 2016.
100. Saha, D.; Hamer, G.; Lee, J.Y. Development of Inter-Leaves Weed and Plant Regions Identification Algorithm using Histogram of Oriented Gradient and K-Means Clustering. In Proceedings of the International Conference on Research in Adaptive and Convergent Systems, Krakow, Poland, 20–23 September 2017; ACM: New York, NY, USA, 2017.
101. Asad, M.H.; Bais, A. Weed detection in canola fields using maximum likelihood classification and deep convolutional neural network. *Inf. Process. Agric.* **2020**, *7*, 535–545. [CrossRef]
102. dos Santos Ferreira, A.; Freitas, D.M.; da Silva, G.G.; Pistori, H.; Folhes, M.T. Weed detection in soybean crops using ConvNets. *Comput. Electron. Agric.* **2017**, *143*, 314–324. [CrossRef]
103. Bakhshipour, A.; Jafari, A. Evaluation of support vector machine and artificial neural networks in weed detection using shape features. *Comput. Electron. Agric.* **2018**, *145*, 153–160. [CrossRef]
104. Forero, M.G.; Herrera-Rivera, S.; Ávila-Navarro, J.; Franco, C.A.; Rasmussen, J.; Nielsen, J. Color Classification Methods for Perennial Weed Detection in Cereal Crops. In *Progress in Pattern Recognition, Image Analysis, Computer Vision, and Applications*; Springer International Publishing: Cham, Switzerland, 2019; pp. 117–123.
105. Wang, S.; Han, Y.; Chen, J.; Pan, Y.; Cao, Y.; Meng, H.; Zheng, Y. A Deep-Learning-Based Low-Altitude Remote Sensing Algorithm for Weed Classification in Ecological Irrigation Area. In *Communications in Computer and Information Science*; Springer: Singapore, 2019; pp. 451–460.
106. Kounalakis, T.; Triantafyllidis, G.A.; Nalpantidis, L. Image-based recognition framework for robotic weed control systems. *Multimed. Tools Appl.* **2017**, *77*, 9567–9594. [CrossRef]
107. Lameski, P.; Zdravevski, E.; Kulakov, A. Weed Segmentation from Grayscale Tobacco Seedling Images. In *Advances in Intelligent Systems and Computing*; Springer International Publishing: Cham, Switzerland, 2016; pp. 252–258.
108. Potena, C.; Nardi, D.; Pretto, A. Fast and Accurate Crop and Weed Identification with Summarized Train Sets for Precision Agriculture. In *Intelligent Autonomous Systems 14*; Springer International Publishing: Cham, Switzerland, 2017; pp. 105–121.
109. Bah, M.; Hafiane, A.; Canals, R. Deep Learning with Unsupervised Data Labeling for Weed Detection in Line Crops in UAV Images. *Remote Sens.* **2018**, *10*, 1690. [CrossRef]
110. Dyrmann, M.; Jørgensen, R.N.; Midtiby, H.S. RoboWeedSupport - Detection of weed locations in leaf occluded cereal crops using a fully convolutional neural network. *Adv. Anim. Biosci.* **2017**, *8*, 842–847. [CrossRef]
111. Dutta, A.; Gitahi, J.M.; Ghimire, P.; Mink, R.; Peteinatos, G.; Engels, J.; Hahn, M.; Gerhards, R. Weed detection in close-range imagery of agricultural fields using neural networks. *Publ. DGPF* **2018**, *27*, 633–645.
112. Ramirez, W.; Achanccaray, P.; Mendoza, L.F.; Pacheco, M.A.C. Deep Convolutional Neural Networks for Weed Detection in Agricultural Crops Using Optical Aerial Images. In Proceedings of the 2020 IEEE Latin American GRSS & ISPRS Remote Sensing Conference (LAGIRS), Santiago, Chile, 22–26 March 2020; IEEE: Piscataway, NJ, USA, 2020.
113. Wang, A.; Xu, Y.; Wei, X.; Cui, B. Semantic Segmentation of Crop and Weed using an Encoder-Decoder Network and Image Enhancement Method under Uncontrolled Outdoor Illumination. *IEEE Access* **2020**, *8*, 81724–81734. [CrossRef]
114. Jin, X.; Che, J.; Chen, Y. Weed Identification Using Deep Learning and Image Processing in Vegetable Plantation. *IEEE Access* **2021**, *9*, 10940–10950. [CrossRef]
115. Beeharry, Y.; Bassoo, V. Performance of ANN and AlexNet for weed detection using UAV-based images. In Proceedings of the 2020 3rd International Conference on Emerging Trends in Electrical, Electronic and Communications Engineering (ELECOM), Balaclava, Mauritius, 25–27 November 2020; IEEE: Piscataway, NJ, USA, 2020.
116. Jogi, Y.; Rao, P.N.; Shetty, S. CNN based Synchronal recognition of Weeds in Farm Crops. In Proceedings of the 2020 4th International Conference on Electronics, Communication and Aerospace Technology (ICECA), Coimbatore, India, 5–7 November 2020; IEEE: Piscataway, NJ, USA, 2020.
117. Hu, K.; Coleman, G.; Zeng, S.; Wang, Z.; Walsh, M. Graph weeds net: A graph-based deep learning method for weed recognition. *Comput. Electron. Agric.* **2020**, *174*, 105520. [CrossRef]
118. Jiang, H.; Zhang, C.; Qiao, Y.; Zhang, Z.; Zhang, W.; Song, C. CNN feature based graph convolutional network for weed and crop recognition in smart farming. *Comput. Electron. Agric.* **2020**, *174*, 105450. [CrossRef]
119. You, J.; Liu, W.; Lee, J. A DNN-based semantic segmentation for detecting weed and crop. *Comput. Electron. Agric.* **2020**, *178*, 105750. [CrossRef]
120. Espejo-Garcia, B.; Mylonas, N.; Athanasakos, L.; Fountas, S.; Vasilakoglou, I. Towards weeds identification assistance through transfer learning. *Comput. Electron. Agric.* **2020**, *171*, 105306. [CrossRef]

121. Olaniyi, O.M.; Daniya, E.; Abdullahi, I.M.; Bala, J.A.; Olanrewaju, E.A. Weed Recognition System for Low-Land Rice Precision Farming Using Deep Learning Approach. In *Advances in Intelligent Systems and Computing*; Springer International Publishing: Cham, Switzerland, 2020; pp. 385–402.
122. Gao, J.; French, A.P.; Pound, M.P.; He, Y.; Pridmore, T.P.; Pieters, J.G. Deep convolutional neural networks for image-based Convolvulus sepium detection in sugar beet fields. *Plant Methods* **2020**, *16*, 29. [CrossRef]
123. Ahmad, A.; Saraswat, D.; Aggarwal, V.; Etienne, A.; Hancock, B. Performance of deep learning models for classifying and detecting common weeds in corn and soybean production systems. *Comput. Electron. Agric.* **2021**, *184*, 106081. [CrossRef]
124. Sivakumar, A.N.V.; Li, J.; Scott, S.; Psota, E.; Jhala, A.J.; Luck, J.D.; Shi, Y. Comparison of Object Detection and Patch-Based Classification Deep Learning Models on Mid- to Late-Season Weed Detection in UAV Imagery. *Remote Sens.* **2020**, *12*, 2136. [CrossRef]
125. Bah, M.D.; Dericquebourg, E.; Hafiane, A.; Canals, R. Deep Learning Based Classification System for Identifying Weeds Using High-Resolution UAV Imagery. In *Advances in Intelligent Systems and Computing*; Springer International Publishing: Cham, Switzerland, 2018; pp. 176–187. [CrossRef]
126. Yu, J.; Schumann, A.W.; Cao, Z.; Sharpe, S.M.; Boyd, N.S. Weed Detection in Perennial Ryegrass With Deep Learning Convolutional Neural Network. *Front. Plant Sci.* **2019**, *10*, 1422. [CrossRef] [PubMed]
127. Ofori, M.; El-Gayar, O. An Approach for Weed Detection Using CNNs And Transfer Learning. In Proceedings of the Annual Hawaii International Conference on System Sciences. Hawaii International Conference on System Sciences, Maui, HI, USA, 1 August 2020–1 January 2021. [CrossRef]
128. Jabir, B.; Falih, N. Deep learning-based decision support system for weeds detection in wheat fields. *Int. J. Electr. Comput. Eng. (IJECE)* **2022**, *12*, 816. [CrossRef]
129. Jin, X.; Sun, Y.; Che, J.; Bagavathiannan, M.; Yu, J.; Chen, Y. A novel deep learning-based method for detection of weeds in vegetables. *Pest Manag. Sci.* **2022**, *78*, 1861–1869. [CrossRef]
130. Nasiri, A.; Omid, M.; Taheri-Garavand, A.; Jafari, A. Deep learning-based precision agriculture through weed recognition in sugar beet fields. *Sustain. Comput. Inform. Syst.* **2022**, *35*, 100759. [CrossRef]
131. García, B.; Mylonas, N.; Athanasakos, L.; Vali, E.; Fountas, S. Combining generative adversarial networks and agricultural transfer learning for weeds identification. *Biosyst. Eng.* **2021**, *204*, 79–89. [CrossRef]
132. Anthoniraj, S.; Karthikeyan, P.; Vivek, V. Weed Detection Model Using the Generative Adversarial Network and Deep Convolutional Neural Network. *J. Mob. Multimed.* **2022**, *18*, 275–292. [CrossRef]
133. Fawakherji, M.; Potena, C.; Prevedello, I.; Pretto, A.; Bloisi, D.D.; Nardi, D. Data Augmentation Using GANs for Crop/Weed Segmentation in Precision Farming. In Proceedings of the 2020 IEEE Conference on Control Technology and Applications (CCTA), Montreal, QC, Canada, 24–26 August 2020; IEEE: Piscataway, NJ, USA, 2020. [CrossRef]
134. Fawakherji, M.; Potena, C.; Pretto, A.; Bloisi, D.D.; Nardi, D. Multi-Spectral Image Synthesis for Crop/Weed Segmentation in Precision Farming. *Robot. Auton. Syst.* **2021**, *146*, 103861. [CrossRef]
135. Giselsson, T.M.; Jørgensen, R.N.; Jensen, P.K.; Dyrmann, M.; Midtiby, H.S. A Public Image Database for Benchmark of Plant Seedling Classification Algorithms. *arXiv* **2017**, arXiv:1711.05458.

Disclaimer/Publisher's Note: The statements, opinions and data contained in all publications are solely those of the individual author(s) and contributor(s) and not of MDPI and/or the editor(s). MDPI and/or the editor(s) disclaim responsibility for any injury to people or property resulting from any ideas, methods, instructions or products referred to in the content.

Review

A Survey on Medical Explainable AI (XAI): Recent Progress, Explainability Approach, Human Interaction and Scoring System

Ruey-Kai Sheu [1] and Mayuresh Sunil Pardeshi [2,*]

1. Department of Computer Science, Tunghai University, No. 1727, Section 4, Taiwan Blvd, Xitun District, Taichung 407224, Taiwan
2. AI Center, Tunghai University, No. 1727, Section 4, Taiwan Blvd, Xitun District, Taichung 407224, Taiwan
* Correspondence: mayuresh@thu.edu.tw

Abstract: The emerging field of eXplainable AI (XAI) in the medical domain is considered to be of utmost importance. Meanwhile, incorporating explanations in the medical domain with respect to legal and ethical AI is necessary to understand detailed decisions, results, and current status of the patient's conditions. Successively, we will be presenting a detailed survey for the medical XAI with the model enhancements, evaluation methods, significant overview of case studies with open box architecture, medical open datasets, and future improvements. Potential differences in AI and XAI methods are provided with the recent XAI methods stated as (i) local and global methods for preprocessing, (ii) knowledge base and distillation algorithms, and (iii) interpretable machine learning. XAI characteristics details with future healthcare explainability is included prominently, whereas the pre-requisite provides insights for the brainstorming sessions before beginning a medical XAI project. Practical case study determines the recent XAI progress leading to the advance developments within the medical field. Ultimately, this survey proposes critical ideas surrounding a user-in-the-loop approach, with an emphasis on human–machine collaboration, to better produce explainable solutions. The surrounding details of the XAI feedback system for human rating-based evaluation provides intelligible insights into a constructive method to produce human enforced explanation feedback. For a long time, XAI limitations of the ratings, scores and grading are present. Therefore, a novel XAI recommendation system and XAI scoring system are designed and approached from this work. Additionally, this paper encourages the importance of implementing explainable solutions into the high impact medical field.

Keywords: eXplainable Artificial Intelligence (XAI); XAI recommendation system; XAI scoring system; medical XAI; survey; approach

1. Introduction

XAI is recently dominating the research field for improving the transparency of the working model with the user. The brief history of AI development relates to statistical analysis, machine learning, natural language processing, computer vision, and data science. Even though such developments were present, it was not able to exceed human intelligence which was later progressed by neural networks, reinforcement learning, and deep learning. Such AI applications advancements were not only beneficial for weather forecasting analysis, self-driving cars, and the AlphaGo game capable of competing with the best humans' skills, but also were found to be of critical importance within the medical domain and its progress [1,2]. Human–Computer Interaction (HCI) research is also progressing to automate many applications and provide solutions [3]. Nevertheless, the improvements within the life expectancy have been recently improved with the use of advanced technologies and still will be beneficial to tackle the problems faced within different categories of the medical domains. Therefore, developments within the medical domain are discussed which focuses mainly on pneumonia status, bloodstream infections (BSI), acute kidney injury (AKI) and hospital mortality (HM) prediction [4]. XAI is necessary to be evaluated with the medical domain progression as it provides complete details of each algorithmic step thought to be trusted within the medical domain, practitioners, and experts. The three stages in XAI can be given as (i) explainable building process for facilitating acceptance, (ii) explainable decisions for enabling trust with users and administrators, and (iii) explainable decision process for the interoperability with business logic [5]. The goal of XAI is to provide machine and deep learning algorithms for better performance with explainability, which further allows ease of user trust, understanding, acceptance, and management.

Even though the drawbacks of the previous AI system including black box models, catastrophic consequences in medical diagnosis were discussed by some reference [6] but later by the progression with the model development, enhancement and tuning high accuracy, quality of work, and speed was achieved. XAI was also found to be the European Union's General Data Protection Regulation (GDPR) standard complaint, as no data is revealed to the outside system/participants by disclosing private medical datasets and providing explanations in the decision process.

1.1. Motivation

The motivation for this work is thought from realizing "Why explainability is necessary in the medical domain?", or it can also be given as the actual motivation is the laws and ethics aspects in the applications of XAI need to be considered before they can be applicable in the medical domain. In various parts of the world, the right to explanation is already enshrined by law, for example by the well-known GDPR, which has huge implications for medicine and makes the field of XAI necessary in the first place [7]. The medical AI is termed as a high-risk AI application in the proposal by European legislation, which is regulated for the fundamental rights of human dignity and privacy protection. In this case, the decision is based solely on real-time AI processing after the decision to assess, which is overcome by the "right to explanation". As the GDPR prohibits decisions solely based on automated processing, the final decision is drawn from the human in the loop approach and informed consent of the data subject. The legal responsibility of medical AI malfunctioning leads to civil liability instead of criminality. Additionally, compulsory insurance is required in the future against the risks of AI applications by the liability law. The ethics in medical AI gives a sustainable development goal for the "good health and well-being" by the United Nations [8]. The bias or flaw in training data due to the societal inclination impact may lead to the limitations in AI performance. Therefore, the factors given by the ethics committee discussions about the contribution of medical AI needs to be given so as to know the specific part decision/action, communication by AI agent, the responsibility taken by the competent person, transparency/explainability, method reference, avoiding manipulation for high accuracy, avoiding discrimination, and the algorithm must not control AI decision and actions. The purpose is to make AI a friend, and combining all of the above responsibilities, it would be termed as XAI. Therefore, the XAI approach provided within this paper constitutes one of the major portions for the directions of future approach and perspective.

1.2. Interpretability

A recent survey on medical XAI focuses completely on interpretability [9]. As the medical field possessess a high level of accountability and transparency, a greater interpretability is needed to be explained by the algorithm. Even though the interpretability is treated equally across all the hospitals, it should be handled with caution; medical practices should be the prime focus for interpretability development, and data based on mathematical knowledge for technical applications are encouraged. The different interpretability categories referenced here are perceptive and mathematical structures. The perceptive interpretability is mostly a visual evidence that can be analyzed using saliency maps, i.e., LIME, Class Activation Map (CAM), Layer-wise Relevance Propagation (LRP), etc.

In signal methods, the stimulation/collection of neurons are detected, i.e., feature maps, activation maximization, etc. The verbal interpretability is the human understandable logical statements based on the predicates, connectives, i.e., disjunctive normal form (DNF) and NLP. The mathematical structure based interpretability is the popular mechanism used through machine learning and neural network algorithms, whereas the predefined models are the relation between variable to output variable that includes logistic regression, Generative Discriminative Machine (GDM), reinforcement learning, etc. Ultimately, the feature extraction from the input source is performed by graphs presentation, clustering, frame singular value decomposition (F-SVD), etc.

1.3. Feedback Loop

A feedback loop designed for the XAI continuous development includes multiple phases, which can be given as follows [10]. The model debugging and visualization is performed first, then model compilation is performed by testing, after which the model is then released based on versioning. During the output phase, the predictions are performed by explainable decisions in which different models are compared for analysis and performance monitoring is performed successively followed by debugging and feedback loop. The model's explainability increases based on how much it supports open box architecture. The deep learning models, i.e., convolutional neural networks (CNN), recurrent neural networks (RNN) are the least explainable and are the predecessor of ensemble model, i.e., random forest, XGB. The statistical models and graphical models are easy to understand and are more straight forward, i.e., SVM, Bayesian brief net, Markov models, etc. The decision trees, linear models, and rule-based models are the most explainable and completely open box architecture models. The different XAI categories explained within this reference include dimension reduction which are presented as most important input features by selecting optimal dimensions, e.g., optimal feature selection, cluster analysis, LASSO, sparse deep learning, and sparse balanced SVM. The feature importance is used to capture characteristics and correlation amongst features for XAI models, e.g., feature weighting, DeepLIFT, SHAP, whereas the attestation mechanism captures the important areas where attention is required by the model, e.g., MLCAM, CAM, GRAD-CAM, Respond-CAM. The XAI well-known knowledge distillation is drawing the knowledge from a complicated model to a more rationalized model, e.g., rule-based system, mimic learning, fuzzy rules, and decision rules. Ultimately, the surrogate models are the locally faithful models and approximate reference models to surrogate model, e.g., LIME, LRP, etc.

1.4. General XAI Process

As the XAI necessity is thought to be effective for improvements within the future system. Therefore, the initial steps required for the XAI process are as follows:

(a) Pre-processing: The data cleaning, recovery/imputation and top feature analysis are described in this phase. The data cleaning refers to the handling of the incorrect, duplicate, corrupted, or incomplete dataset, whereas the data imputation refers to the substitute values for replacing missing data. In case of SHapley Additive exPlanations (SHAP), which is a part of game theoretic approach for identifying the top dominating features to help achieve better prediction results [8].

(b) Methodology: The model specifically designed for the effective implementation of the machine or deep learning construction and tuning. There are many machine learning algorithms, i.e., naïve bayes, linear regression, decision trees, support vector machine (SVM), etc., whereas neural networks are used to mimic human brains by providing a series of algorithms for recognizing relationships within the dataset [9,10]. The interpretable deep learning refers to the similar concepts except inspecting data processing at each layer and thus helping the designer to control the data movement and mathematical operations within it. Furthermore, the layers can also be configured by setting the feature learning by convolution, max pooling, and classification by fully connected, activation functions, etc.

(c) Explanation: This phase provides the explanation for each decision transparently to know the importance and action taken by the algorithm. The explanation provides detailed reasoning for all the decisions taken within the model from preprocessing, algorithm for prediction, classification, evaluation, and conclusion. As the explanations form the crucial content of XAI, it improves the acceptance of the deployed system to the end user, domain experts, or clients.

(d) Re-evaluation: Feedback system designed to understand limitations as the difference in choices made by the users and the algorithms. At the end of the algorithm, the end user can interact with the system by providing the necessary feedback for each decision and parameters used, which later can be evaluated effectively by re-configuration in the successive version. Therefore, it not only promotes ease of usage but also makes the end user as the part of the system, which can improve the next version of the training data and weights enhancement.

1.5. Objectives

The objectives for this survey can be given as follows:
- Determine the current progress within the different infection/diseases based on AI algorithms and their respective configurations.
- Describe the characteristics, explainability, and XAI methods for tackling design issues in the medical domain.
- Discuss the future of medical XAI, supported by explanation measures by human-in-the-loop process in the XAI based systems with case studies.
- Demonstrate a novel XAI Recommendation System and XAI Scoring System applicable to multiple fields.

A paper plan for this survey is given as follows: related works describe the various infections/diseases-based references, methods, and evaluations in Section 2; the difference between AI and XAI methods is given in Section 3; and recent XAI methods usage with its importance in Section 4. Afterwards, the characteristics of XAI0-based explanation in Section 5; future of XAI explainability in Section 6; and prerequisite for AI and XAI explainability in Section 7. Lastly, details about the case study for application usage in Section 8; XAI limitations in Section 9; XAI Recommendation System in Section 10; and XAI Scoring System in Section 11, followed by the conclusion and references.

2. Related Works

In this section, we are going to present the background for the medical domain with respect to the various infection or diseases related works, which are recently presented as a solution using AI or XAI. The research work presented in medical fields is mostly evaluated using mathematical statistics and machine learning algorithms as given in Tables 1–4. Therefore, it presents several opportunities to provide XAI-based implementation and improve the current understanding with better evaluation using classification.

The highly affecting acute respiratory disease syndrome (ARDS) or pneumonia-based evaluation supports various features such as vital signs and chest X-rays (CXR) [11]. The classification in this case can be mostly performed within the combination or independent data sources of vital signs and/or CXR. Usually the patients within this case are required to be first identified with specific symptoms of cough, fever, etc., and then the vital signs and/or CXR are used by the medical examiners to diagnose and know the healing progress of the pneumonia status. Later, the discharge is predicted using this work, and also more detailed configuration can help to understand the algorithm behavior. The mechanism for local determines a single decision system, whereas for global it determines multiple decisions.

Figure 1 presents the mindmap diagram for the literature survey analysis. The explanation type ante-hoc is for open/human understandable models and post hoc for black boxes and Deep Neural Networks (DNN). One of the commonly occurring infections within patients is bloodstream infection (BSI), which can be identified by the presence of bacterial or fungal microorganisms within the blood samples [21]. It is also popularly known as sepsis/septic shock and has severe symptoms. The most common symptoms include fever, increase in heart rate, high blood pressure, chills, and gastrointestinal issues. In the previous studies, the BSI was studied in detail with vital signs and laboratory variables with ICU admission data. The preprocessing is mostly done to recover the missing data in the BSI and Non-BSI cases, which is later evaluated using the machine learning model. The BSI once detected then later can be cured using medicine treatment.

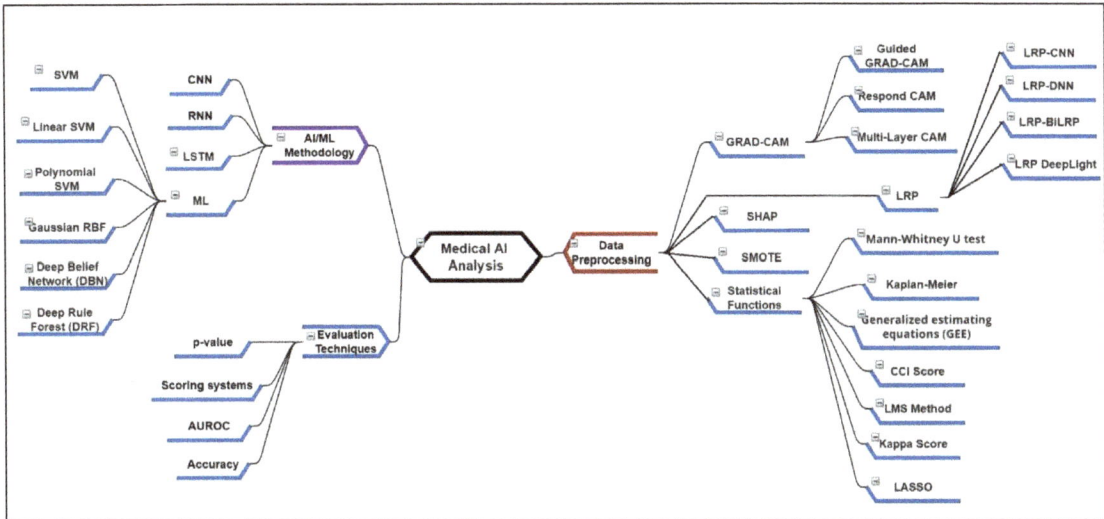

Figure 1. MindMap diagram for the medical AI analysis.

Table 1. General analysis for the explanation-based preprocessing.

Ref. #	Reference	Reference Paper/Mechansim	Data Preprocessing	Evaluation Methods/Algorithms	Outcome/Explanation Type
[11]	Selvaraju, R.R. et al. (2017)	GRAD-CAM/Global	GRAD-CAM	VGG, Structured CNN, Reinforcement Learning comparisons.	Textual explanations and AUROC/post-hoc
[12]	Tang, Z. et al. (2019)	Guided GRAD-CAM/Global	GRAD-CAM and feature occlusion analysis.	Segmentation on heatmaps and CNN scoring.	AUROC, PR curve, t-test and p-value/post-hoc
[13]	Zhao, G. et al. (2018)	Respond CAM/Global	GRAD-CAM, weighted feature maps and contours.	Sum to score property on 3D images by CNN.	Natural images captioning by prediction/post-hoc
[14]	Bahdanau et al. (2014)	Multi-Layer CAM/Global	Conditional probability	Encoder–decoder, neural machine translation and bidirectional RNN.	BLEU score, language translator and confusion matrix/post-hoc
[15]	Lapuschkin, S. et al. (2019)	LRP 1/Local (Layer-wise relevance propagation).	Relevance heatmaps.	Class predictions by classifier, Eigen-based clustering, LRP, spectral relevance analysis.	Detects source tag, elements and orientations. Atari breakout/ante-hoc
[16]	Samek, W. et al. (2016)	LRP 2/Local	Sensitivity	LRP, LRP connection to the Deep Taylor Decomposition (DTD).	Qualitative and quantitative sensitivity analysis. importance of context measured/post-hoc
[17]	Thomas, A. et al. (2019)	LRP DeepLight/Local	Axial brain slices and brain relevance maps.	Bi-directional long short-term memory (LSTM) based DL models for fMRI.	Fine-grained temporo-spatial variability of brain activity, decoding accuracy and confusion matrix/post-hoc

Table 1. *Cont.*

Ref. #	Reference	Reference Paper/Mechansim	Data Preprocessing	Evaluation Methods/Algorithms	Outcome/Explanation Type
[18]	Arras, L. et al. (2016)	LRP CNN/Local	Heatmap visualizations/PCA projections.	Vector-based document representations algorithm.	Classification performance and explanatory power index/ante-hoc
[19]	Hiley, L. et al. (2020)	LRP DNN/Local	Sobel filter and DTD selective relevance (temporal/spatial) maps	A selective relevance method for adapting the 2D explanation technique	Precision is the percentage overlap of pixels, std. and Avg. precision comparison/ante-hoc
[20]	Eberle, O. et al. (2020)	LRP BiLRP/Global	DTD to derive BiLRP propagation rules.	Systematically decompose similarity scores on pairs of input features (nonlinear)	Average cosine similarity to the ground truth, similarity matrix/ante-hoc

Table 2. Analysis for the BSI-affected patients treatment research.

Ref. #	Reference Paper	Dataset	Data Preprocessing/Mechanism	Evaluation Methods/Algorithms	Outcome/Explanation Type
[21]	Burnham, J.P. et al. (2018)	430 patients	Chi-squared/Fisher exact test, Student t test /Mann–Whitney U/Global	Multivariate Cox proportional hazards models	Kaplan–Meier curves and p-values/ante-hoc
[22]	Beganovic, M. et al. (2019)	428 patients	Chi-square/ Fisher exact test for categorical variables, and t test/ Wilcoxon rank for continuous variables./Global	Propensity scores (PS) using logistic regression with backward stepwise elimination and Cox proportional hazards regression model.	p-values./ante-hoc
[23]	Fiala, J. et al. (2019)	757 patients	Generalized estimating equations (GEE) and Poisson regression models/Global	Logistic regression models, Cox proportional hazards (PH) regression models	p-value before and after adjustment/ante-hoc
[24]	Fabre, V. et al. (2019)	249 patients	χ^2 test and Wilcoxon rank sum test/Local	multivariable logistic regression for propensity scores	Weighted by the inverse of the propensity score and 2-sided p-value/ante-hoc
[25]	Harris, P.N.A. et al. (2018)	391 patients	Charlson Comorbidity Index (CCI) score, multi-variate imputation/Global	Miettinen–Nurminen method (MNM) or logistic regression.	A logistic regression model, using a 2-sided significance level
[26]	Delahanty, R.J. et al. (2018)	2,759,529 patients	5-fold cross validation/Local	XGboost in R.	Risk of Sepsis (RoS) score, Sensitivity, Specficity and AUROC/post-hoc
[27]	Kam, H.J. et al. (2017)	5789 patients	Data imputation and categorization./Local	Multilayer perceptron's (MLPs), RNN and LSTM model.	Accuracy and AUROC/post-hoc
[28]	Taneja, I. et al. (2017)	444 patients	Heatmaps, Riemann sum, categories and batch normalization/Global	Logistic regression, support vector machines (SVM), random forests, adaboost, and naïve Bayes.	Sensitivity, Specificity, and AUROC/ante-hoc
[29]	Oonsivilai, M. et al. (2018)	243 patients	Z-score, the Lambda, mu, and sigma (LMS) method. 5-fold cross-validated and Kappa based on a grid search/Global	Decision trees, Random forests, Boosted decision trees using adaptive boosting, Linear support vector machines (SVM), Polynomial SVMs, Radial SVM and k-nearest neighbours (kNN)	Comparison of perfor-mance rankings, Calibration, Sensitiv-ity, Specificity, p-value and AUROC/ante-hoc
[30]	García-Gallo, J.E. et al. (2019)	5650 patients	Least Absolute Shrinkage and Selection Operator (LASSO)/Local	Stochastic Gradient Boosting (SGB)	Accuracy, p-values and AUROC/post-hoc

Table 3. Analysis for the AKI-affected patients treatment research.

Ref. #	Reference	Dataset	Criteria	Data Preprocessing/Mechanism	Evaluation Methods/Algorithms	Outcome/Explanation Type
[31]	Lee, H-C. et al. (2018)	1211	Acute kidney injury network (AKIN)	Imputation and hot-deck imputation/Global	Decision tree, random forest, gradient boosting machine, support vector machine, naïve Bayes, multilayer perceptron, and deep belief networks.	AUROC, accuracy, p-value, sensitivity and specificity/ante-hoc
[32]	Hsu, C.N. et al. (2020)	234,867	KDIGO	Least absolute shrinkage and selection operator (LASSO), 5-fold cross validation/Local	Extreme gradient boost (XGBoost) and DeLong statistical test.	AUROC, Sensitivity, and Specificity/ante-hoc
[33]	Qu, C. et al. (2020)	334	KDIGO	Kolmogorov–Smirnov test and Mann–Whitney U tests/Local	Logistic regression, support vector machine (SVM), random forest (RF), classification and regression tree (CART), and extreme gradient boosting (XGBoost).	Feature importance rank, p-value and AUROC/ante-hoc
[34]	He, L. et al. (2021)	174	KDIGO	Least absolute shrinkage and selection operator (LASSO) regression, Bootstrap resampling and Harrell's C statistic/Local	Multivariate Cox regression model and Kaplan-Meier curves.	p-value, Accuracy, Sensitivity, Specificity, and AUROC/ante-hoc
[35]	Kim, K. et al. (2021)	482,467	KDIGO	SHAP, partial dependence plots, individual conditional expectation, and accumulated local effects plots/Global	XGBoost model and RNN algorithm	p-value, AUROC/post-hoc
[36]	Penny-Dimri, J.C. et al. (2021)	108,441	Cardiac surgery-associated (CSA-AKI)	Five-fold cross-validation repeated 20 times and SHAP/Global	LR, KNN, GBM, and NN algorithm.	AUC, sensitivity, specificity, and risk stratification/post-hoc
[37]	He, Z.L. et al. (2021)	493	KDIGO	Wilcoxon's rank-sum test, Chi-square test and Kaplan–Meier method/Local	LR, RF, SVM, classical decision tree, and conditional inference tree.	Accuracy and AUC/ante-hoc
[38]	Alfieri, F. et al. (2021)	35,573	AKIN	Mann–Whitney U test/Local	LR analysis, stacked and parallel layers of convolutional neural networks (CNNs)	AUC, sensitivity, specificity, LR+ and LR-/post-hoc
[39]	Kang, Y. et al. (2021)	1 million.	N.A.	conjunctive normal form (CNF) and Disjunctive normal form (DNF) rules/Global	CART, XGBoost, Neural Network, and Deep Rule Forest (DRF).	AUC, log odd ratio and rules based models/post-hoc
[40]	S. Le et al. (2021)	2347	KDIGO	Imputation and standardization/Global	XGBoost and CNN.	AUROC and PPV/post-hoc

Table 4. Analysis for the hospital mortality prediction research.

Ref. #	Reference	Dataset	Ventilator	Data Preprocessing/Mechanism	Evaluation Methods/Algorithms	Outcome/Explanation Type
[41]	Mamandipoor, B. et al. (2021)	Ventila dataset with 12,596	Yes	Mathews correlation coefficient (MCC)/Global	LR, RF, LSTM, and RNN.	AUROC, AP, PPV, and NPV/post-hoc
[42]	HU, C.A. et al. (2021)	336	Yes	Kolmogorov–Smirnov test, Student's t-test, Fisher's exact test, Mann–Whitney U test, and SHAP/Global	XGBoost, RF, and LR.	p-value, AUROC/ante-hoc
[43]	Rueckel, J. et al. (2021)	86,876	Restricted ventilation (atelectasis)	Fleischner criteria, Youden's J Statistics, Nonpaired Student t-test/Global	Deep Neural Network.	Sensitivity, Specificity, NPV, PPV, accuracy, and AUROC/post-hoc
[44]	Greco, M. et al. (2021)	1503	Yes	10-fold cross validation, Kaplan–Meier curves, imputation and SVM-SMOTE/Global.	LR and Supervised machine learning models	AUC, Precision, Recall, F1 score/ante-hoc
[45]	Ye, J. et al. (2020)	9954	No	Sequential Organ Failure Assessment (SOFA) score, Simplified Acute Physiology Score II (SAP II), and Acute Physiology Score III (APS III)./Global	Majority voting, XGBoost, Gradient boosting, Knowledge- guided CNN to combine CUI features and word features.	AUC, PPV, TPR, and F1 score/ante-hoc
[46]	Kong, G. et al. (2020)	16,688	Yes	SOFA and SAPS II scores./Local	Least absolute shrinkage and selection operator (LASSO), RF, GBM, and LR.	AUROC, Brier score, sensitivity, specificity, and calibration plot/ante-hoc
[47]	Nie, X. et al. (2021)	760	No	Glasgow Coma Scale (GCS) score, and APACHE II/Global	Nearest neighbors, decision tree, neural net, AdaBoost, random forest, and gcForest.	Sensitivity, specificity, accuracy, and AUC/ante-hoc
[48]	Theis, J. et al. (2021)	2436	N.A.	SHAP, SOFA, Oxford Acute Severity of Illness Score (OASIS), APS-III, SAPS-II score, and decay replay mining/Global	LSTM encoder–decoder, Dense Neural Network.	AUROC, Mean AUROC and 10-FOLD CV AUROC/post-hoc
[49]	Jentzer, J.C. et al. (2021)	5680	Yes	The Charlson Comorbidity Index, individual comorbidities, and severity of illness scores, including the SOFA and APACHE-III and IV scores/Global	AI-ECG algorithm	AUC/post-hoc
[50]	Popadic, V. et al. (2021)	160	Yes	N.A./Local	Univariate and multivariate logistic regression models	p-values, ROC curves/ante-hoc

A severe type of infection or condition, which can be caused by multiple factors affecting blood flow to the kidney or medications side effects is known as acute kidney injury [31]. The symptoms can be basically seen in the lab tests, which include urine output and serum creatinine levels. In case of ventilation support, additional parameters are considered for the features. The preprocessing could help to improve data quality and provide promising results. Machine learning has shown to identify the stage and level of AKI, which has helped to apply proper medication treatment, recovery for the mild and control the severe conditions. In case of comorbidities or critical conditions, the hospital mortality is thought to be an important prediction [41]. There are more features available for such cases, as it involves distinct ICU parameters. Additionally, the medication courses and its related effects are available. The criteria for considering critical cases is the first filter for preprocessing and later data imputation can be added, if necessary. Previously, many works have provided such predictions using time windows before 48, 72 h, etc. by using either statistical, machine learning, and/or CNN methods. Such work is important in case of shifts in the medical treatment department or medication course.

3. Potential Difference between the AI and XAI Methods

Figure 2 presents various factors responsible for the difference in AI and XAI methods.

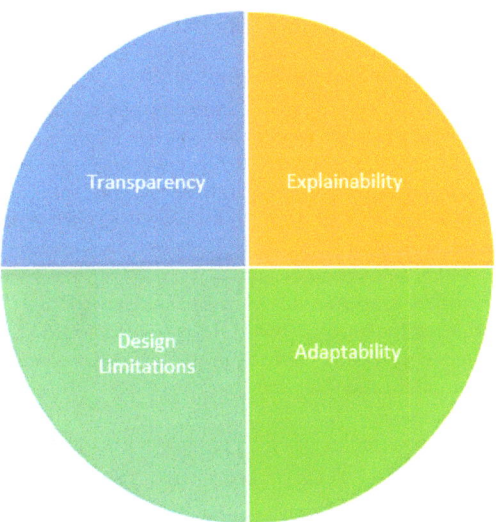

Figure 2. Difference in AI and XAI Methods.

Researchers are required to select XAI methods for the benefits as discussed below:

3.1. Transparency in the System Process

The use of conventional black-box-based AI models have limited its use and transparency of the system. Therefore, XAI methods are known for their transparent systems that provide the details of the data preprocessing, model design, detail implementation, evaluation, and conclusion. Transparency provides the user with complete system design that can be later configured, improved, versioned, and audited effectively.

3.2. Explainability of the System

The AI model lacks explainability for the system process. Therefore, the user's trust can be gained by a highly explained XAI-based decision process. The decision taken on every step of the system process and its supporting explanation makes it more effective. In case of model design issues, the explainability can also help to identify at which process step the erroneous decision was made and thus later can be resolved. The explainability is crucial for the initial data analysis, decision, and action for the whole XAI model.

3.3. Limitations on the Model Design

The AI models are usually black box and are not accessible to the end users. In comparison, the XAI provides models more interpretability at each structural layer, which is used to know the data quality, feature distribution, categorization, severity analysis, comparison, and classification. Thus, the acceptance of the XAI models is more due to interpretability. The user is more confident and has trust in the system. Nevertheless, false positive values can also be caught and analyzed in detail to avoid system failure and better treatment.

3.4. Adaptability to the Emerging Situations

The XAI models are known for high adaptability by using the feedback technique. The domain experts/medical examiners may be interested in applying/modifying a new feature. In severe cases, ICU parameters can also be adopted for better discharge and mortality classification. Due to recent infections, the cases of comorbidities are on the rise and such complex cases need high adaptability and explainability for the treatment. Ultimately, the model quality can be kept consistent and will be applicable to long-term usage.

4. Recent XAI Methods and Its Applicability

The recent reference papers show the approach of providing interpretability and transparency of using the models as shown in Figure 3. Even though the models, dataset, criteria, and outcome are specified in detail in many medical domain papers, still the explainability and justifiability needs to be provided for every case. In the future, interactive AI systems will be in more demand for providing such explainability and interaction with the domain experts to continuously improve the outcome, which is adapted to various situations such as changes in human, weather, and medical conditions. The tables from 1 to 4 are probable approaches for the respective infection/disease and are deemed to be appropriate for the hospital-based recovery prediction. For this section, we are going to discuss the preprocessing methods used for the recent paper, algorithms used within their respective models, and outcome.

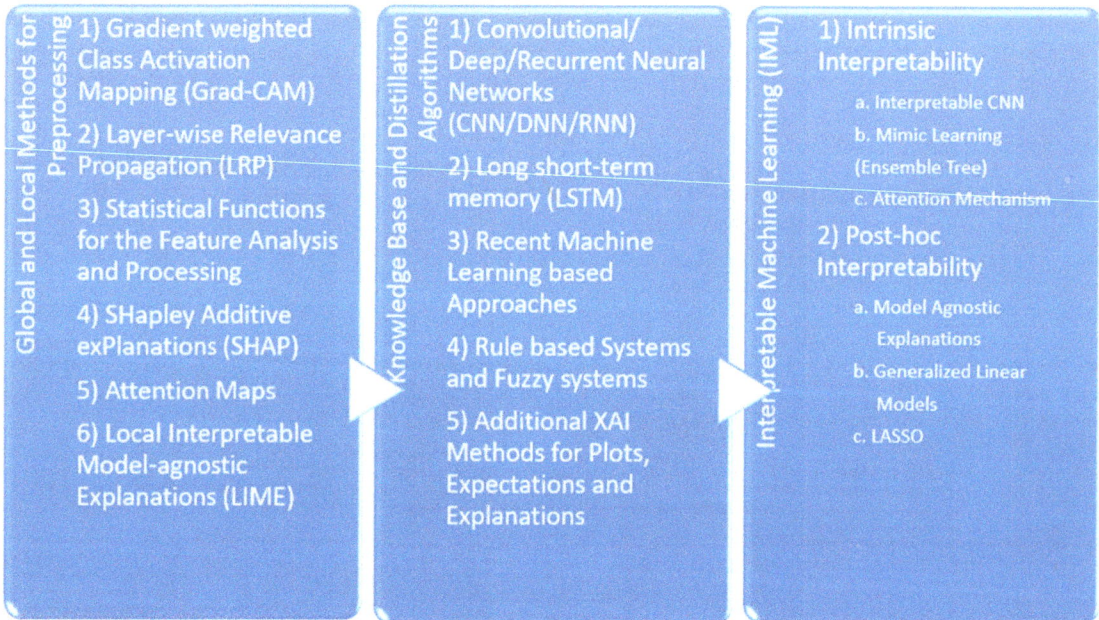

Figure 3. XAI Methods Categorization.

4.1. Local and Global Methods for the Preprocessing

4.1.1. Gradient Weighted Class Activation Mapping (Grad-CAM)

Grad-CAM [11] is used for prediction of the respective concept by referring to the gradients of the target, which is passed to the final convolutional layer. The important regions are highlighted using the coarse localization mapping. It is also known to be a variant of heat map, which can be used by image registration to identify the different image sizes and scales for the prediction. Grad-CAM is a propagation method, easy to visualize and provides user-friendly explanations. It is one of the popular methods in object detection and is recently used frequently within the medical domain to identify different diseases and affected areas of the patient. The chest X-ray (CXR), CT-scan, brain tumors, fractures in the different human/animal parts can be easily highlighted by such application. As the accuracy with sensitive domain is not recommended, there are several other versions for the CAM supported analysis include Guided Grad-CAM [12], Respond-CAM [13], Multi-layer CAM [14], etc. The Guided Grad-CAM is used to check models prediction by identifying salient visual features. Thus, the interest class relevant features are highlighted by the saliency maps. The Grad-CAM and guided backpropagation pointwise multiplication is known as saliency maps. The Guided Grad-CAM is known to generate class specific maps, which are the last convolutional layers feature map dot product and neurons combining to a predicted class score by partial derivatives. The Respond CAM is used to operate on the 3D images having complex structures of macromolecular size from the cellular electron cryo-tomography (CECT). The Respond-CAM has a sum to score property for better results than Grad-CAM and is used to highlight 3D images' class discriminative parts using weighted feature maps.

The Respond-CAM's sum-to-score property can be given as $y^{(c)}$ as the class score, $b^{(c)}$ is the last layer CNN parameter, $\sum_{i,j,k}(L_A^{(c)})$ is the class c sum for Grad-CAM/Respond-CAM and C as the number of classes given in Equation (1).

$$y^{(c)} = b^{(c)} + \sum_{i,j,k}(L_A^{(c)})_{i,j,k} \qquad (1)$$

The Multi-layer Grad-CAM is used to compute conditional probability of the selected feature with a single maxout hidden layer. It is based on maxout units, a single hidden layer with a softmax function to normalize output probability.

4.1.2. Layer-Wise Relevance Propagation (LRP)

It is also one of the popularly used propagation methods, which operates by using the propagation rules for propagating the prediction backward in the neural network. The LRP can flexibly operate on input such as images, videos, and texts. The relevance scores can be recorded in each layer by applying different rules. The LRP is based and justified using a deep taylor decomposition (DTD). It can be set on a single or set of layers in the neural network and can be scaled in the complex DNN by providing high explanation quality. It is also popularly used in the medical domain consisting of CXR, axial brain slices, brain relevance maps, and abnormalities, etc. The versions available in LRP are LRP CNN, LRP DNN, LRP BiLRP, LRP DeepLight for the heatmap visualizations. The LRP relevance is higher as compared to other visualization/sensitivity analysis. The input representations are forward-propagated using CNN until the output is reached and back-propagated by the LRP until the input is reached. Thus, the relevance scores for the categories are yielded in LRP CNN [18]. For the LRP DNN [19], the CNN is tuned with initial weights for the activity recognition with pixel intensity. In LRP BiLRP [20], the input features pairs having similarity scores are systematically decomposed by this method. The high nonlinear functions are scaled and explained by using composition of LRP. Thus, the BiLRP provides a similarity model for the specific problem by verifiability and robustness.

The BiLRP is presented as a multiple LRP combined procedure and recombined on input layer. Here, x and x' are input which are to be compared for similarity, $\varnothing_x$ as a group of network layer with $\{\varnothing_1$ to $\varnothing_L\}$, and y(x, x') as the combined output given in Equation (2).

$$\text{BiLRP}(y, x, x') = \sum_{m=1}^{h} \text{LRP}([\varnothing_L \circ \cdots \circ \varnothing_1]_m, x) \bigotimes \text{LRP}([\varnothing_L \circ \cdots \circ \varnothing_1]_m, x') \qquad (2)$$

The DeepLight LRP [17] performs decoding decision decomposition, which is used to analyze the dependencies between multiple factors on multiple levels of granularity. It is used to study the fine-grained temporo-spatial variability of the high dimension and low sample size structures.

4.1.3. Statistical Functions for the Feature Analysis and Processing

The statistical analysis [21] of survivors and non-survivor's comparison for categorical variables is performed by chi-square test/Fisher's exact test and reported as interquartile range (IQRs) and standard deviation/medians. Whereas, the continuous variables by Mann–Whitney U test or Student's *t*-test and expressed as frequencies. The Kaplan–Meier is used for graphical analysis of the relationship between two features with a significance log rank test. The hazard model of multivariate cox proportional regulation regulates the risk factor for the outcome and is analyzed graphically by the log-log prediction plot. In such cases, a significant *p*-value is less than 0.05 for single variate and 0.10 for bi-variate analysis. The generalized estimating equation (GEE) [23] is used to present the correlations between the feature matched sets. The incidence difference between the feature inheritance with GEE matching is within pre and post data adjustments. The Charlson comorbidity index score [25] is used to determine the comorbidities affected hospitalized patient life span risk within one year by a weighted index. The multivariate imputation is performed by the multiple imputation for the post-hoc sensitivity analysis for discrete and continuous data using chained equations. The lambda, mu, and sigma (LMS) method [29] is used to calculate the spirometric values for the normal lower limits in the z-scores. The kappa is an account chance agreement, where measurement agreement produces output as kappa 1.0 else 0. The least absolute shrinkage and selection operator (LASSO) [32] is a method of variable selection and regularization for improving prediction accuracy as a regression analysis. The imbalance classification problem is popularly solved by using Synthetic Minority Oversampling Technique (SMOTE) [44]. The cause of imbalance is usually due to the minority class, which are later duplicated in the training set before fitting the model. Such duplication helps to balance class duplication but does not provide any additional information.

4.1.4. SHapley Additive exPlanations (SHAP)

The SHAP [35] uses ranking based algorithms for feature selection. The best feature is listed in the descending values by using SHAP scores. It is based on the features attribution magnitude and is an additive feature attribution method. SHAP is a framework that uses shapley values to explain any model's output. This idea is a part of game theoretic approach which is known for its usability in optimal credit allocation. SHAP can compute well on the black box models as well as tree ensemble models. It is efficient to calculate SHAP values on optimized model classes but can suffer in equivalent settings of model-agnostic settings. Individual aggregated local SHAP values can also be used for global explanations due to their additive property. For deeper ML analysis such as fairness, model monitoring, and cohort analysis, SHAP can provide a better foundation.

4.1.5. Attention Maps

Popularly used to be applied on the LSTM RNN model, which highlights the specific times when predictions are mostly influenced by the input variables and has a high interpretability degree for the users [51]. In short, the RNN's predictive accuracy, disease state, decomposition for performance, and interpretability is improved. The attention vector learns feature weights, to relate the next model's layer with certain features mostly used with LSTM for forwarding attention weights at the end of the network.

$$a_k = softmax\,(W_k x_k) \tag{3}$$

Here, the W_k learned weights are used for calculating a_k for every k feature of x_k. A feature on every time step is x_k weighted with a learned attention vector, which is later given as y_k in Equation (4).

$$y_k = a_k \odot x_k \qquad (4)$$

An ICU critical task to capture individual physiological data that is time sensitive is demonstrated in DeepSOFA [52]. The attention mechanism is used to highlight variables in time series, which are crucial for mortality prediction outcome. Successively, the time step is assigned with more weights thought to be more influential for outcome.

4.1.6. Local Interpretable Model-Agnostic Explanations (LIME)

The LIME is a feature-scoring method, which performs the input data samples perturbation and checks for prediction change for understanding the model. In SurvLIME [53], cox proportional hazard is used to approximate a survival model within the range of the test area. The cox uses covariates coefficient of linear combination for the prediction impact of solving unconstrained optimization problems and other applications. The black-box-based human understandable explanations are given by medical examiner XAI [54], which is a LIME-like with rule-based XAI. In this case, a model agnostic technique is used for handling sequential, multi-labelled, ontology-linked data. This model trains a decision tree on labeled synthetic neighbors and the decision rules help to extract the explanations. The applications are used to predict the next diagnosis visit of the patient based on EHR data using RNN. The Lime based super-pixel generation is given in Appendix A.1.

4.2. Knowledge Base and Distillation Algorithms

4.2.1. Convolutional/Deep/Recurrent Neural Networks (CNN/DNN/RNN)

CNN is a deep learning method, which is used to depict the human brain for higher performance and solving of complex tasks. It basically takes an input data/image, assigns weights and biases to its various factors, and later differentiates them from each other. The filters used here act as a relevant converter for spatial and temporal dependencies. The CNNs designed for structured output are used for image captioning [11]. To improve this captioning, the local discriminative image regions are found to be better with the CNN + LSTM models. The CNN scoring [12] provides precise localization. Later, based on some categories and thresholds, the scores are calculated. The DNN [43] is termed on the network consisting of multiple hidden layers. The DNN, once trained, can provide better performance for the suspicious image findings, which can be used to identify faults and status. The RNN is mostly used in the natural language processing applications as they are sequential data algorithms. It is usually preferred for remembering its input by its internal memory structure and thus is mostly suitable for machine learning methods involving sequential data. The bi-directional RNN [14] is designed to function as an encoder and decoder, which emulates searching through sequences at the time of its decoding. Thus, the sequences of forward and backward hidden states can be accessed.

4.2.2. Long Short-Term Memory (LSTM)

The advancement for processing, classifying and making predictions on time series data is achieved by using LSTM. The vanishing gradient problem is popularly solved by using LSTM. The bi-directional LSTM [17] is used to model the within and across multiple structures with the spatial dependencies. Deeplight also uses a bi-directional LSTM, which contains a pair of independent LSTM iterating in the reverse order and later forwarding their output to the fully-connected softmax output layer. The LSTM encoder takes n-sized embedded sequences with dual layer, n cells, and outputs dense layers. The second LSTM is the reverse architecture known as a decoder to reconstruct the input. The dropout layer can be used in between encoder and decoder to avoid overfitting.

In this LRP, the linear/non-linear classifier f is used with input a having dimension d, positive prediction $f(a) > 0$, and R_d is having a single dimension of relevance.

$$f(a) \approx \sum_{d=1}^{D} R_d \tag{5}$$

Here, $R_j^{(l)}$ with a j neuron at l network layer, $R_{i \leftarrow j}^{(l-1,l)}$ defined by deep light where $Z_{ij} = a_i^{(l-1)} w_{ij}^{(l-1,l)}$ having coefficient weight w, a as input, and ϵ as stabilizer given in Equation (6).

$$R_j^{(l)} = \sum_{i \epsilon(l)} R_{i \leftarrow j}^{(l-1,l)}$$

$$R_{i \leftarrow j}^{(l-1,l)} = \frac{Z_{ij}}{Z_j + \epsilon \cdot \text{sign}(Z_j)} R_j^{(l)} \tag{6}$$

4.2.3. Recent Machine Learning-Based Approaches

The support vector machines (SVMs) are used for regression, classification, and outlier detection, which are supervised learning algorithms. It is more popularly used in high-dimensional spaces, which can be even greater than sample size. The linear SVM [29] is used in ultra large datasets for solving multiclass classification problems, which is the version of the cutting plane algorithm. The polynomial SVM is also known as polynomial kernel, which shows the polynomial having feature space with a training set focusing on the similarity vectors. The decision boundary flexibility is controlled by degree parameter. Hence, the decision boundary can increase based on the higher degree kernel. The SVM also uses one more kernel function known as Gaussian RBF (Radial Basis Function). The value calculated on the basis of some point or origin distance is RBF kernel. In machine learning, a deep neural network class or generative graphical model is known as deep belief network (DBN) [31]. It is constructed with latent variables of multiple layers having interconnected layers excepts for the units in each layer. The deep rule forest (DRF) [39] are multilayer tree models, which uses rules as the combination of features to outcome interaction. The DRF are based on the random forest and deep learning based algorithms for identifying interactions. Validation errors can be effectively reduced by DRFs hyper-parameters fine tuning.

The DBN [55] consists of the following evolution of a restricted boltzmann machine (RBM) having posterior probability of each node with values 1 or 0.

$$P(h_i = 1|v) = f(b_i = W_i v) \tag{7}$$

$$P(h_i = 1|h) = f(a_i = W_i h) \tag{8}$$

Here, the $f(x) = 1/(1 + e^{-x})$, which has energy and distribution function as:

$$E(v, h) = -\sum_{i \in v} a_i v_i - \sum_{j \in h} b_j h_j - \sum_{i,j} v_i h_j w_{ij} \tag{9}$$

$$p(v, h) = \frac{1}{z} e^{-E(v,h)} \tag{10}$$

The RBM follows unsupervised learning with pdf $p(v)$, $\theta \epsilon \{W, a, b\}$ as likelihood function, and v as input vector given as $p(v, \theta)$, where the gradient method has $\log p(v, \theta)$ as likelihood function and higher learning can be achieved by revising gradient parameters as $\frac{\partial p(v,\theta)}{\partial \theta}$.

$$\theta(n+1) = \theta(n) + a \times \left(-\frac{\partial p(v, \theta)}{\partial \theta} \right), \quad \theta \epsilon \{W, a, b\}$$

$$-\frac{\partial logp(v, w_{ij})}{\partial w_{ij}} = E_v\left[p(h_i|v) \times v_j\right] - v_j^{(i)} \times f\left(W_i \times v^{(i)} + b_i\right)$$

$$-\frac{\partial logp(v, b_i)}{\partial b_i} = E_v\left[p(h_i|v) \times v_j\right] - f\left(W_i \times v^{(i)}\right)$$

$$-\frac{\partial logp(v, a_j)}{\partial a_i} = E_v\left[p(h_i|v) \times v_j\right] - v_j^{(i)} \quad (11)$$

4.2.4. Rule-Based Systems and Fuzzy Systems

A rule-based system uses knowledge representation rules for obtaining the knowledge coded in systems. They are completely dependent on the expert systems, which solves the knowledge-intensive problem by reasoning similar to human experts. It is used in stroke prediction models by interpretable classifiers using Bayesian analysis [56]. The interpretability of decision statements is simplified by the high dimensional and multivariate feature space by the discretization of if-then conditions. The decision list has posterior distribution yielded by Bayesian rule list. The structure used here to support sparsity has a highly accurate medical scoring system. The interpretable mimic learning uses gradient boosting trees and has high prediction performance as a knowledge distillation approach [57]. Mimic learning uses a teacher and student model, where the teacher model eliminates training data noise/error and soft labels are passed to the student model as regularization to avoid overfitting. It is applied in the medical domain of acute lung injury and achieves high prediction results. It is also known to be applicable in speech processing, multitask learning, and reinforcement learning. Fuzzy rules are a form of if-then conditional statements that are yielding truth to a certain degree instead of complete true/false. A deep rule-based fuzzy system is used to predict ICU patient's mortality which consists of a heterogeneous dataset combining categorical and numeric attributes in hierarchical manner [58]. The interpretable fuzzy rules can be found in each unit of hidden layer within this model. Also to gain interpretability, a supervised random attribute shift is added in the stack approach.

The supervised clustering has fuzzy partition matrix and cluster centers. Here, β_{dp} is the output weight vectors having a building unit as dp-th, where the partition matrix is U_{dp} and output set as T given in Equation (12).

$$\beta_{dp} = \left(\frac{1}{Const}I + U_{dp}^T U_{dp}\right)^{-1} U_{dp} T \quad (12)$$

The interpretability is the layer's prediction with random projections for higher linear separability, where α' is the sub constants of α, Z_{dp} as random projection matrix, and Y_{dp} as the last unit's output vector.

$$X_{dp} = X + \alpha' Y_{dp} Z_{dp}$$

$$Y_{dp} = U_{dp} \beta_{dp} \quad (13)$$

4.2.5. Additional XAI Methods for Plots, Expectations, and Explanations

The partial dependence plot (PDP) in machine learning presents a marginal effect between input of one or multiple features on the final prediction, which is usually having a partial dependency. The PDP algorithm performs the average of all input variables except for PDP computed variable n [59]. This variable n is then checked in relation to the change in target variable for the purpose of recording and plotting. In comparison to the PDP, individual conditional expectations focus on specific instances that disclose variations in the recovery of the patient's subgroup [60]. The XAI-based explanation to the classifier prediction is best achieved by the Local Interpretable Model-agnostic Explanations (LIME) as an interpretable model approximating black box model to the instance under consideration [61]. The artifacts are user defined interpretable modules and are used later to generate local black boxes for instance neighbors. The user intervention and artifact limits are overcome by Semantic LIME (S-LIME) for possessing semantic features which are independently generated using unsupervised learning.

The fidelity function is given below consisting of model g with the instance x and y for feature characterizing agreement and the function π having exponential kernel with weighted σ with a distance D.

$$\mathcal{F}(x, f, g, \pi) = \sum_{y \in X} \pi(x, y).(f(y) - g(y))^2 \tag{14}$$

$$D(x, y) = \sum_{x_i = 1} |x_i - y_i| \tag{15}$$

LIME is popular to highlight the important features and provides explanation based on its coefficient but suffers due to randomness in sampling step, making it unacceptable in medical applications. To gain trust, safeguard stakes, and avoid legal issues, a high stability and an acceptable adherence level system is proposed known as optimized LIME explanations (OptiLIME) for diagnostics [62]. The mathematical properties are clearly highlighted and are kept stable across several runs in OptiLIME to search for the best kernel width in an automated way. As per the formula given below in Equation (16), the declining R^2 is converted into $l(kw, \widetilde{R}^2)$ a global maximum to get the best width. Here, the $\widetilde{R}^2$ is the expected adherence with random kw values.

$$l(kw, \widetilde{R}^2) = \begin{cases} R^2(kw), & if\ R^2(kw) \leq \widetilde{R}^2 \\ 2\widetilde{R}^2 - R^2(kw), & if\ R^2(kw) > \widetilde{R}^2 \end{cases} \tag{16}$$

In the classical ROC plot and AUC, the alterable threshold leads to the changes in false positive and false negative errors types [63]. As the partial part of ROC and AUC are useful in imbalanced data, then optional methods include partial AUC and the area under precision recall (PR) curve but are still insufficient to be trusted completely. Therefore, a new method known as partial AUC (pAUC) and c statistics of ROC are present, maintaining characteristics of AUC which are continuous and discrete measures, respectively. For the horizontal partial AUC, where x = 1 for the AUC integration border and other parts as true negative. Integration with baseline as x-axis and baseline x = 0 in case of swapping x and y axis. Thus, by transforming x (FPR) to 1 − x (TNR) then TNR can be received as required and x = 0 changes to 1.

$$pAUC_x \triangleq \int_{y_1}^{y_2} 1 - r^{-1}(y) dy \tag{17}$$

The partial c statistic (c_Δ) for ROC data is given in the normalized form as below in Equation (18). The c_Δ can be expressed as J out of positive's P and the k as a subset out of negative's N.

$$\hat{C}_\Delta \triangleq \frac{2PN.c_\Delta}{J.N + K.P} \tag{18}$$

The partial c statistic can be summed up as shown by the whole curve having q disjoint partial curves.

$$c = \sum_{i=1}^{q} (c_\Delta)_i \tag{19}$$

4.3. Interpretable Machine Learning (IML)

Machine learning has made phenomenal progress recently in a wide variety of applications including movie recommendation, language translation, speech recognition, self-driving cars, etc. [64,65]. IML aims to provide human-friendly explanations with the combined efforts from computer science, social science, and human–computer interaction. As self-driving cars need to make decisions by themselves in real time, the black box model would not be feasible and acceptable. Therefore, an open box model with explainability will convey the decision to the user and its choice based on the related reason [66], i.e., why the daily route was changed is due to the traffic congestion in the upcoming lane. The two categories of IML can be given as:

4.3.1. Intrinsic Interpretability

Inherently interpretable models consist of self-explanatory features within their structure. It has more accurate explanations with the slight trade-off of prediction performance. The global interpretable models can be either made by interpretable constraint or by complex model extraction. In interpretability constraints, the pruning of decision trees is performed to replace subtrees with leaves for deep trees instead of balanced structure. In the case of interpretable CNN, natural objects are identified accurately by adding regularization loss for learning disentangled representations, whereas in interpretable model extraction, also known as mimic learning, the trade-off of explanation is not substantial. In this case, the complex model is converted into a simple interpretable models, i.e., decision trees, linear model. The obtained model has better prediction and explainability performance, e.g., the ensemble tree or DNN is transformed into a decision tree, where the overfitting is handled by active learning.

The local interpretable models are more focused on providing specific prediction by a more justified model architecture. The attention mechanisms used by the RNNs as sequential models are interpreted by the attention weight matrix for explaining individual predictions. Attention mechanism is mostly used in image captioning with CNN for image vectorization and RNN for descriptions. Additionally, it can also be used in neural machine translation applications.

4.3.2. Post-Hoc Interpretability

These are the independent model, which requires supporting models to provide explanation. The post-hoc global explanation consists of machine learning models that capture several patterns from the training data and retain knowledge into the model. Here, the knowledge within the pre-trained models are presented to the end user understanding. In machine learning, the data is converted to features, which are interpretable and are mapped to output, i.e., feature importance. Model agnostic explanations are known to be a black box model with no transparency, whereas in permutation feature importance, the n features are shuffled to check the model's average prediction score and is known to be an efficient and robust strategy. The model-specific explanation is based on internal model structure for its explanation. The generalized linear models (GLM) consist of linear model combinations for features transformation, e.g., linear regression, logistic regression, etc. GLM has limitations when the feature dimensions become too large. In tree-based ensemble models, i.e., random forests, gradient-boosting machines (GBM), XGBoost, which measure feature contribution by accuracy, feature coverage, or split data count. In case of DNN explanation, the representations are given by the neurons at the intermediate layers for detail analysis. The activation maximization is utilized for iterative optimization of the image interpretations at different layers. Even though some noise and errors can be faced during classification, generative models are found to provide better visualization. Therefore, the CNN can capture better visualization from object corners, textures to object parts, and then whole objects or scenes. The RNN are better known for abstract knowledge where language modeling is required for learning representations. The RNN are good at capturing complex characteristics such as dependencies, syntax, and semantics. RNNs can capture hierarchical representations from different hidden layers, whereas the multi-layer LSTM are used to construct bi-directional language models with context aware understanding of words.

The post-hoc local explanations are focused on individual predictions based on the features supporting it and are also known as attribution methods. In model-agnostic explanations, the predictions from different machine learning models as black boxes are explained without guarantee, whereas the local approximations explanation supports interpretable white box based explanation in an adjacent part of the input, e.g., attribution methods, sparse linear models such as LASSO. The perturbation-based methods, the feature contribution, determines the prediction score. Thus, if the input part can change the prediction, then it is known as counterfactual explanation. The model specific explanations refer to white-boxes such as back-propagation method, deep representations, and perturbation methods. For perceptive interpretability, refer to Appendix A.2.

5. Characteristics of Explainable AI in Healthcare

In this section, a complete aspect of the medical XAI system is given in detail. Considering the hospital situation, the interaction, explanation, and transparency detail of the system will be disclosed. The characteristics will provide a complete overview about the new generation of XAI healthcare system, equipped with enhanced capabilities [66] as shown in Figure 4.

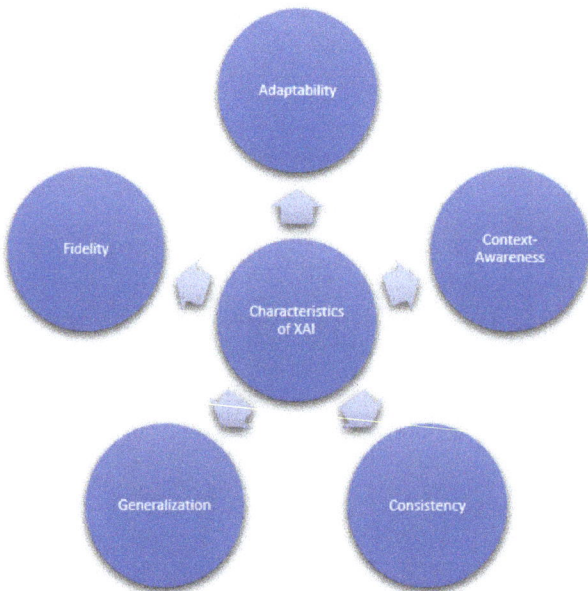

Figure 4. XAI Characteristics.

5.1. Adaptability

The transparency needs to be provided equally amongst all the healthcare system users. In case of medical examiners, the details of complete dataset preprocessing, algorithm function and decision analysis at each step should be provided. The medical examiners should be familiar with the system usage by training provided earlier to consultation and the protocol followed by the hospital treatment standards. The detailed decisions based on the model training on the previous year's consultations can help the medical examiners to check on multiple factors and then provide a final decision.

Early prediction systems can help medical examiners to take immediate actions to avoid severe conditions. In case of nurses, the statistics of the patient's health can be displayed to help them record the patient's health recovery and administer the required procedure. The history records of the patients should be accessible and should provide reminders about the emergency and regular scheduled procedure to be achieved. For the administrators, the patient's record, clinical tests, previous history of payments, and alerts for the future treatment possibility as decided by the medical examiners can be predicted. The patients connected to the hospital system can receive the daily reminders of the personal treatment, doses, warnings about the diets, alerts for the improvements and updates from the system in case of major infections spread, etc.

5.2. Context-Awareness

An XAI system should be complete in every sense. In case of diagnosis, the system should provide detailed vital signs, CXR, clinical tests are given as the affected patient's conditions and disease. Hence, a prediction or classification system is used to provide the discharge status after one week or month based on the patient's history records. For the surgical/ICU department, the features would vary as the oxygenation, ventilators, supporting instruments based on the alternative mechanical methods, etc. Therefore, the algorithms used in this XAI model should be adaptable to the new feature depending on the patient's case-to-case basis. In the drug development/consultation process, the XAI algorithm can predict the required dose schedule, the weighted contents of the drug, the combination of drug should be suggested for the comorbidities case, etc. The risk associated with the different types of cases should also be disclosed in the drug usage case. Exceptional circumstances can also be made for high risk patients and the supporting dietary or treatment with different age and continental patients should be constituted.

5.3. Consistency

The healthcare dataset and model should be consistent during the continuing patient's treatment. Also there should be consistency between multiple evaluations for the same patient. Therefore, versioning is required to be maintained and report the updates as per the module changes. A version reports the updates as per the module enhancements. A version report can be made available to know the updated module details and the changes affecting the past patient's records. Log records should be maintained for every patient that can display complete history with health status and respective time series records. The system log records should be immutable and must store the versioning information with the updates and fixes. A database maintained with such rules must also include the patient's medication course applied, clinical test report, ICU/emergency facilities details and some special treatment applied based on some exceptional circumstances. The comorbidities are related to complex cases that may require careful treatment and dependency factors to be analyzed. Consistency is an important aspect of the hospital's quality control and research department.

5.4. Generalization

In the healthcare system, every patient's data consists of vital signs, CXR, clinical tests and comorbidities. In case of ICU treatment, additional features are present. The designed model must be able to distinguish between multiple patients based on the features with high accuracy and less error rate. Thus, if many instances have similar explanations, then the generalization is not acceptable for the treatment and operating process. The XAI model should be adaptable to different features and must be effective to provide distinct explanations based on the case-to-case basis. The XAI algorithm must be able to provide high transparency of every category of the patient's data, i.e., vital signs, CXR, clinical test, and comorbidities. It will be useful to distinguish between patients' affected status in different categories. These explanations will be helpful to the medical examiners and medical staff for knowing about the patient's current health status, i.e., slight/mild/severely affected and to take appropriate further actions.

5.5. Fidelity

A designed XAI model should be configured as per the available dataset categories and must be specific to the objective application, i.e., healthcare. To provide a more effective explanation, the model must be interpretable. Thus, the benefit of having interpretable models is to analyze the processing of the input data at each level. Considering the CXR images, the interpretable model will provide analysis by CXR image quality as high, average, and low. Additionally, to know whether the CXR processing for the chest cage edge identification is aligned or not. The feature identification for the different categories of diseases/infection as infiltrate, cardiomegaly, effusion, COVID-19, etc. The level of severity analysis of the patient's condition either as normal, slight, mild, and severe infection are some of the factors. Furthermore, interpretation must be aligned with the XAI model prediction to enable the patient's discharge and/or mortality status prediction with high transparency.

6. Future of Explainability in Healthcare

In this section, we have identified and provided the four key aspects for the future of explainability in healthcare. The human-in-the-loop (HITL) enhances the classification capability of XAI, human–computer interaction (HCI) provides the deep understanding of a patient's condition, explanation evaluation provides key insights for personalized outcome, and explainable intelligent systems (EIS) significantly improves the medical XAI system. The demands of the medical system in healthcare are always at priority. The future of XAI shows promising solutions that can improve the healthcare facilities as shown in Figure 5.

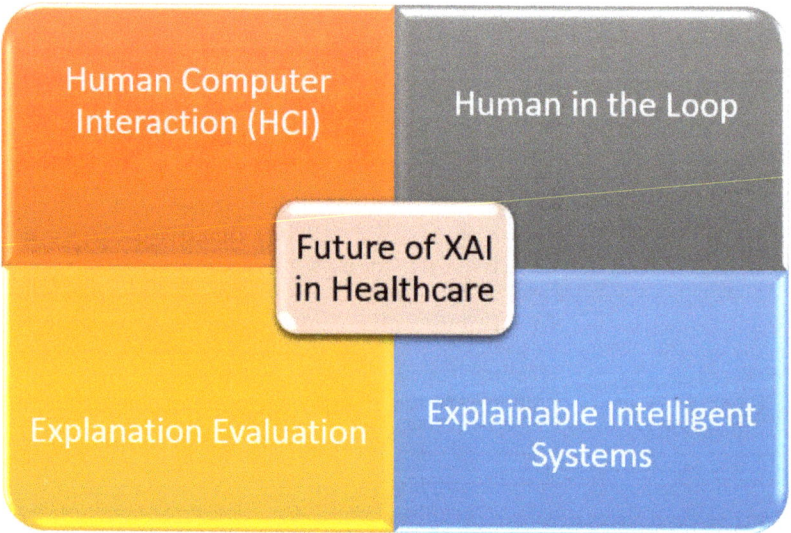

Figure 5. XAI in Healthcare.

6.1. Human–Computer Interaction (HCI)

The concept of HCI refers to the interaction between the real world and augmented reality [67]. The human subject here is the patient whose interaction with the computer is recorded for the symptoms feature identification purposes. Here, the computer sensors are used to record the human movements, e.g., sneezing, coughing, chest pain, stress level, lack of focus, etc. The HCI then provides the output based on machine learning algorithms for the predictions of the results. The HCI is also a crucial aspect in the future of XAI as it will add the symptoms feature for disease identification. The HCI has further applications to detect human poses, body structure, movement discontinuities, speech recognition, object handling using motion detection, psychological response, etc. Even though the recent AI is thought to be progressing, with the future XAI, a complete human body functioning is thought to be a progressive step towards the goal.

6.2. Human in the Loop

Applying the XAI concept in the healthcare domain is thought to be incomplete without the human in the loop process [68]. Considering an infection/disease, there can be several symptoms including EHR, CXR, clinical tests, etc. In recent works, it can be noticed that multimodal data analysis is a challenge for the machine learning algorithms because of trade-offs, less domain knowledge, high false positives, etc. To effectively solve such a challenge, the domain expert should be continuously involved within the interpretable model implementation to set the required hyper-parameters at each level, manage the trade-off, add/remove features manually, decision-based system, manual labeling of data, handling exceptional conditions, etc. A versioning-based system or feedback evaluation system should be used for continuous improvement so that the final system will be used in the hospital evaluation with trust. Human in the loop is hence necessary to manage the identification/diagnosis or prediction system for the new category of infection/diseases without replacing the whole XAI model and by adapting to the current scenario.

6.3. Explanation Evaluation

The XAI explanation for the final results evaluation is one of the most crucial aspects in healthcare. During the peak hours for patient's diagnosis and health prediction, medical examiners prefer to only check the final result as an expert opinion. Therefore, the final explanation provided by the system should be effective and acceptable. Nevertheless, recent works have discussed the selection of the explanation from multiple robots [69]. For the different robots the explanation may vary, so during the initial phase of the model deployment in the hospital center, the medical examiners are asked to choose the sentence type from the multiple explanation options as best suitable to the respective medical examiner/user. The type of explanation selection determines which robot is most suitable to the medical examiner and is thus finalized to that specific medical examiner's personal account. Therefore, both the system transparency of the evaluation and the explainability are achieved. The detailed explanation of the results provides model interpretability and helps to gain the user's trust.

6.4. Explainable Intelligent Systems

Modern healthcare is being strengthened and revolutionized by the development in AI [70]. The XAI-based system can improvise the previous analysis, learning, predict, and perform actions with explainability for the surgery-assisted robots, relationship within genetic codes to detect, and evaluate minor patterns. The XAI intelligent system is aimed at explaining the AI-led drug discovery, so that faster, cheaper, and effective drug development is performed, e.g., COVID-19, cancer treatments, etc. Healthcare robotics are used for assisting certain patients in paralysis, smart prosthesis, assistive limbs, spinal cord injuries and can explain how much recovery in the patient is recorded. Additionally, during the surgery process, the robots can explain the decision taken and necessary actions. The AI-powered stethoscope can be used in remote areas where medical personnel shortage is present and can analyze high clinical data for discovering disease patterns and abnormalities. Ultimately, the intelligent systems can treat and provide better explanations for transparent and trustable processes.

7. Prerequisite for the AI and XAI Explainability

A user is recommended to choose complete XAI explainability categories of preprocessing, methodology and healthcare as shown in Figure 6, as a part of the human-in-the-loop approach with the discussions provided in the following subsections:

Explainable AI (XAI) Approach		
XAI PREPROCESSING	**XAI METHODOLOGY**	**XAI EVALUATION**
Dataset Consistency	Feature Validation	Model Metrics
Imputation Functions	Novel Approach	Classification Cases
Data Distribution	Method Inefficiency	Recursion
Image Registration	Multi-modal Data Analysis	Fusion Algorithm
Feature Scoring	Model Synchronization	Graphs
Feature Priority	Feature Analysis	Manual Features Effect
Equal Feature Scoring	Severity Level Analysis	Feedback Model
Threshold for Feature Selection	Feature Effectiveness	Model Re-Training
Manual Feature Selection	Feature Averaging	Model Design
Binary/Multiclass Features	Feature Improvements	Feedback Versioning

Figure 6. Explainable AI Approach Planning.

7.1. Discussion for the Initial Preprocessing

- Whether the dataset is consistent?

The dataset is the input given to the model for its processing. In practical aspects, the dataset is not always complete, as it may include missing data, incomplete data, etc. Thus, consistency within the dataset is very crucial. Therefore, the dataset should always need to be checked prior to the utilization, as it may lead to miscalculation for predictions.

- Which data imputation functions are required for data consistency?

In case of an inconsistent dataset, which is usually encountered by the researchers, an appropriate selection of data imputation techniques is quite necessary. This process can also be known as cleaning, which performs fixing inaccurate values by deleting, modifying, or replacing the records. The imputation operations include missing/non-missing at random, mean, median, replace by zero or constant, multivariate imputation by chained equation (MICE), stochastic regression, interpolation/extrapolation, hot deck, data augmentation, etc.

- Presentation of analysis of the data distributions?

The dataset can be analyzed by its distribution in detail. The distribution is used to present the relationship between observations within the sample space. There can be various types of distribution, i.e., normal, uniform, exponential, Bernoulli, binomial, poisson, etc. The distribution will provide an idea by the analysis with the graphical presentation.

- Image registration techniques required for the image dataset?

In medical image processing, the input given for the chest X-rays (CXR) is not always consistent. For the alignment, the scene/object must be aligned in the correct angle. Thus, the issues of image scaling, rotation and skew needs are addressed by using image translation.

- Whether some feature scoring techniques are used prior?

Recently a need for high accuracy and productivity is present within the medical domain. Thus, feature engineering and scoring helps us to achieve this goal. The feature scoring is calculated based on the relevant features obtained by local explanation which has optimal credit allocation using SHAP. Several other methods include GRAD-CAM, saliency maps, LRP, Deep LIFT, LIME, etc.

- Is there any priority assigned to some features by domain experts?

    In case, the expert will be supportive for achieving better prediction accuracy.

- What actions are taken in case of equal feature scores?

    In the complex cases of feature scores showing equal values, the domain experts have to take the decision as to which features need to be considered on priority. There can be top 20 or 30 features shown by the feature scoring techniques, but the least important features having high variability needs to be eliminated. In such a case, manual selection of features on a case by case basis would be applicable.

- Is there some threshold assigned for feature selection?

    Possessing several features is usually not effective, as it may lead to high imbalance within the dataset. Thus, applying thresholds to sort features as priorities can also be thoughtful to better prediction. In case of general ward patients, the thresholds applied are on the age, pulse rate, body temperature, respiratory rate, oxygenation (SaO2), etc. are considered to be beneficial.

- Are the features selected based on domain expert's choice?

    Comorbidities may cause complications in some rare patient cases. To handle such a situation, the domain expert/medical examiner can select a set of features from a particular sub-category including the ICU features. The categories of severe patient or critically ill can be given as slight, mild, or severe.

- How are binary or multiclass output based features used?

    There can be binary or multi-class based output that can be managed effectively to provide considerate prediction. The domain expert in binary case can select either a class 0 or class 1 for priority, whereas for multiclass, a specific priority listing can be assigned to the features with that multi-class features.

7.2. Discussion for the Methodology Applicability in XAI

- What feature aspects make the method selection valid?

    The machine learning algorithms are divided into multiple categories, i.e., unsupervised/supervised, regression, clustering, classification, etc. For a small dataset, principal component analysis, singular value decomposition, k-means, etc. can be applied. In the case of a large dataset, where speed and accuracy is important then classification algorithms experimented are SVM, random forest, XGBoost, neural networks, etc.

- Is the approach genuine for the system model?

    A good survey paper reference will be useful to know the recent models and their respective results. Therefore, selecting an appropriate method for the preprocessing for feature scoring then using a suitable algorithm based on the available dataset by performing multiple experiments based on the shortlisted/recent models with hyper-tuning can yield better results. A good sense of data behavior will be useful for selecting the suitable model and configuring neural network architecture with parameters.

- Why would some methods be inefficient? Are references always useful for literature?

    A recent literature works before or during the initial work of the XAI-based project would be crucial in this case as given in Tables 1–4. It is recommended that instead of implementing all the methods, a reference from several books and papers can save time and help to understand different model behavior based on the dataset availability, thus helping us to know which methods can be better from the survey paper.

- In case of multi-modal data analysis is the model suitable? Will it be efficient to use such a model?

A multi-modal dataset includes data from different categories/formats (i.e., raw data, images, speech) required to be evaluated by a machine learning/artificial intelligence model with a binary or multiclass output. Such multiple data is hard to be evaluated by a single model. Thus, a hybrid model consisting of a combined independent model is made to ensure appropriate processing for the respective format data, which is later combined by regression, voting, ensemble, classification predictions, etc. Appropriate combination methods used will have efficient performance.

- How is the model synchronization made for input?

Multi-modal data has different input provided to the respective same/different models. A separate algorithm is present, which collects the output of both the models and gives the prediction/classification. Thus, the synchronization is achieved in this process in parallel.

- Are the features of the current patient affected more or less than the average dataset?

The processing model must provide the details of the patient's condition by his features. During the prediction of the results, it is expected prior by the XAI to provide the patient's condition in detail comparison to the population for the domain expert analysis and acceptance with trust.

- What is the current patient's stage the methods have classified?

The XAI informs in advance about the patient's affected stages, i.e., stage 1, stage 2, and stage 3. These disease-affected stages depict the critical condition the patient is at present. The patient affected stage is useful for the medical examiner to provide the required medical treatment.

- What is the current patient's medication course assigned and its results?

Upon assigning a medication course, the medical examiner can check the patient's recovery progress and can change/update it accordingly. In case of comorbidities, the patient medication course may vary. A medication may have different recovery progress based on case to case basis.

- How much of a percentage of a patient's clinical features are affected?

An affected patient's data such as vital signs, clinical laboratory features, intensive care unit (ICU) features, and medication courses are crucial for the complete status overview. The overall recovery of the patient can be expressed in percentage, which must provide detailed patient's features for the confirmation.

- Are the features showing positive or negative improvement status?

With the hospital admission, a medical course in the standard operating procedure (SOP) improves the patient's condition which shows a positive improvement. In rare cases of speciality treatment requirements, negative improvements can also be seen in the patient's status. Thus, the patients are required to be shifted to the specialty care to the different section or hospital ICU.

- Which output metrics are suitable for the model evaluation?

A learning model is evaluated either on a statistical function, machine learning, or AI. The statistical function usage provides numerical p-value or graphical results, whereas machine learning models provide prediction with the metric of accuracy and deep learning by classification. In fact, all the metrics are suitable but the medical examiner can select the one which is more accurate and easy to interpret.

7.3. Discussion for the Evaluation Factors in XAI

- What cases are important for the output classification?

In case of binary output, the learning model must provide the classification as either infected or not infected (Yes/No), whereas for multi-class output, the learning model must classify clearly about the infection, current stage and improving/deteriorating condition by handling the false alarms carefully. The XAI explainability in such a case will play a crucial role.

- Is the output improving based on recursion?

The AI expert/architect can design the model carefully with the necessary parameters or layer configuration. The model can either be back-propagation, epoch based or a feedback model. It is a best practice to update the training at regular intervals with auto-weights adjustments. In the case of a feedback model, the domain expert's suggestions are considered for the feature engineering.

- How are the multi-model outputs combined?

As discussed previously, the multi-modal data handled by the multiple models are combined as approximated by an algorithm. The output can then be represented either by a profitability value/prediction/graph. The limitation for multiple model systems is to overcome for handling a single model output.

- How are the bar graphs compared and evaluated?

The bar graphs are usually drawn by the iteration value and its respective prediction. The graphs are given for AUROC curve, PR curve, sensitivity vs. specificity, NPV, PPV, etc. The graphs are crucial in any system for the performance analysis as well as its effectiveness.

- Whether the user/domain expert likes to manually select features for evaluation?

In some special cases of comorbidities, there can be many false positive alarms that can cause panic. To handle such a situation, the domain expert can select manually a group of features by his choice and can take an appropriate treatment decision for the affected patient's ahead.

- Is the system designed to record feedback from the domain experts?

As none of the system is considered perfect but it is supposed to continuously update itself for improvement. The feedback from the domain experts can resolve major issues about the new infections or its variants, which are not known by the trained system before and may compromise on performance.

- Whether the model updates training features with every evaluation?

The model re-training takes high processing time, which is a major issue in the ML/DL models. Therefore, an appropriate schedule is planned by the domain experts and research team to update the model based on the couple of weeks/months interval. Re-training is important for the system adaptation for the future tasks, and alternatively it can be done by proxy system for transferring weights later.

- Which model is suitable for such medical cases, machine learning (ML) or deep learning (DL)?

A white box model such as a decision tree is easy to understand but cannot be easily applied on complex human body features, as it may react differently than one another. Some ML models are known to work with high accuracy for some PPG disease cases, whereas for several other major infections/diseases, an interpretable deep learning-based model is required to provide explainability for every step of the multi-class output with the growth analysis.

- How the feedback suggestions by domain experts are adapted in the current model?

In case of data imbalance, the domain expert may decide to remove some features from the classification for high accuracy. However, if the feedback consists of adding scoring functions features, i.e., CCI, SOFA, OASIS, APS, SPSS, etc. then it is updated in the new version. The system adaptability is crucial for its progress.

Table 5 provides the problems addressed by the references as given for the Section 6, even though there exist some concepts that still need to be worked on by the future XAI research, including multi-modal data analysis [5], model synchronization, recursion, fusion algorithm, manual features effect, feedback model, model re-training [15], model design [15,16], and feedback design.

Table 5. XAI Human-In-The-Loop References.

XAI Category	Sub-Section	References
Pre-Processing	Dataset Consistency	[15,18,21–23,25,26]
	Imputation	[4,22,25,26,28,29]
	Data Distribution	[15,18,21,22,25,26,28,29]
	Image Registration	[4,5,17,19]
	Feature Scoring	[11–13,17,18,25,28]
	Feature Priority	[9,11,16,23–25,28]
	Equal Feature Scoring	[11,15,19,20,29]
	Threshold (Feature Selection)	[12,17,20,23–26]
	Manual Feature Selection	[12,24–26,28]
	Binary/Multi-Class Feature	[4,23,25,29]
Methodology	Feature Validation	[14,18,25,26,28]
	Novel Approach	[4,11,16,20]
	Method Inefficiency	[22,23,25,26,28]
	Feature Analysis	[11,18,25–29]
	Severity Level Analysis	[9,23,25,29]
	Feature Effectiveness	[18,20,23,25,28]
	Feature Averaging	[4,28,29]
	Feature Improvements	[16,17,24,25,29]
Evaluation	Model Metrics	[4,17,26–29]
	Classification	[11,12,14,17,27]
	Graphs	[12,16,18,25,27–29]

8. Case Study

The recent progress in XAI have further advanced the research with higher accuracy and explainability. Table 6 shows some of the medical datasets with references. The following discussion will help to understand the influential score (I-score) for the affected pneumonia patients [71] in Figure 7. The EHR data is known to possess many challenges, which would be very interesting when the supporting decisions taken for predictions are explainable.

Table 6. Publicly available dataset for medical experiments.

Dataset Source	Medical Domain	Category	Size
RSNA(Radiological Society of North America) and NIH [72]	Pneumonia	NIH chest X-ray dataset with initial annotation.	26,601 CXR Images
Kermany [73]	Pneumonia	Chest X-rays	5856 CXR Images
Chest radiographs (SCR) dataset X-ray images [74]	Pneumonia	Chest radiographs	247 frontal viewed posterior-anterior (PA)
Central Line-Associated Bloodstream infections (CLABSI) in California Hospitals [75]	Blood Stream Infections (BSI)	The CLABSI (text/csv) dataset contains reported infections, baseline data predictions, days count for central line, standard infection ratio (SIR), associated confidence interval of 95%, and grading with respect to national baseline.	Details from 461 hospitals.
MIMIC Clinical Database [76,77]	Epidemics (HER Data)	The MIMIC dataset consists of ICU data with high patient's count including vital signs, laboratory test, and medication courses.	The MIMIC-III database has 26 relational tables containing patient's data (SUBJECT_ID), hospital admissions (HADM_ID), and ICU admissions(ICUSTAY_ID).
ICES Data Repository [78]	EHR Data	EHR Data recorded from the health services of Ontario.	13 million people.
Veterans Health Administration [79,80]	EHR Data	EHR data from US Veteran's Affairs (VA) dataset	1293 health care facilities with 171 medical center and 1112 outpatient sites.

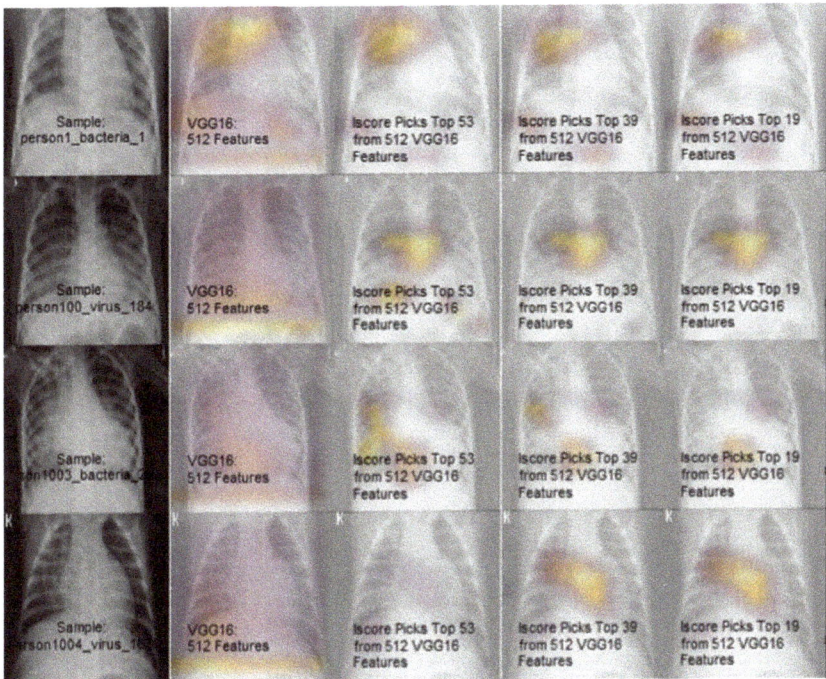

Figure 7. I-Score Model shown by reducing the top features.

An interaction-based methodology is proposed for the image noise and non-informative variables elimination using I-score for feature prediction. The explainable and interpretable features are used to demonstrate its feature prediction, which has an interactive effect.

Even though there is a tradeoff observed with the learning performance and effectiveness of explainability, it can still be overcome by providing new features scoring methods. An explanation factor is determined clearly by the prediction, which is related to the feature selection technique known to be interpretable with explainable. In this case, I-score variable used for explainability is processed in discrete form, which may be converted from the continuous variable as required.

The largest marginal I-score is also specified as a random variable drawn from a normal distribution. The I-score is maximized by the optimal subsets of variables which are searched by the backward dropping algorithm (BDA). The BDA achieves its goal by variable elimination in a stepwise manner from an initial subset in a variable space. In this research, a pre-trained CNN is applied before I-score and/or alternative BDA, which is later evaluated using a feed-forward neural network requiring less number of parameters. The predicted 512 features from pneumonia affected CXR are further reduced to top 19 features which are explained to provide warning about the disease location.

The EHR used for the prediction of acute critical illness (ACI) in the hospitalized patients needs to be explained by an open box model [81]. The open box model is usually an interpretable neural network model with explainable AI (XAI). The drawbacks of AI models are known to lack correct results, i.e., identifying false positives or true negatives for critical care. To overcome this problem, the XAI-EWS is designed to provide visual explanations for sepsis, AKI and ACI. The architecture of XAI-EWS consists of deep Taylor decomposition (DTD) with temporal convolutional network (TCN) in the prediction module and a separate explanation module. The explanation module is used to support prediction relevant to the clinical parameters. These clinical parameters are listed in the form of top 10 parameters with high weights to recent values. The XAI-EWS model provides the user with the transparency of the system model and helps to earn trust by giving explanation of every key decision within the algorithm. The XAI-EWS has an individual and population based perspective for the model based explanation. The DTD has been beneficial for predicting the development of ACI from the individual perspective. Back-propagation is used for the relevance output processing with a global parameter having mean relevance scores and the correlations in the clinical parameters in local explanation of the population perspective.

The Influence score (I-score) and the backward dropping algorithm are used in combination for the Figure 7 output demonstration. This proposed methodology includes selecting high potential variables for influential modules, which is filtered by inter-activeness and later combined for the prediction. This I-score is known to work better with discrete variables. In case of random variables taken from normal distribution, then optimal cut-off is set to the highest marginal I-score. It supports limited categories to avoid classification error rates.

$$I = \sum_{j \in \mathcal{P}} n_j^2 (\overline{Y}_j - \overline{Y})^2 \qquad (20)$$

Here, $\overline{Y}_j$ is the average of Y observations over the local average jth partition and global average $\mathcal{P}$. Y is the response variable (binary 0/1) and all explanatory variables are discrete. $\mathcal{P}_K$ is a subset of K explanatory variables $\{x_{b1} \ldots x_{bk}\}$. $n_1(j)$ is the number of observations with $Y = 1$ in partition element j as given in Equation (20). The BDA algorithm is used as a greedy algorithm, which selects optimal subsets of variables having highest I-score. The architecture consists of an interaction based convolutional neural network (ICNN).

The explainable deep CNN (DCNN) is used for the classification of normal, virus infection (pneumonia), and COVID-19 pneumonia-infected patients [82]. A fine distinguishing criteria is set by designing application specific DCNN for different infection categories with high accuracy.

The training set consists of a gold-standard diagnosis set by a radiologist by confirmation. The training set provided is quite balanced in this system consisting of healthy, pneumonia-infected, other virus-infected and COVID-19-infected, thus providing high accuracy by avoiding trade-off of features within infection categories belonging to the same patients.

The base model is adapted from VGG-19, where its convolution kernel is set as per the requirements. The final values are set with bright colors to identify the region of interest for medical analysis. The hyper-parameters are trained using the grid search to find the best settings. The two CNN models include CNN1 for training samples with category labels, test samples to partition the standard set, and CNN2 for the virus infection output. Finally, the CNN2 generates the output by classifying the infection in Figure 8 detail.

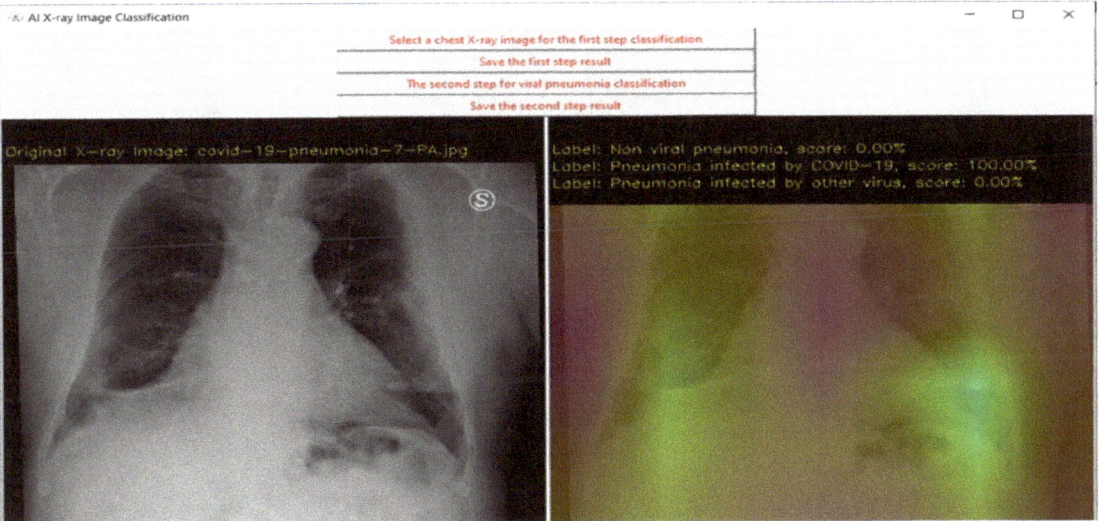

Figure 8. Explainable DCNN for CXR analysis and COVID-19 pneumonia classification.

The graph diffusion pseudo-labelling by deep network for CXR-based COVID-19 identification is presented [83]. The Figure 9 GraphXCovid is used for COVID-19 identification by using a deep semi-supervised framework. It performs pseudo-labelling generation using dirichlet energy by a novel optimization model. Thus with minimal labels a high sensitivity in COVID-19 is generated. An iterative scheme in a deep net and attention maps are the highlights of this model. The work is considered to be the successor of deep SSL technique [84–87] combining generalization and feature extraction of deep neural networks. Therefore, the process can be given in detail as optimizing epochs for deep net extraction for graph construction, diffuse labelled sets to un-labelled data. Thus, pseudo-labels are generated, which optimizes the model parameter by regular updates, which is later iterated until completion. In this case, the medical data imbalance problem is handled during the diffusion process.

The feedback system is designed to evaluate which robot explanation is more suitable for an explainable AI system [69,88]. In the initial implementation which consists of having multiple feature evaluators such as CAM, Grad-CAM, and network dissection are used to support explanation by the robot as shown in Figure 10. The feature engineering pre-processing uses top 20% by the heat-maps, which are labeled with the respective concepts. The classification models then provide detailed accuracy for explanation by Resnet-18, Resnet-50, VGG 16, and AlexNet.

Even though the outcomes are the same by best accuracy, the explanations are distinct. In such a case the user can decide which robot explanation is more suitable for his understanding, and based on that the further outcomes are planned to be explained. The selection is taken from multiple questionnaires at the beginning, which are later evaluated based on five points Likert scale. In parallel, the suggestions given by the user in the feedback box are also collected for the advancements in the new version [89].

Nevertheless, multiple surveys for computer vision [90], deep learning [91], AI imaging techniques [92], and explainable AI [93] are present, which are informative and provide basic as well as in-depth knowledge about the concept and applications.

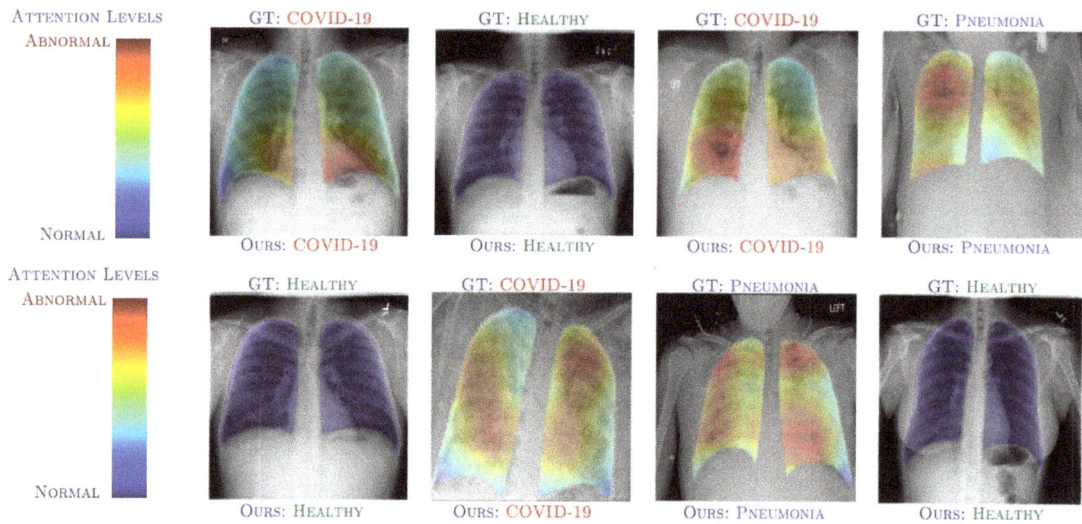

Figure 9. GraphXCovid pneumonia CXR results.

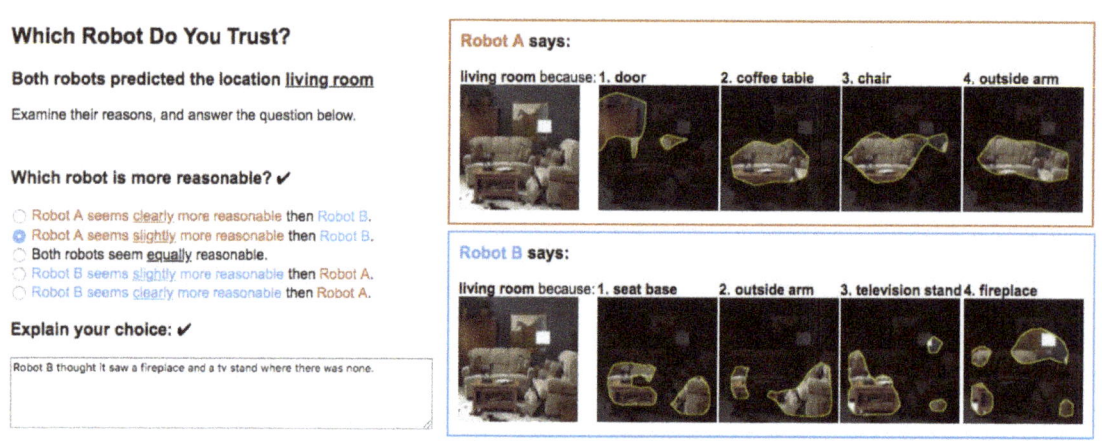

Figure 10. Feedback system for the human rating-based evaluation.

9. XAI Limitations

In order to claim the research work to be XAI-compatible, the designed model should be rated based on its explainability level and XAI evaluation. In practice, many XAI systems are not adaptable to the new challenges of model tuning and training [71]. Even though some models are well-designed but are not correctly trained and classified, which usually suffer from performance issues. Therefore, such challenges are hidden and are discussed in more detail as follows.

9.1. Explainability Ratings for the New Models

Several recent works claim to be XAI compliant but no specific rating based standard is present [71,82,83]. Therefore, at the base level a system can be checked for preprocessing, methods, and post evaluation explanation of the model which can still be further improved by point-based ratings. The explainability also depends on the ease of understanding and the prediction/classification of the XAI system usage. The language compatibility and parameter settings provide more transparency. The data statistics provided at every level and interpretability at every DNN layer can further improve the transparency with more explanation [94]. The decision transparency is rarely dealt with, and the user has no exact knowledge about the sudden change in decision [71]. Therefore, the decision taken at every step should be disclosed to the user, which can be helpful in the scenario of having false alarms and leading to chaos within the hospital staff.

9.2. Measurement System for the XAI Evaluation

During the XAI evaluation phase, the results comparison is not measurable [81]. Recently, robot-based explanations popularly known as bots have different classification explanation depending upon distinct models [88], i.e., CAM, GRAD-CAM, Network Dissection, etc. Subsequently, the CXR feature highlighting may differ by distinct models. Therefore, in such a case, the user's profile account having different priorities can make the required settings for such preferences optionally by using multiple XAI analysis techniques known as intrinsic and post-hoc, which are used everywhere [83].

It is still unknown how to apply an automatically adapting system that can select different optimal techniques to provide the best classification/prediction. Ensemble algorithm can be one of the optional solutions but it uses a brute force method and leads to a performance trade-off. Nevertheless, a local dataset does tend to bias the classification and is not effective enough on the global data. A severe security threat for making the model training biased is made by adversarial attacks [95]. Therefore, identifying such a bias issue is also a challenge.

9.3. XAI System Adaptation to the Continuous Improvements

Many of the recent works have designed a specific hyper-tuned classification model that may have shortcomings/over-fitting on the different dataset or feature modification [82,83]. Thus, a model will not perform significantly for the transfer learning cases too. The XAI system must also be open to adapt to the new feature set, i.e., vitals signs, CXR image, ICU parameters, clinical tests, etc. [96]. Modification of such features by using a user feedback system is crucial for the system's upgrade. A user feedback compatible system improves the model scope and is thought to be continuously adaptable. In case of model with input of single or multi-modal data source, the user must be able to choose either of the option as per data available, e.g., in a multi-modal system, if the CXR is not available for the patient's disease diagnosis, then it must be able to classify on single source of vital sign or different data source of ICU parameters/clinical tests, whereas the model parameter tuning is also important for optimal performance, which needs to be performed automatically without sacrificing performance.

9.4. Human in the Loop Approach Compatibility

The saliency map analysis is not perfectly designed to identify certain features with vital signs data or CXR image data [71]. Therefore, the medical examiner's based data labeling is required at the initial training phase of the model for better quality and to achieve high classification output [97]. Higher expectations from XAI has subsequently led to an intelligent system that can discuss and convince its classification to the medical examiners, so that an effective decision about the medical treatment can be made. Achieving such a highly capable human in the loop can greatly benefit the XAI progress. In addition to the previous discussion about user profile management for priority-based feature selection, XAI must be able to serve also in the multi-specialty hospitals concerning different departments.

Interest for the single patient's health analysis is one of the major challenges. Considering the XAI transparency and explainability at every DNN layer, the user must also be able to configure the layer size and weights for the effective analysis of diagnosis or severity analysis by feature highlighting is necessary.

10. XAI Recommendation System (XAI-RS)

Table 7 for XAI post-treatment recommendation system is beneficial to the hospitalized/treated patients for a group of diseases. As discussed earlier, the hospitalized patients were treated for a group of diseases i.e., pneumonia, bloodstream infections, acute kidney injury, mortality prediction, etc. have different features and symptoms. Therefore, the XAI-RS can evaluate the results as per the recent health condition of the discharged patient. For every patient, the XAI-RS will be personally evaluated. Thus, the patient's recovered from AKI will have a default recommendation set with additional suggestions for personalized evaluation.

Table 7. XAI post-treatment recommendation chart.

1. Diet	2. Medicine/Treatment	3. Exercise	4. Regular Checkup	5. Side Effects
a. Fruits	a. Morning dose 1	a. Walking/Running	a. Daily	a. Vomiting
b. Vegetables	b. Afternoon dose 2	b. Yoga	b. Alternate day	b. Dizziness
c. Seafood	c. Evening dose 3	c. Cycling	c. Weekly	c. Headache
d. Meat	d. Lotions/Drops	d. Swimming	d. Bio-sensors/Remote health monitoring	d. Loss of Appetite
e. Grains	e. Physiotherapy	e. Sports	e. Monthly	e. Skin rashes
f. Soup	f. Injections		f. Quarterly/Year	e. Palpitations
g. Milk Products	g. Dialysis			

Table 8 presents the XAI-RS for the AKI-affected patient. These recommendations are default to every AKI-discharged patient but if the patient is addicted to smoking and/or consuming alcohol, then an additional recommendation needs to be added as shown by the orange highlighted box. The purpose of the XAI recommendation system is to provide best continuous treatment to the post-discharged patients to live a healthy lifestyle.

Table 8. XAI Personal Post-Treatment Recommendation Chart for AKI Patient.

1. Diet	2. Medicine/Treatment	3. Exercise	4. Regular Checkup	5. Side Effects
a. Fruits	a. Morning dose 1	a. Walking/Running	a. Daily	a. Vomiting
b. Vegetables	b. Afternoon dose 2	b. Yoga	b. Monthly Report	b. Loss of Appetite
c. Seafood	c. Evening dose 3	c. Swimming		c. Skin rashes
d. Soup	d. Injections			
	e. Dialysis			

11. XAI Scoring System (XAI-SS)

The XAI-SS determines the standard grade for the newly designed and in use XAI systems. This scoring system in Table 9 can be extended to multiple areas, i.e., industrial, finance, sensor communications, etc. The scores can be assigned based on international (10 points), group (8 points), and local (6 points) policy achievements. As each of the 10 XAI factors are assigned equal scores with balanced/equal weightage, the final evaluation grade is assigned as Class I ($\geq$90%), Class II ($\geq$70% and <90%), and Class III ($\geq$60% and <70%). Designing a new XAI system must involve experts for setting the objectives for a high quality work plan.

The training provided to the XAI system must involve an international dataset for its effectiveness (Table 6), whereas the detail for the XAI factors can be referred for preprocessing (Sections 4.1 and 7.1), model selection (Section 4.2), model re-configuration

(Section 7.2), interpretability (Sections 1.2 and 4.3), explainability (Section 7), evaluation (Sections 1.3 and 7.3), human-in-the-loop (Section 6.2), and XAI-RS (Section 10). It is recommended that the XAI should be evaluated every year to maintain the XAI system's quality and validity.

Table 9. Checklist for XAI scoring system.

Serial No.	XAI Scoring Factor	Description	Checklist (10 pts Each)
1	XAI based Objectives	To serve the evaluation purpose for effective problem solving following laws and ethics.	
2	Dataset for Training Model	Whether the dataset has global and local scope?	
3	Data Pre-processing	Manage the data consistency and imbalance issue.	
4	Model Selection	To perform feature analysis and select an appropriate model with a novel approach.	
5	Model Reconfiguration	The model's hyper-parameter tuning for better prediction by handling bias and variance.	
6	Interpretability	How much does the model support intrinsic and post-hoc interpretability?	
7	Explainability	Transparency in every step and decision of the model should be given by the algorithm.	
8	Evaluation, Feedback loop and Post-evaluation	The outcome should provide meaningful results. Graphs, prediction, and classification should be cross-verifiable. The feedback loop consisting of interacting with domain experts is helpful for post-evaluation.	
9	Human-in-the-Loop Process	Continuously involve the domain expert for improving multi-modal data and feature management.	
10	XAI Recommendation System	To maintain the discharged patient's health conditions.	

The Table 10 Grades for the XAI scoring system in Figure 11 provides the evaluation for the recent XAI medical references [98]. It is helpful to analyze the applicability of XAI in the recent works and for the future works mapping.

Table 10. Grades for the XAI Scoring System.

Reference	XAI Scores	Grades
W. Qiu et al. 2022 [99]	90	Class I
Y. Yang et al. 2022 [100]	86	Class II
L. Zou et al. 2022 [101]	88	Class II
C. Hu et al. 2022 [102]	86	Class II
L. Zhang et al. 2022 [103]	84	Class II

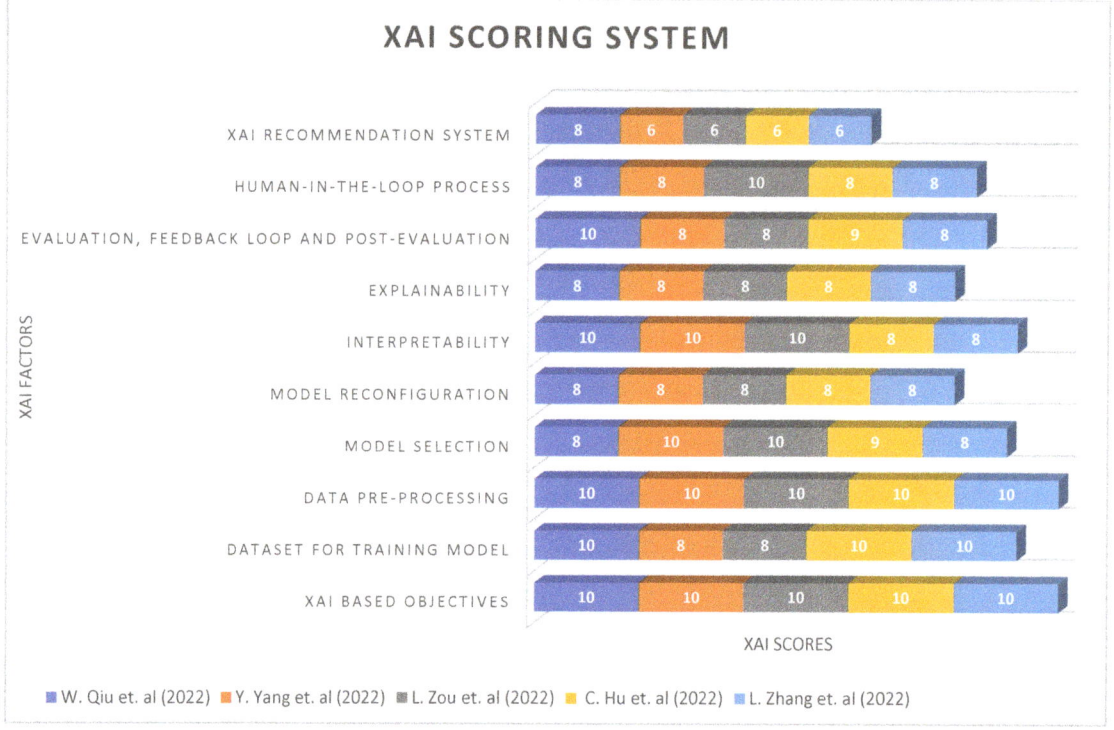

Figure 11. XAI Scoring System Evaluation.

12. Conclusions

The XAI survey presents a detailed approach for the XAI research development. Legal and ethical XAI aspects are presented, as well as some important areas within the medical field that need attention to improve it further and gain the user's trust by providing transparency within the process. The contribution of XAI Recommendation System and XAI Scoring System will be suitable for overall development of XAI in the future. The future work will be focused on presenting the enhancements by XAI and further contributions as the recent progress is quite impressive.

Author Contributions: Conceptualization, R.-K.S. and M.S.P.; methodology, R.-K.S. and M.S.P.; software, M.S.P.; validation, R.-K.S. and M.S.P.; formal analysis, R.-K.S. and M.S.P.; investigation, R.-K.S. and M.S.P.; resources, R.-K.S.; data curation, M.S.P.; writing—original draft preparation, R.-K.S. and M.S.P.; writing—review and editing, R.-K.S. and M.S.P.; visualization, R.-K.S. and M.S.P.; supervision, R.-K.S.; project administration, R.-K.S.; funding acquisition, R.-K.S. All authors have read and agreed to the published version of the manuscript.

Funding: This work is funded by the Ministry of Science and Technology (MOST), Taiwan with the project code "MOST 111-2321-B-075A-001".

Institutional Review Board Statement: Not applicable.

Informed Consent Statement: Not applicable.

Data Availability Statement: Not applicable.

Acknowledgments: We: the authors, would like to thank Chieh-Liang Wu, Department of Critical Care Medicine, Taichung Veterans General Hospital, Taichung City 40705, Taiwan. Wu has helped to understand the internal working/challenges within the medical field and involve us within multiple healthcare projects, which we have implemented in the team. In collaboration, we have implemented the project for adverse event prediction using deep learning algorithm (AEP-DLA), pneumonia status prediction, XAI-based blood stream infection classification, etc. It has become possible to achieve publication sharing of this medical survey with the efforts, discussion, and working with many hospital staff from the past 4 years. We hope this survey will be helpful to new researchers and present tremendous opportunities to overcome the upcoming challenges in XAI healthcare.

Conflicts of Interest: The authors declare no conflict of interest.

Appendix A

Appendix A.1 LIME-Based Super-Pixel Generation

Pixels present the image in grid format and as a part of the original picture. Pixels are also known as artifacts that are meant to create digital images. In contrast, the originating source and semantic meaning of a specific super-pixel can be estimated [104,105]. A super-pixel is usually a pixel group or combination based on a set of common properties including pixel color value. The benefits of using super-pixel are given as: (a) Less complexity: The grouping of pixels based on distinct properties reduces the complexity and requires less computations. (b) Significant entities: A group of super-pixels having texture properties achieves an expressiveness through embedding. (c) Marginal information loss: In case of over-segmentation, crucial areas are highlighted with a minor deficit of less valuable data. Figure A1 shows the four types of super-pixel classification algorithms are explained as given below:

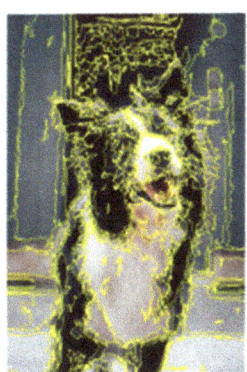   

(**a**) Felzenszwalb (**b**) Quick-Shift (**c**) SLIC (**d**) Compact-Watershed

Figure A1. Comparison of super-pixel approaches.

(i) Felzenszwalb and Huttenloch (FSZ) algorithm: The FSZ is a well-known algorithm for graph based approach and utilized for image edge operation with a O(M logM) complexity. The FSZ algorithm takes a weighted gradient with the same properties between two adjacent pixels. Successively, a future super-pixel seed is administered per pixel for obtaining the shortest gradient difference and largest for adjacent segments.

(ii) Quick-Shift (QS): QS is the default algorithm used by LIME. The QS algorithm generates super-pixels by mode seeking segmentation scheme. It then moves every point towards higher density leading to increased density.

(iii) Simple Linear Iterative Clustering (SLIC): The SLIC belongs to a cluster based super-pixel algorithm. It operates on the basis of k-means algorithm with a search space proportional size ($S \times S$) for reducing distance calculations significantly.

$$D = \sqrt{d_c^2 + \left(\frac{d_s}{S}\right)^2 m^2}$$

The spatial d_s and color d_c proximity is combined by weighted distance measure having complexity of $O(N)$. The size and compactness of super-pixel is given by m as given in the above equation. The algorithm initiates k cluster process as centers, which is grid scanned with S pixel's distance with $S = \sqrt{\frac{N}{K}}$ approximately. To avoid super-pixel placement on the edge, the centers are shifted towards the smallest gradient. Later, each pixel is allotted to proximate clusters whose search area is overlaid on a super-pixel. Successively, the new cluster center is taken as pixels' average vector for every center's update. Finally, the residual error E is minimized until the threshold value and independent pixels are added to proximate super-pixel.

(iv) Compact-Watershed (CW): The CW algorithm is the optimized version of watershed super-pixel algorithm. The gradient image is used as input, where the altitude is the gray tone of each pixel that can be depicted as a topographical surface. The watershed with catchment basins are resulted from the continuous surface flooding, which may lead to over-segmentation and can be avoided by markers. The algorithm includes: (i) Flooding avoidable markers for each label, (ii) Marked areas neighboring pixels are collected in priority queue and graded by gradient magnitude equivalent to its priority level. (iii) Pixels with highest priority are pulled out and labeled according to their neighboring pixel. Later, labeled pixels are added in this queue. (iv) Step 3 is repeated until the priority queue is null.

The CW in terms of size and extensions has more compact super-pixels than watershed algorithms. To obtain this, the use of Euclidean distance for the difference with a pixel from super-pixel seed point by weighted distance measure and the gray value comparison within pixel and gray pixel's value is performed.

Appendix A.2 Perceptive Interpretability Methods

The XAI based perceptive interpretable methods [106,107] are LIME and GRAD-CAM, which are the CNN architecture-based decision explanation methods. The input given is a trained model for this interpretable process known as post-hoc analysis. The details of the XAI perceptive models for the shapes detection of cancerous masses in the breast imaging by computer-aided diagnosis (CAD) are as given below:

(a) Local Interpretable Model-Agnostic Explanations (LIME): The LIME interpretation is provided by highlighting the top contributing of class S, which is evaluated on an image classification by observing a ground truth prospect. Figure A2 shows the S perturbations executed by LIME, which are similar to GRAD-CAM, whereas Figure A3 GRAD-CAM is using the same image for comparison with LIME in Figure A2. The ground truth class is used for the CNN based prediction is the same class used for LIME perturbations. The green color indicates positive correlation of regions with the CNN decision, and red color indicates negative correlation. Figure A2 shows the saliency scheme is used for the presentation of saliency zones, where higher intensity red color focuses more for classification and lower intensity blue color has less focus. Several classes are distinguished by header bars with distinct colors.

(b) Class Activation Mapping (CAM): The Figure A3 shows the GRAD-CAM methods based pictorial presentation of eight fine-tuned networks for visual explanation. The figure includes for every class, two sample images are presented and every network's respective saliency maps for the images. The links within lesion area and network performance are highlighted for which accurate prediction of images is given by the network. The ground truth is considered for generating saliency maps of the approximate features. Subsequently, the lesion areas which are incorrectly identified by CN architectures are also affected during classification tasks. In case of SqueezeNet, the AUC is not significant and AUC trade-offs with the number of parameters obtained from VGG-16 are incapable of highlighting the lesion area, whereas, the lesion of the images is accurately highlighted by the ResNet-50, DenseNet-121, and DenseNet-161. Therefore, the GRAD-CAM is distinct from LIME as it emphasizes the color intensity closer to the center of the lesion area. Figure A3 presents the super-pixels with red color for negative contribution and green color otherwise. Similarly, several classes are distinguished by header bars with distinct colors. Samples belonging to the same class are represented by similar images header color.

Thus, it can be noticed that graphical reasoning has good understanding in the CAM-based interpretations, whereas the evaluation of LIME and GRAD-CAM activation maps are completely different which can be noticed from Figures A2 and A3. Ultimately, it is recommended to consider both interpretations and models evaluation to gain a broader view of this process.

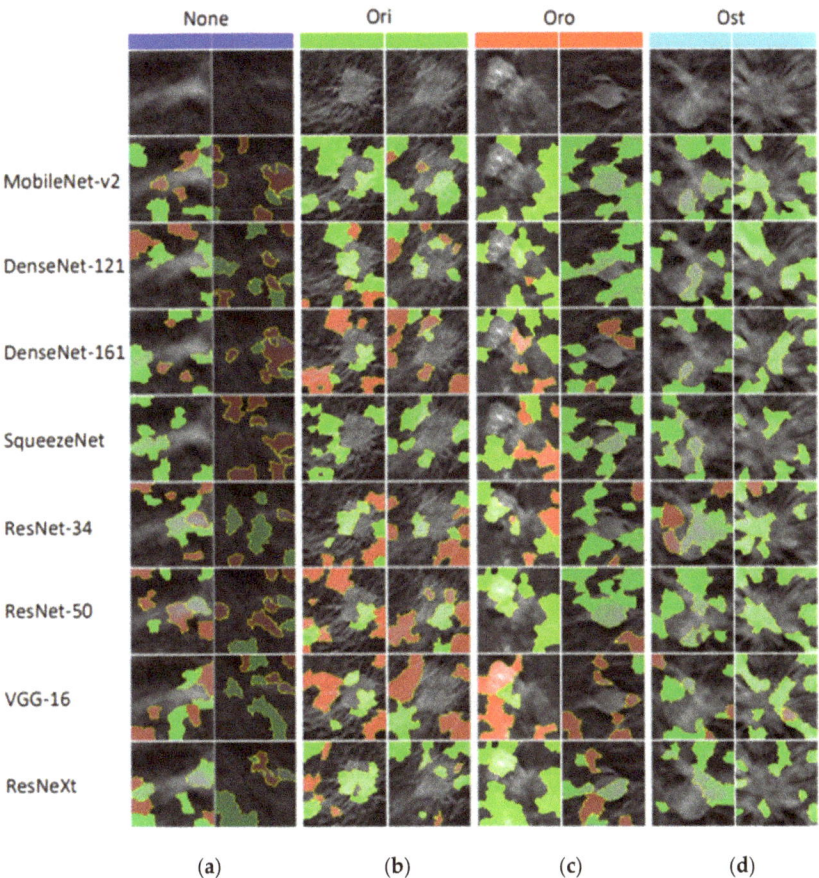

Figure A2. LIME Region of Interest (ROI) images for (**a**) No lesions (None); (**b**) Irregular opacity (Ori); (**c**) Regular opacity (Oro); and (**d**) Stellar opacity (Ost).

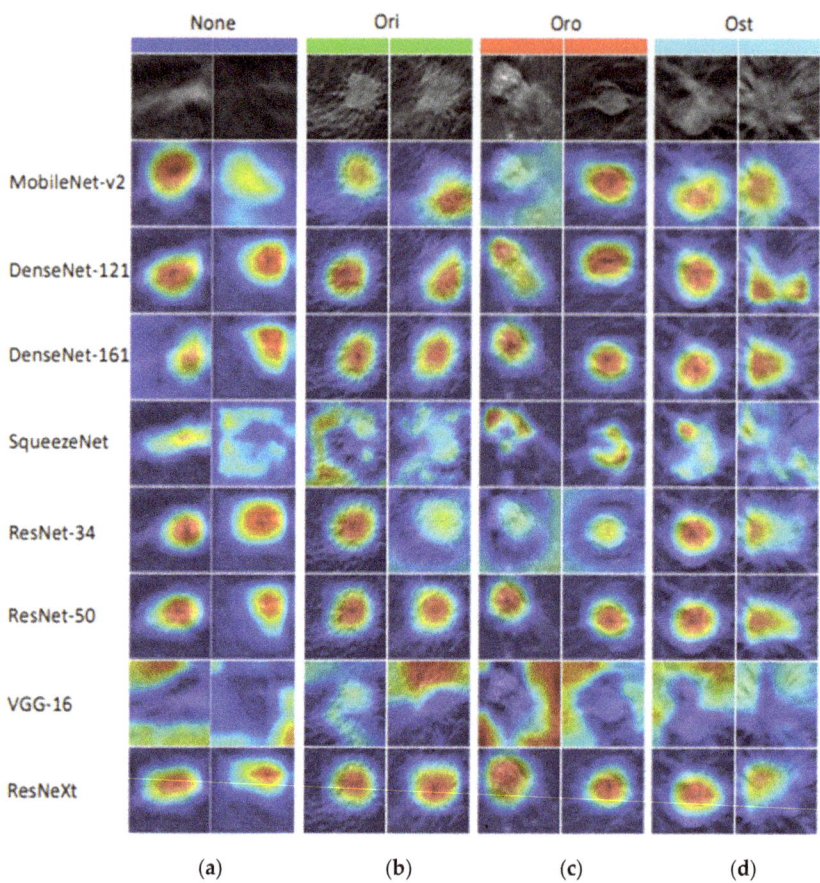

Figure A3. GRAD-CAM Region of Interest (ROI) images for (**a**) No lesions (None); (**b**) Irregular opacity (Ori); (**c**) Regular opacity (Oro); and (**d**) Stellar opacity (Ost).

References

1. Esmaeilzadeh, P. Use of AI-based tools for healthcare purposes: A survey study from consumers' perspectives. *BMC Med. Inform. Decis. Mak.* **2020**, *20*, 1–19. [CrossRef] [PubMed]
2. Houben, S.; Abrecht, S.; Akila, M.; Bär, A.; Brockherde, F.; Feifel, P.; Fingscheidt, T.; Gannamaneni, S.S.; Ghobadi, S.E.; Hammam, A.; et al. Inspect, understand, overcome: A survey of practical methods for ai safety. *arXiv* **2021**, arXiv:2104.14305.
3. Juliana, J.F.; Monteiro, M.S. What are people doing about XAI user experience? A survey on AI explainability research and practice. In *International Conference on Human-Computer Interaction*; Springer: Cham, Switzerland, 2020.
4. Xie, S.; Yu, Z.; Lv, Z. Multi-disease prediction based on deep learning: A survey. *CMES-Comput. Modeling Eng. Sci.* **2021**, *127*, 1278935. [CrossRef]
5. Clodéric, M.; Dès, R.; Boussard, M. The three stages of Explainable AI: How explainability facilitates real world deployment of AI. *Res. Gate* **2020**.
6. Li, X.H.; Cao, C.C.; Shi, Y.; Bai, W.; Gao, H.; Qiu, L.; Wang, C.; Gao, Y.; Zhang, S.; Xue, X.; et al. A survey of data-driven and knowledge-aware explainable ai. *IEEE Trans. Knowl. Data Eng.* **2020**, *34*, 29–49. [CrossRef]
7. Schneeberger, D.; Stöger, K.; Holzinger, A. The European legal framework for medical AI. In Proceedings of the International Cross-Domain Conference for Machine Learning and Knowledge Extraction, Dublin, Ireland, 25–28 August 2020; Springer: Cham, Switzerland, 2020; pp. 209–226.
8. Muller, H.; Mayrhofer, M.; Van Veen, E.; Holzinger, A. The Ten Commandments of Ethical Medical AI. *Computer* **2021**, *54*, 119–123. [CrossRef]
9. Erico, T.; Guan, C. A survey on explainable artificial intelligence (xai): Toward medical xai. *IEEE Trans. Neural Netw. Learn. Syst.* **2020**, *32*, 4793–4813.

10. Guang, Y.; Ye, Q.; Xia, J. Unbox the black-box for the medical explainable ai via multi-modal and multi-centre data fusion: A mini-review, two showcases and beyond. *Inf. Fusion* **2022**, *77*, 29–52.
11. Selvaraju, R.R.; Cogswell, M.; Das, A.; Vedantam, R.; Parikh, D.; Batra, D. Grad-cam: Visual explanations from deep networks via gradient-based localization. In Proceedings of the IEEE international conference on computer vision, Venice, Italy, 22–29 October 2017; pp. 618–626.
12. Tang, Z.; Chuang, K.V.; DeCarli, C.; Jin, L.W.; Beckett, L.; Keiser, M.J.; Dugger, B.N. Interpretable classification of Alzheimer's disease pathologies with a convolutional neural network pipeline. *Nat. Commun.* **2019**, *10*, 2173. [CrossRef]
13. Zhao, G.; Zhou, B.; Wang, K.; Jiang, R.; Xu, M. RespondCAM: Analyzing deep models for 3D imaging data by visualizations. In *Medical Image Computing and Computer Assisted Intervention—MICCAI 2018*; Frangi, A.F., Schnabel, J.A., Davatzikos, C., Alberola-López, C., Fichtinger, G., Eds.; Springer: Cham, Switzerland, 2018; pp. 485–492.
14. Bahdanau, D.; Cho, K.; Bengio, Y. Neural machine translation by jointly learning to align and translate. *arXiv* **2014**, arXiv:1409.0473.
15. Lapuschkin, S.; Wäldchen, S.; Binder, A.; Montavon, G.; Samek, W.; Müller, K.-R. Unmasking clever hans predictors and assessing what machines really learn. *Nat. Commun.* **2019**, *10*, 1096. [CrossRef]
16. Samek, W.; Montavon, G.; Binder, A.; Lapuschkin, S.; Müller, K. Interpreting the predictions of complex ML models by layer-wise relevance propagation. *arXiv* **2016**, arXiv:1611.08191.
17. Thomas, A.W.; Heekeren, H.R.; Müller, K.-R.; Samek, W. Analyzing neuroimaging data through recurrent deep learning models. *Front. Neurosci.* **2019**, *13*, 1321. [CrossRef] [PubMed]
18. Arras, L.; Horn, F.; Montavon, G.; Müller, K.; Samek, W. 'What is relevant in a text document?': An interpretable machine learning approach. *arXiv* **2016**, arXiv:1612.07843. [CrossRef]
19. Hiley, L.; Preece, A.; Hicks, Y.; Chakraborty, S.; Gurram, P.; Tomsett, R. Explaining motion relevance for activity recognition in video deep learning models. *arXiv* **2020**, arXiv:2003.14285.
20. Eberle, O.; Buttner, J.; Krautli, F.; Mueller, K.-R.; Valleriani, M.; Montavon, G. Building and interpreting deep similarity models. *IEEE Trans. Pattern Anal. Mach. Intell.* **2020**, *44*, 1149–1161. [CrossRef] [PubMed]
21. Burnham, J.P.; Rojek, R.P.; Kollef, M.H. Catheter removal and outcomes of multidrug-resistant central-line-associated bloodstream infection. *Medicine* **2018**, *97*, e12782. [CrossRef]
22. Beganovic, M.; Cusumano, J.A.; Lopes, V.; LaPlante, K.L.; Caffrey, A.R. Comparative Effectiveness of Exclusive Exposure to Nafcillin or Oxacillin, Cefazolin, Piperacillin/Tazobactam, and Fluoroquinolones Among a National Cohort of Veterans With Methicillin-Susceptible Staphylococcus aureus Bloodstream Infection. *Open Forum Infect. Dis.* **2019**, *6*, ofz270. [CrossRef] [PubMed]
23. Fiala, J.; Palraj, B.R.; Sohail, M.R.; Lahr, B.; Baddour, L.M. Is a single set of negative blood cultures sufcient to ensure clearance of bloodstream infection in patients with Staphylococcus aureus bacteremia? The skip phenomenon. *Infection* **2019**, *47*, 1047–1053. [CrossRef]
24. Fabre, V.; Amoah, J.; Cosgrove, S.E.; Tamma, P.D. Antibiotic therapy for Pseudomonas aeruginosa bloodstream infections: How long is long enough? *Clin. Infect. Dis.* **2019**, *69*, 2011–2014. [CrossRef] [PubMed]
25. Harris, P.N.A.; Tambyah, P.A.; Lye, D.C.; Mo, Y.; Lee, T.H.; Yilmaz, M.; Alenazi, T.H.; Arabi, Y.; Falcone, M.; Bassetti, M.; et al. Effect of piperacillin-tazobactam vs meropenem on 30-day mortality for patients with E coli or Klebsiella pneumoniae bloodstream infection and ceftriaxone resistance: A randomized clinical trial. *JAMA* **2018**, *320*, 984–994. [CrossRef] [PubMed]
26. Delahanty, R.J.; Alvarez, J.; Flynn, L.M.; Sherwin, R.L.; Jones, S.S. Development and Evaluation of a Machine Learning Model for the Early Identification of Patients at Risk for Sepsis. *Ann. Emerg. Med.* **2019**, *73*, 334–344. [CrossRef]
27. Kam, H.J.; Kim, H.Y. Learning representations for the early detection of sepsis with deep neural networks. *Comput. Biol. Med.* **2017**, *89*, 248–255. [CrossRef]
28. Taneja, I.; Reddy, B.; Damhorst, G.; Dave Zhao, S.; Hassan, U.; Price, Z.; Jensen, T.; Ghonge, T.; Patel, M.; Wachspress, S.; et al. Combining Biomarkers with EMR Data to Identify Patients in Different Phases of Sepsis. *Sci. Rep.* **2017**, *7*, 10800. [CrossRef]
29. Oonsivilai, M.; Mo, Y.; Luangasanatip, N.; Lubell, Y.; Miliya, T.; Tan, P.; Loeuk, L.; Turner, P.; Cooper, B.S. Using machine learning to guide targeted and locally-tailored empiric antibiotic prescribing in a children's hospital in Cambodia. *Open Res.* **2018**, *3*, 131. [CrossRef]
30. García-Gallo, J.E.; Fonseca-Ruiz, N.J.; Celi, L.A.; Duitama-Muñoz, J.F. A machine learning-based model for 1-year mortality prediction in patients admitted to an Intensive Care Unit with a diagnosis of sepsis. *Med. Intensiva Engl. Ed.* **2020**, *44*, 160–170. [CrossRef]
31. Lee, H.-C.; Yoon, S.B.; Yang, S.-M.; Kim, W.H.; Ryu, H.-G.; Jung, C.-W.; Suh, K.-S.; Lee, K.H. Prediction of Acute Kidney Injury after Liver Transplantation: Machine Learning Approaches vs. Logistic Regression Model. *J. Clin. Med.* **2018**, *7*, 428. [CrossRef] [PubMed]
32. Hsu, C.N.; Liu, C.L.; Tain, Y.L.; Kuo, C.Y.; Lin, Y.C. Machine Learning Model for Risk Prediction of Community-Acquired Acute Kidney Injury Hospitalization From Electronic Health Records: Development and Validation Study. *J. Med. Internet Res.* **2020**, *22*, e16903. [CrossRef]
33. Qu, C.; Gao, L.; Yu, X.Q.; Wei, M.; Fang, G.Q.; He, J.; Cao, L.X.; Ke, L.; Tong, Z.H.; Li, W.Q. Machine learning models of acute kidney injury prediction in acute pancreatitis patients. *Gastroenterol. Res. Pract.* **2020**, *2020*, 3431290. [CrossRef]

34. He, L.; Zhang, Q.; Li, Z.; Shen, L.; Zhang, J.; Wang, P.; Wu, S.; Zhou, T.; Xu, Q.; Chen, X.; et al. Incorporation of urinary neutrophil gelatinase-Associated lipocalin and computed tomography quantification to predict acute kidney injury and in-hospital death in COVID-19 patients. *Kidney Dis.* **2021**, *7*, 120–130. [CrossRef]
35. Kim, K.; Yang, H.; Yi, J.; Son, H.E.; Ryu, J.Y.; Kim, Y.C.; Jeong, J.C.; Chin, H.J.; Na, K.Y.; Chae, D.W.; et al. Real-Time Clinical Decision Support Based on Recurrent Neural Networks for In-Hospital Acute Kidney Injury: External Validation and Model Interpretation. *J. Med. Internet Res.* **2021**, *23*, e24120. [CrossRef] [PubMed]
36. Penny-Dimri, J.C.; Bergmeir, C.; Reid, C.M.; Williams-Spence, J.; Cochrane, A.D.; Smith, J.A. Machine learning algorithms for predicting and risk profiling of cardiac surgery-associated acute kidney injury. In *Seminars in Thoracic and Cardiovascular Surgery*; WB Saunders: Philadelphia, PA, USA, 2021; Volume 33, pp. 735–745.
37. He, Z.L.; Zhou, J.B.; Liu, Z.K.; Dong, S.Y.; Zhang, Y.T.; Shen, T.; Zheng, S.S.; Xu, X. Application of machine learning models for predicting acute kidney injury following donation after cardiac death liver transplantation. *Hepatobiliary Pancreat. Dis. Int.* **2021**, *20*, 222–231. [CrossRef] [PubMed]
38. Alfieri, F.; Ancona, A.; Tripepi, G.; Crosetto, D.; Randazzo, V.; Paviglianiti, A.; Pasero, E.; Vecchi, L.; Cauda, V.; Fagugli, R.M. A deep-learning model to continuously predict severe acute kidney injury based on urine output changes in critically ill patients. *J. Nephrol.* **2021**, *34*, 1875–1886. [CrossRef]
39. Kang, Y.; Huang, S.T.; Wu, P.H. Detection of Drug–Drug and Drug–Disease Interactions Inducing Acute Kidney Injury Using Deep Rule Forests. *SN Comput. Sci.* **2021**, *2*, 1–14.
40. Le, S.; Allen, A.; Calvert, J.; Palevsky, P.M.; Braden, G.; Patel, S.; Pellegrini, E.; Green-Saxena, A.; Hoffman, J.; Das, R. Convolutional Neural Network Model for Intensive Care Unit Acute Kidney Injury Prediction. *Kidney Int. Rep.* **2021**, *6*, 1289–1298. [CrossRef]
41. Mamandipoor, B.; Frutos-Vivar, F.; Peñuelas, O.; Rezar, R.; Raymondos, K.; Muriel, A.; Du, B.; Thille, A.W.; Ríos, F.; González, M.; et al. Machine learning predicts mortality based on analysis of ventilation parameters of critically ill patients: Multi-centre validation. *BMC Med. Inform. Decis. Mak.* **2021**, *21*, 1–12. [CrossRef]
42. Hu, C.A.; Chen, C.M.; Fang, Y.C.; Liang, S.J.; Wang, H.C.; Fang, W.F.; Sheu, C.C.; Perng, W.C.; Yang, K.Y.; Kao, K.C.; et al. Using a machine learning approach to predict mortality in critically ill influenza patients: A cross-sectional retrospective multicentre study in Taiwan. *BMJ Open* **2020**, *10*, e033898. [CrossRef] [PubMed]
43. Rueckel, J.; Kunz, W.G.; Hoppe, B.F.; Patzig, M.; Notohamiprodjo, M.; Meinel, F.G.; Cyran, C.C.; Ingrisch, M.; Ricke, J.; Sabel, B.O. Artificial intelligence algorithm detecting lung infection in supine chest radiographs of critically ill patients with a diagnostic accuracy similar to board-certified radiologists. *Crit. Care Med.* **2020**, *48*, e574–e583. [CrossRef] [PubMed]
44. Greco, M.; Angelotti, G.; Caruso, P.F.; Zanella, A.; Stomeo, N.; Costantini, E.; Protti, A.; Pesenti, A.; Grasselli, G.; Cecconi, M. Artificial Intelligence to Predict Mortality in Critically ill COVID-19 Patients Using Data from the First 24h: A Case Study from Lombardy Outbreak. *Res. Sq.* **2021**. [CrossRef]
45. Ye, J.; Yao, L.; Shen, J.; Janarthanam, R.; Luo, Y. Predicting mortality in critically ill patients with diabetes using machine learning and clinical notes. *BMC Med. Inform. Decis. Mak.* **2020**, *20*, 1–7. [CrossRef]
46. Kong, G.; Lin, K.; Hu, Y. Using machine learning methods to predict in-hospital mortality of sepsis patients in the ICU. *BMC Med. Inform. Decis. Mak.* **2020**, *20*, 1–10. [CrossRef]
47. Nie, X.; Cai, Y.; Liu, J.; Liu, X.; Zhao, J.; Yang, Z.; Wen, M.; Liu, L. Mortality prediction in cerebral hemorrhage patients using machine learning algorithms in intensive care units. *Front. Neurol.* **2021**, *11*, 1847. [CrossRef]
48. Theis, J.; Galanter, W.; Boyd, A.; Darabi, H. Improving the In-Hospital Mortality Prediction of Diabetes ICU Patients Using a Process Mining/Deep Learning Architecture. *IEEE J. Biomed. Health Inform.* **2021**, *26*, 388–399. [CrossRef] [PubMed]
49. Jentzer, J.C.; Kashou, A.H.; Attia, Z.I.; Lopez-Jimenez, F.; Kapa, S.; Friedman, P.A.; Noseworthy, P.A. Left ventricular systolic dysfunction identification using artificial intelligence-augmented electrocardiogram in cardiac intensive care unit patients. *Int. J. Cardiol.* **2021**, *326*, 114–123. [CrossRef]
50. Popadic, V.; Klasnja, S.; Milic, N.; Rajovic, N.; Aleksic, A.; Milenkovic, M.; Crnokrak, B.; Balint, B.; Todorovic-Balint, M.; Mrda, D.; et al. Predictors of Mortality in Critically Ill COVID-19 Patients Demanding High Oxygen Flow: A Thin Line between Inflammation, Cytokine Storm, and Coagulopathy. *Oxidative Med. Cell. Longev.* **2021**, *2021*, 6648199. [CrossRef] [PubMed]
51. Kaji, D.A.; Zech, J.R.; Kim, J.S.; Cho, S.K.; Dangayach, N.S.; Costa, A.B.; Oermann, E.K. An attention based deep learning model of clinical events in the intensive care unit. *PLoS ONE* **2019**, *14*, e0211057. [CrossRef]
52. Shickel, B.; Loftus, T.J.; Adhikari, L.; Ozrazgat-Baslanti, T.; Bihorac, A.; Rashidi, P. DeepSOFA: A Continuous Acuity Score for Critically Ill Patients using Clinically Interpretable Deep Learning. *Sci. Rep.* **2019**, *9*, 1–12. [CrossRef]
53. Maxim, K.; Lev, U.; Ernest, K. SurvLIME: A method for explaining machine learning survival models. *Knowl.-Based Syst.* **2020**, *203*, 106164. [CrossRef]
54. Panigutti, C.; Perotti, A.; Pedreschi, D. Medical examiner XAI: An ontology-based approach to black-box sequential data classification explanations. In Proceedings of the 2020 Conference on Fairness, Accountability, and Transparency (FAT* '20). Association for Computing Machinery, New York, NY, USA, 27–30 January 2020; pp. 629–639. [CrossRef]
55. Hua, Y.; Guo, J.; Zhao, H. Deep Belief Networks and deep learning. In Proceedings of the 2015 International Conference on Intelligent Computing and Internet of Things, Harbin, China, 17–18 January 2015; pp. 1–4. [CrossRef]
56. Letham, B.; Rudin, C.; McCormick, T.H.; Madigan, D. Interpretable classifiers using rules and Bayesian analysis: Building a better stroke prediction model. *Ann. Appl. Stat.* **2015**, *9*, 1350–1371. [CrossRef]

57. Che, Z.; Purushotham, S.; Khemani, R.; Liu, Y. Interpretable Deep Models for ICU Outcome Prediction. *AMIA Annu. Symp. Proc.* **2017**, *2016*, 371–380.
58. Davoodi, R.; Moradi, M.H. Mortality prediction in intensive care units (ICUs) using a deep rule-based fuzzy classifier. *J. Biomed. Inform.* **2018**, *79*, 48–59. [CrossRef]
59. Johnson, M.; Albizri, A.; Harfouche, A. Responsible artificial intelligence in healthcare: Predicting and preventing insurance claim denials for economic and social wellbeing. *Inf. Syst. Front.* **2021**, 1–17. [CrossRef]
60. Xu, Z.; Tang, Y.; Huang, Q.; Fu, S.; Li, X.; Lin, B.; Xu, A.; Chen, J. Systematic review and subgroup analysis of the incidence of acute kidney injury (AKI) in patients with COVID-19. *BMC Nephrol.* **2021**, *22*, 52. [CrossRef] [PubMed]
61. Angiulli, F.; Fassetti, F.; Nisticò, S. Local Interpretable Classifier Explanations with Self-generated Semantic Features. In Proceedings of the International Conference on Discovery Science, Halifax, NS, Canada, 11–13 October 2021; Springer: Cham, Switzerland, 2021; pp. 401–410.
62. Visani, G.; Bagli, E.; Chesani, F. OptiLIME: Optimized LIME explanations for diagnostic computer algorithms. *arXiv* **2020**, arXiv:2006.05714.
63. Carrington, A.M.; Fieguth, P.W.; Qazi, H.; Holzinger, A.; Chen, H.H.; Mayr, F.; Manuel, D.G. A new concordant partial AUC and partial c statistic for imbalanced data in the evaluation of machine learning algorithms. *BMC Med. Inform. Decis. Mak.* **2020**, *20*, 1–12. [CrossRef]
64. Du, M.; Liu, N.; Hu, X. Techniques for interpretable machine learning. *Commun. ACM* **2020**, *63*, 68–77. [CrossRef]
65. Murdoch, W.J.; Singh, C.; Kumbier, K.; Abbasi-Asl, R.; Yu, B. Definitions, methods, and applications in interpretable machine learning. *Proc. Natl. Acad. Sci. USA* **2019**, *116*, 22071–22080. [CrossRef]
66. Adadi, A.; Berrada, M. Explainable AI for healthcare: From black box to interpretable models. In *Embedded Systems and Artificial Intelligence*; Springer: Singapore, 2020; pp. 327–337.
67. Nazar, M.; Alam, M.M.; Yafi, E.; Mazliham, M.S. A Systematic Review of Human-Computer Interaction and Explainable Artificial Intelligence in Healthcare with Artificial Intelligence Techniques. *IEEE Access* **2021**, *9*, 153316–153348. [CrossRef]
68. Srinivasan, R.; Chander, A. Explanation perspectives from the cognitive sciences—A survey. In Proceedings of the Twenty-Ninth International Conference on International Joint Conferences on Artificial Intelligence, Yokohama, Japan, 7–15 January 2021; pp. 4812–4818.
69. Zhou, B.; Sun, Y.; Bau, D.; Torralba, A. Interpretable basis decomposition for visual explanation. In Proceedings of the European Conference on Computer Vision (ECCV), Munich, Germany, 8–14 September 2018; pp. 119–134.
70. Mohseni, S.; Zarei, N.; Ragan, E.D. A Multidisciplinary Survey and Framework for Design and Evaluation of Explainable AI Systems. *ACM Trans. Interact. Intell. Syst.* **2021**, *11*, 1–45. [CrossRef]
71. Lo, S.H.; Yin, Y. A novel interaction-based methodology towards explainable AI with better understanding of Pneumonia Chest X-ray Images. *Discov. Artif. Intell.* **2021**, *1*, 1–7. [CrossRef]
72. RSNA Pneumonia Detection Challenge Dataset. Available online: https://www.kaggle.com/c/rsna-pneumonia-detection-challenge (accessed on 20 September 2022).
73. Dataset by Kermany et al. Available online: https://www.kaggle.com/paultimothymooney/chest-xray-pneumonia (accessed on 20 September 2022).
74. van Ginneken, B.; Stegmann, M.; Loog, M. Segmentation of anatomical structures in chest radiographs using supervised methods: A comparative study on a public database. *Med. Image Anal.* **2006**, *10*, 19–40. Available online: http://www.isi.uu.nl/Research/Databases/SCR/ (accessed on 20 September 2022). [CrossRef] [PubMed]
75. Central Line-Associated Bloodstream Infections (CLABSI) in California Hospitals. Available online: https://healthdata.gov/State/Central-Line-Associated-Bloodstream-infections-CLA/cu55-5ujz/data (accessed on 20 September 2022).
76. Johnson, A.; Pollard, T.; Mark, R. MIMIC-III Clinical Database (version 1.4). *PhysioNet* **2016**. [CrossRef]
77. Johnson, A.E.W.; Pollard, T.J.; Shen, L.; Lehman, L.H.; Feng, M.; Ghassemi, M.; Moody, B.; Szolovits, P.; Celi, L.A.; Mark, R.G. MIMIC-III, a freely accessible critical care database. *Sci. Data* **2016**, *3*, 160035. [CrossRef]
78. ICES Data Repository. Available online: https://www.ices.on.ca/Data-and-Privacy/ICES-data (accessed on 20 September 2022).
79. Department of Veterans Affairs, Veterans Health Administration: Providing Health Care for Veterans. Available online: https://www.va.gov/health/ (accessed on 9 November 2018).
80. Tomasev, N.; Glorot, X.; Rae, J.W.; Zielinski, M.; Askham, H.; Saraiva, A.; Mottram, A.; Meyer, C.; Ravuri, S.; Protsyuk, I.; et al. A clinically applicable approach to continuous prediction of future acute kidney injury. *Nature* **2019**, *572*, 116–119. [CrossRef]
81. Lauritsen, S.M.; Kristensen, M.; Olsen, M.V.; Larsen, M.S.; Lauritsen, K.M.; Jørgensen, M.J.; Lange, J.; Thiesson, B. Explainable artificial intelligence model to predict acute critical illness from electronic health records. *Nat. Commun.* **2020**, *11*, 1–11. [CrossRef]
82. Hou, J.; Gao, T. Explainable DCNN based chest X-ray image analysis and classification for COVID-19 pneumonia detection. *Sci. Rep.* **2021**, *11*, 16071. [CrossRef]
83. Berthelot, D.; Carlini, N.; Goodfellow, I.; Papernot, N.; Oliver, A.; Raffel, C.A. Mixmatch: A holistic approach to semisupervised learning. *Adv. Neural Inf. Process. Syst.* **2019**, *32*, 14.
84. Tarvainen, A.; Valpola, H. Mean teachers are better role models: Weight-averaged consistency targets improve semi-supervised deep learning results. *Adv. Neural Inf. Process. Syst. NIPS* **2017**, *30*, 1195–1204.
85. Verma, V.; Lamb, A.; Kannala, J.; Bengio, Y.; Lopez-Paz, D. Interpolation consistency training for semi-supervised learning. *Int. Jt. Conf. Artif. Intell. IJCAI* **2019**, *145*, 3635–3641.

86. Raghu, M.; Zhang, C.; Kleinberg, J.; Bengio, S. Transfusion: Understanding transfer learning for medical imaging. *Neural Inf. Process. Syst.* **2019**, *32*, 3347–3357.
87. Aviles-Rivero, A.I.; Papadakis, N.; Li, R.; Sellars, P.; Fan, Q.; Tan, R.; Schönlieb, C.-B. Graphx-net—Chest x-ray classification under extreme minimal supervision. In Proceedings of the International Conference on Medical Image Computing and Computer-Assisted Intervention, Shenzhen, China, 13–17 October 2019; pp. 504–512.
88. Aviles-Rivero, A.I.; Sellars, P.; Schönlieb, C.B.; Papadakis, N. GraphXCOVID: Explainable deep graph diffusion pseudo-labelling for identifying COVID-19 on chest X-rays. *Pattern Recognit.* **2022**, *122*, 108274. [CrossRef] [PubMed]
89. Napolitano, F.; Xu, X.; Gao, X. Impact of computational approaches in the fight against COVID-19: An AI guided review of 17,000 studies. *Brief. Bioinform.* **2022**, *23*, bbab456. [CrossRef] [PubMed]
90. Esteva, A.; Chou, K.; Yeung, S.; Naik, N.; Madani, A.; Mottaghi, A.; Liu, Y.; Topol, E.; Dean, J.; Socher, R. Deep learning-enabled medical computer vision. *NPJ Digit. Med.* **2021**, *4*, 1–9. [CrossRef] [PubMed]
91. Zhou, S.K.; Greenspan, H.; Davatzikos, C.; Duncan, J.S.; Van Ginneken, B.; Madabhushi, A.; Prince, J.L.; Rueckert, D.; Summers, R.M. A review of deep learning in medical imaging: Image traits, technology trends, case studies with progress highlights, and future promises. *Proc. IEEE* **2021**, *109*, 820–838. [CrossRef]
92. Tellakula, K.K.; Kumar, S.; Deb, S. A survey of ai imaging techniques for covid-19 diagnosis and prognosis. *Appl. Comput. Sci.* **2021**, *17*, 40–55.
93. Fábio, D.; Cinalli, D.; Garcia, A.C.B. Research on Explainable Artificial Intelligence Techniques: An User Perspective. In Proceedings of the 2021 IEEE 24th International Conference on Computer Supported Cooperative Work in Design (CSCWD), IEEE, Dalian, China, 5–7 May 2021.
94. Neves, I.; Folgado, D.; Santos, S.; Barandas, M.; Campagner, A.; Ronzio, L.; Cabitza, F.; Gamboa, H. Interpretable heartbeat classification using local model-agnostic explanations on ECGs. *Comput. Biol. Med.* **2021**, *133*, 104393. [CrossRef]
95. Selvaganapathy, S.; Sadasivam, S.; Raj, N. SafeXAI: Explainable AI to Detect Adversarial Attacks in Electronic Medical Records. In *Intelligent Data Engineering and Analytics*; Springer: Singapore, 2022; pp. 501–509.
96. Payrovnaziri, S.N.; Chen, Z.; Rengifo-Moreno, P.; Miller, T.; Bian, J.; Chen, J.H.; Liu, X.; He, Z. Explainable artificial intelligence models using real-world electronic health record data: A systematic scoping review. *J. Am. Med. Inform. Assoc.* **2020**, *27*, 1173–1185. [CrossRef]
97. Van Der Velden, B.H.; Kuijf, H.J.; Gilhuijs, K.G.; Viergever, M.A. Explainable artificial intelligence (XAI) in deep learning-based medical image analysis. *Med. Image Anal.* **2022**, *79*, 102470. [CrossRef]
98. Antoniadi, A.M.; Du, Y.; Guendouz, Y.; Wei, L.; Mazo, C.; Becker, B.A.; Mooney, C. Current challenges and future opportunities for XAI in machine learning-based clinical decision support systems: A systematic review. *Appl. Sci.* **2021**, *11*, 5088. [CrossRef]
99. Qiu, W.; Chen, H.; Dincer, A.B.; Lundberg, S.; Kaeberlein, M.; Lee, S.I. Interpretable machine learning prediction of all-cause mortality. *medRxiv* **2022**. [CrossRef]
100. Yang, Y.; Mei, G.; Piccialli, F. A Deep Learning Approach Considering Image Background for Pneumonia Identification Using Explainable AI (XAI). *IEEE/ACM Trans. Comput. Biol. Bioinform.* **2022**, 1–12. [CrossRef] [PubMed]
101. Zou, L.; Goh, H.L.; Liew, C.J.; Quah, J.L.; Gu, G.T.; Chew, J.J.; Kumar, M.P.; Ang, C.G.; Ta, A. Ensemble image explainable AI (XAI) algorithm for severe community-acquired pneumonia and COVID-19 respiratory infections. *IEEE Trans. Artif. Intell.* **2022**, 1–12. [CrossRef]
102. Hu, C.; Tan, Q.; Zhang, Q.; Li, Y.; Wang, F.; Zou, X.; Peng, Z. Application of interpretable machine learning for early prediction of prognosis in acute kidney injury. *Comput. Struct. Biotechnol. J.* **2022**, *20*, 2861–2870. [CrossRef]
103. Zhang, L.; Wang, Z.; Zhou, Z.; Li, S.; Huang, T.; Yin, H.; Lyu, J. Developing an ensemble machine learning model for early prediction of sepsis-associated acute kidney injury. *Iscience* **2022**, *25*, 104932. [CrossRef] [PubMed]
104. Schallner, L.; Rabold, J.; Scholz, O.; Schmid, U. Effect of superpixel aggregation on explanations in LIME—A case study with biological data. In *Joint European Conference on Machine Learning and Knowledge Discovery in Databases*; Springer: Cham, Switzerland, 2019; pp. 147–158.
105. Wei, Y.; Chang, M.C.; Ying, Y.; Lim, S.N.; Lyu, S. Explain black-box image classifications using superpixel-based interpretation. In Proceedings of the 2018 24th International Conference on Pattern Recognition (ICPR), Beijing, China, 20–24 August 2018; IEEE: Piscataway, NJ, USA, 2018; pp. 1640–1645.
106. Hussain; Mehboob, S.; Buongiorno, D.; Altini, N.; Berloco, F.; Prencipe, B.; Moschetta, M.; Bevilacqua, V.; Brunetti, A. Shape-Based Breast Lesion Classification Using Digital Tomosynthesis Images: The Role of Explainable Artificial Intelligence. *Appl. Sci.* **2022**, *12*, 6230. [CrossRef]
107. Zhang, Y.; Tiňo, P.; Leonardis, A.; Tang, K. A survey on neural network interpretability. *IEEE Trans. Emerg. Top. Comput. Intell.* **2021**, *5*, 726–742. [CrossRef]

MDPI AG
Grosspeteranlage 5
4052 Basel
Switzerland
Tel.: +41 61 683 77 34

Sensors Editorial Office
E-mail: sensors@mdpi.com
www.mdpi.com/journal/sensors

Disclaimer/Publisher's Note: The statements, opinions and data contained in all publications are solely those of the individual author(s) and contributor(s) and not of MDPI and/or the editor(s). MDPI and/or the editor(s) disclaim responsibility for any injury to people or property resulting from any ideas, methods, instructions or products referred to in the content.

www.ingramcontent.com/pod-product-compliance
Lightning Source LLC
LaVergne TN
LVHW070504100526
838202LV00014B/1784